Computational Aspects of Algebraic Curves

LECTURE NOTES SERIES ON COMPUTING

Editor-in-Chief: D T Lee (*Academia Sinica, Taiwan*)

Published

Vol. 1: Computing in Euclidean Geometry
Eds. D-Z Du & F Hwang

Vol. 2: Algorithmic Aspects of VLSI Layout
Eds. D T Lee & M Sarrafzadeh

Vol. 3: String Searching Algorithms
G A Stephen

Vol. 4: Computing in Euclidean Geometry (Second Edition)
Eds. D-Z Du & F Hwang

Vol. 5: Proceedings of the Conference on
Parallel Symbolic Computation — PASCO '94
Ed. H Hong

Vol. 6: VLSI Physical Design Automation: Theory and Practice
S M Sait & H Youssef

Vol. 7: Algorithms: Design Techniques and Analysis
Ed. M H Alsuwaiyel

Vol. 8: Computer Mathematics
Proceedings of the Fourth Asian Symposium (ASCM 2000)
Eds. X-S Gao & D Wang

Vol. 9: Computer Mathematics
Proceedings of the Fifth Asian Symposium (ASCM 2001)
Eds. K Yokoyama & K Shirayanagi

Vol. 10: Computer Mathematics
Proceedings of the Sixth Asian Symposium (ASCM 2003)
Eds. Z Li & W Sit

Vol. 11: Geometric Computation
Eds. F Chen & D Wang

Vol. 12: Planar Graph Drawing
T Nishizeki & S Rahman

Vol. 13: Computational Aspects of Algebraic Curves
ed. T Shaska

Lecture Notes Series on Computing – Vol. 13

Computational Aspects of Algebraic Curves

University of Idaho, USA 26 – 28 May 2005

Editor

Tanush Shaska

Oakland University, USA

World Scientific

NEW JERSEY • LONDON • SINGAPORE • BEIJING • SHANGHAI • HONG KONG • TAIPEI • CHENNAI

Published by

World Scientific Publishing Co. Pte. Ltd.

5 Toh Tuck Link, Singapore 596224

USA office: 27 Warren Street, Suite 401-402, Hackensack, NJ 07601

UK office: 57 Shelton Street, Covent Garden, London WC2H 9HE

British Library Cataloguing-in-Publication Data
A catalogue record for this book is available from the British Library.

COMPUTATIONAL ASPECTS OF ALGEBRAIC CURVES
Lecture Notes Series on Computing — Vol. 13

ISBN 981-256-459-4

Printed in Singapore.

CONTENTS

PREFACE

Over the years algebraic curves have found applications in many areas, including cryptography, primality testing, integer factorization, error-correcting codes, computer vision, etc. The development of new computational techniques has made it possible to attack some classical problems of algebraic geometry related to algebraic curves. In this new light we organized a conference at the University of Idaho, where the focus was on such computational techniques. This continued the trend of such conferences with the same theme: i) *Special Session on Computational Aspects of Algebraic Curves*, ACA 03, North Carolina State (2003), ii) *Special Session on Computational Aspects of Algebraic Curves*, ACA 04, Lamar University (2004), and iii) *Special Session on Algorithmic Algebraic and Analytic Geometry*, Annual AMS meeting (2005), organized by this editor in collaboration with others.

The conference on *Computational Aspects of Algebraic Curves* was held at the University of Idaho in Moscow, Idaho, May 26th-28th, 2005. The goal of this conference was to look at some of the problems related to algebraic curves from a computational point of view. The participants in the conference included D. Joyner, A. Bialostocki, R. Pries, J. Voight, I. Burhanuddin, J. Achter, A. Wootton, B. Zyka, N. Bialke, R. Vogeler, D. Manz, S. Zheng, C. Shor, M. van Opstall, J. Son, et al. The conference was fully supported by NSA under the grant R1-05-0129. On behalf of all the participants I want to thank the NSA for its support.

I want to thank in particular Dr. Wagner and her staff at NSA for their patience during the duration of this grant and the dean of College of Science at the University of Idaho, Dr. Parrish, and her staff for sharing the burden of the organizational tasks of the conference.

Further, I want to thank the Applications of Computer Algebra (ACA) international advisory board for allowing me to organize a special session on algebraic curves both in ACA 03 and ACA 04. Also thanks go to Saugata Basu (Georgia Institute of Technology), Victoria A. Powers (Emory University), Mika K. Sepälä (Florida State University), and Emil J. Volcheck (National Security Agency) for being great co-organizers of the special session on *Algorithmic Algebraic and Analytic Geometry* during the annual AMS meeting in Atlanta 2005. Without those previous meetings the current conference would not have happened.

Special thanks go to Emil Volchek for encouraging me to pursue the idea of organizing such a conference and to Arie Bialostocki for helping me with the organizational tasks.

Tanush Shaska
Rochester, Michigan

FOREWORD BY THE EDITOR

The area of research of algebraic curves and their applications is broad and fast developing and has implications in many branches of science. With the development of new computational methods and better computing power, the area is growing in interest from the mathematics community and also from engineers and computer scientists. In this book we have tried to give a glimpse of some of the computational aspects of algebraic curves and its connections to group theory, computational number theory, cryptography, coding theory, etc.

While no book can include all aspects of such a broad area, we have tried to include papers from a wide variety of topics. In putting together this book, I have tried to include a few survey articles which give an overall view of certain topics.

Elliptic curves are still a big part of the research that goes on in the area of algebraic curves. Papers related to elliptic curves occupy the first part of the book. Schaefer [§3] presents a new proof for the non-degeneracy of the Frey-Rück pairing and a connection to isogenies over the base field. Burhanuddin and Huang [§4] present algorithms to determine the group of rational torsion points on an elliptic curve. Achter [§5] gives an efficient, deterministic algorithm to decide if two abelian varieties over a number field are isogenous. Seppälä [§6] studies the numerical uniformatization of elliptic curves.

The next several articles study the subloci of the moduli space $\mathcal{M}_g$ of curves. D. Glass and R. Pries [§7] study the sublocus $\mathcal{H}_{g,n}$ of $\mathcal{M}_g$ whose points correspond to curves which are $(\mathbb{Z}/2\mathbb{Z})^n$ covers of the projective line. In the case that $n = 2$, the components of $\mathcal{H}_{g,2}$ are discussed and all of the intersections between these components. Unlike the case of the hyperelliptic locus $\mathcal{H}_{g,1}$, which is well-known to be connected, it is shown that the space of $\mathcal{H}_{g,n}$ is not connected when $n = 2$ and $g \geq 4$ or when $n = 3$ and $g \geq 5$. Cardona and Quer [§8] show that for every genus 2 curve with extra automorphisms there is a rational model over its field of moduli. This complements the work of Mestre, who gave an algorithm of determining whether such a model exists for genus 2 curves with automorphism group $\mathbb{Z}$.

Algebraic curves with many automorphisms have been studied for a long time. Hurwitz curves especially have attracted a lot of attention over the years. Vogeler [§9] describes an algorithmic method of computing the length spectrum of the $(2,3,7)$ triangle group G in exact arithmetic form. From G descend all the Hurwitz surfaces, and their spectra can in turn be computed. Wootton [§10] shows that there are only a small number of non-normal Belyi p-gonal surfaces, most of which are well known. For each such surface, we find the full automorphism group and a defining equation. A consequence is the discovery of an explicit family of surfaces which show that Accola's bound is sharp for p-gonal surfaces for all primes $p \geq 5$. Gutierrez, Sevilla, and Shaska [§11] study genus 3 hyperelliptic curves which have an extra involution. The locus of these curves is a 3-dimensional subvariety in the genus 3 hyperelliptic moduli $\mathcal{H}_3$. A birational parametrization of this locus by affine 3-space is given. For every moduli point $\mathfrak{p} \in \mathcal{H}_3$ such that $|\text{Aut}\,(\mathfrak{p})| > 2$, it is shown the field of moduli is a field of definition. A rational model of the curve over its field of moduli for all moduli points $\mathfrak{p} \in \mathcal{H}_3$ such that $|\text{Aut}\,(\mathfrak{p})| > 4$ is provided.

The next couple of articles study algebraic curves over a field of positive characteristic. In his introductory article Voight [§12] surveys what is known about the maximum number of points on a curve of genus g defined over $\mathbb{F}_q$, including an exposition of upper bounds, lower bounds, known values of this maximum, and briefly indicates some methods of constructing curves with many points, providing many references to the literature. Demirbas [§13] computes the automorphism groups of hyperelliptic curves of genus 3 and 4, in characteristic 2, and displays the generators and the automorphisms explicitly.

Joyner and Ksir [§14] compute the $PSL(2,N)$-module structure of the Riemann-Roch space $L(D)$, where D is an invariant non-special divisor on the modular curve $X(N)$, with $N \geq 7$ prime. This depends on a computation of the ramification module, which is given explicitly. These results hold for characteristic p if $X(N)$ has good reduction mod p and p does not divide the order of $PSL(2,N)$. Examples are given for the cases $N = 7, 11$, which are computed using GAP [4]. Applications to AG codes associated to this curve are considered, and specific examples are computed using GAP [4] and MAGMA [6].

In [§15] the author gives a survey of a problem that goes back to Legendre and Jacobi, namely the genus 2 curves with elliptic subcovers. Computational conditions are provided for a curve to have (n, n)-split Jacobians for small n. The paper summarizes previous results of the author and provides some of the details of the proofs that were skipped in those other papers. Furthermore, the author provides a computer algebra package that allows computation with genus 2 curves and with some special classes of higher genus hyperelliptic curves.

The paper of Goldstein, R. Guralnick, and D. Joyner [§16] addresses the following question: Let X be an irreducible non-singular projective variety defined over a field F, and let G be a finite subgroup of ${}_F(X)$: Is it true that every G-invariant divisor class on X is the class of a G-invariant divisor on X? A special case of this question was posed by Lonsted. The authors show that the answer depends only on G and not on the curve. The same question is addressed for degree-0 divisor classes.

The paper by Bialostocki and Shaska [§17] proves that if G is a doubly transitive subgroup of S_p where p is prime, then G isomorphic to one of the following: (i) A_p, S_p, (ii) $p = 11, G = L_2(11)$ or M_{11}, (iii) $p = 23,\ G = M_{23}$, or (iv) $p = \frac{(q^k-1)}{(q-1)}$ and $L_k(q) \leq G \leq Aut(L_k(q))$. This extends a simple condition on finding Galois groups of polynomials. Let $f(x) \in \mathbb{Q}[x]$ be an irreducible polynomial of prime degree $p \geq 5$ and $r = 2s$ be the number of non-real roots of $f(x)$. If s satisfies $s\,(s \log s + 2 \log s + 3)\ \leq p$ then $Gal(f) = A_p, S_p$. We all know a similar condition from elementary algebra when the number of nonreal roots is three.

Staszewski, Völklein, and Wiesend [§18] describe an algorithm (and its implementation in GAP and MAGMA) for computing the number of tuples $(g_1, ..., g_r)$ generating G such that $g_1 \cdots g_r = 1$ and $g_i \in C_i$. This is a new ingredient of the BRAID package, a collection of programs for computing braid orbits of such tuples.

I would like to thank all the people who helped me put together this book. My thanks go to the authors who contributed papers to this volume. Special thanks go to the referees of the papers for all their time and effort. Thanks to the World Scientific staff for their efforts and patience. Finally, I would like to thank the National Security Agency (NSA) for its support through the grant R1-05-0129. It was this grant that made possible the conference on *Computational Aspects of Algebraic Curves* which was held at the University of Idaho in Moscow, Idaho, in May of 2005, from which most of these papers originated.

Tanush Shaska

A new proof for the non-degeneracy of the Frey-Rück pairing and a connection to isogenies over the base field

Edward F. Schaefer*

Department of Mathematics and Computer Science,
Santa Clara University,
Santa Clara, CA 95053, USA.
eschaefe@math.scu.edu

Frey and Rück have described a non-degenerate bilinear pairing on the Jacobians of curves; this includes elliptic curves. We present a new mathematical foundation for this pairing and use it to give a different proof of its non-degeneracy. We then present yet another method of looking at this pairing using isogenies.

1. Introduction

In this article, we will provide a new mathematical foundation for a pairing, first described by Frey and Rück in [2], on the Jacobians of curves over finite fields. This pairing has cryptographic applications. It is used to give a fast translation of the discrete logarithm problem in a Jacobian to the discrete logarithm problem in the multiplicative group of a finite field. It is also used in pairing-based cryptography. We note that an elliptic curve is an example of a Jacobian.

In Section 4, we present a new proof of the non-degeneracy of this pairing. We will use Galois cohomology and Weil reciprocity over finite fields. The proof of non-degeneracy will stem from the non-degeneracy of the Weil pairing.

The pairing of Frey and Rück pairs two groups of prime exponent l. In order to use this pairing, the base field, over which the curve is defined, is

*The author is grateful to Joseph Wetherell for useful conversations. The author is supported by a National Security Agency Standard MDA904-03-1-0030.

always enlarged so as to include the l-th roots of unity. We pair the kernel and co-kernel of the multiplication by l map on the points of the Jacobian over this extended field.

For encryption and decryption, the group in which the computations take place is a subgroup of the Jacobian of a curve over the base field. Typically the base field does not include the l-th roots of unity. So it should not be necessary to always have to consider the Jacobian over the extension of the base field gotten by adjoining the l-th roots of unity.

In Section 5, we will use isogenies other than the multiplication by l map in order to consider the Jacobian over the base field. Perhaps this reinterpretation will lead to new insights into this pairing, as in Section 6.

2. Overview of mathematical methods

In this section, we will review the Galois cohomology and arithmetic geometry that will be used throughout this article. References for the results on cohomology that we simply state are [4], [11], [6]. A reference for results on curves that we simply state is [6].

Let G be a Galois group, finite or infinite, and let M be a G-module. A cocycle from G to M is a map $\xi : G \to M$ with the property that $\xi(\sigma\tau) = {}^{\sigma}\xi(\tau) + \xi(\sigma)$ for all $\sigma, \tau \in G$. If $m \in M$ then we can associate to it the cocycle $\sigma \mapsto {}^{\sigma}m - m$; such a cocycle is called a coboundary. The quotient of the cocycles by the coboundaries is a group denoted $H^1(G, M)$. If G is an infinite Galois group, we just use continuous cocycles. If $0 \to \ker(f) \to M_1 \xrightarrow{f} M_2 \to 0$ is an exact sequence of G-modules then we get a long exact sequence $0 \to \ker(f)^G \to M_1^G \to M_2^G \xrightarrow{\delta_f} H^1(G, \ker(f)) \to H^1(G, M_1) \xrightarrow{\hat{f}} H^1(G, M_2)$. Here M^G denotes the G-invariants of M. The maps between H^1's are induced by the maps between the modules. Let us define δ_f. If $m \in M_2^G$ and $n \in M_1$ is a preimage, then $\delta_f(m)$ is the class including the cocycle $\sigma \mapsto {}^{\sigma}n - n$.

Let K be a field and $\overline{K}$ a separable closure. If M is a $\mathrm{Gal}(\overline{K}/K)$-module, then we use $H^1(K, M)$ to denote $H^1(\mathrm{Gal}(\overline{K}/K), M)$ and $M(K)$ to denote $M^{\mathrm{Gal}(\overline{K}/K)}$. Let K^* denote the multiplicative group of units in K; as a set, $K^* = K \setminus 0$. Hilbert's Theorem 90 tells us $H^1(K, \overline{K}^*) = 0$. Let $l \in \mathbb{Z}$, $l \geq 2$. Raising to the l-th power gives us a short exact sequence of $\mathrm{Gal}(\overline{K}/K)$-modules: $1 \to \mu_l \to \overline{K}^* \xrightarrow{l} \overline{K}^* \to 1$, where μ_l is the group of l-th roots of unity in $\overline{K}$. From above, we have a long exact sequence $1 \to \mu_l(K) \to K^* \xrightarrow{l} K^* \xrightarrow{\delta_l} H^1(K, \mu_l) \to 0$. So we get the Kummer isomorphism $K^*/K^{*l} \xrightarrow{\delta_l} H^1(K, \mu_l)$ by $k \in K^* \mapsto [\sigma \mapsto {}^{\sigma}\sqrt[l]{k}/\sqrt[l]{k}]$.

Similarly, let A be an abelian variety defined over the field K. Abelian varieties include Jacobian varieties. By A we mean $A(\overline{K})$, the set of points with coordinates in $\overline{K}$. Let $A[l]$ denote the kernel of the multiplication by l map. This is the l-torsion on A. We have the short exact sequence of $\mathrm{Gal}(\overline{K}/K)$-modules $0 \to A[l] \to A(\overline{K}) \stackrel{l}{\to} A(\overline{K}) \to 0$. This gives us the long exact sequence $0 \to A[l](K) \to A(K) \stackrel{l}{\to} A(K) \stackrel{\delta_l}{\to} H^1(K, A[l]) \to H^1(K, A)$. From this we get an injection $A(K)/lA(K) \stackrel{\delta_l}{\hookrightarrow} H^1(K, A[l])$. If $P \in A(K)$ and $Q \in A(\overline{K})$ has the property that $lQ = P$ then $\delta_l(P) = [\sigma \mapsto {}^{\sigma}Q - Q]$.

If L is a normal extension of K and M is a $\mathrm{Gal}(\overline{K}/K)$-module then there is a restriction map $H^1(K, M) \stackrel{\mathrm{res}}{\to} H^1(L, M)$ induced by $\mathrm{Gal}(\overline{K}/L) \subseteq \mathrm{Gal}(\overline{K}/K)$. If $[\xi] \in H^1(\mathrm{Gal}(L/K), M(L))$ then ξ induces a cocycle

$$\inf(\xi) : \mathrm{Gal}(\overline{K}/K) \to \mathrm{Gal}(L/K) \stackrel{\xi}{\to} M(L) \subseteq M,$$

where the first map is simply the quotient map on Galois groups. This induces an inflation map from $H^1(\mathrm{Gal}(L/K), M(L))$ to $H^1(K, M)$. We get an exact sequence

$$\begin{aligned} 0 \to H^1(\mathrm{Gal}(L/K), M(L)) \stackrel{\inf}{\to} H^1(K, M) \stackrel{\mathrm{res}}{\to} \\ \to H^1(L, M)^{\mathrm{Gal}(L/K)} \to H^2(\mathrm{Gal}(L/K), M(L)) \end{aligned} \tag{1}$$

For the purposes of this article, we will not need to understand the structure of nor the map to $H^2(\mathrm{Gal}(L/K), M(L))$.

If G is a group with exponent n and M has exponent m and $\gcd(m, n) = 1$, then $H^i(G, M) = 0$ for all i.

Proposition 1: *Let K be a finite field of characteristic p and let $l \neq p$ be a prime. Let M be a finite dimensional $\mathbb{F}_l$-vector space on which* $\mathrm{Gal}(\overline{K}/K)$ *acts. Then* $\dim_{\mathbb{F}_l} H^1(K, M) = \dim_{\mathbb{F}_l} M(K)$.

Proof: Let L be the field of definition of all elements of M. Let L_l be the extension of L of degree l. We have the exact sequence $0 \to H^1(\mathrm{Gal}(L_l/K), M) \stackrel{\inf}{\to} H^1(K, M) \stackrel{\mathrm{res}}{\to} H^1(L_l, M)$. The restriction map from $H^1(K, M)$ to $H^1(L_l, M)$ can be factored as $H^1(K, M) \stackrel{\mathrm{res}}{\to} \mathrm{Hom}(\mathrm{Gal}(\overline{K}/L), M) \stackrel{\mathrm{res}}{\to} \mathrm{Hom}(\mathrm{Gal}(\overline{K}/L_l), M) = H^1(L_l, M)$. Since M is an $\mathbb{F}_l$-vector space and L_l is the degree l extension of L, we see that the latter restriction map is trivial. Thus the inflation map induces an isomorphism of $H^1(\mathrm{Gal}(L_l/K), M)$ and $H^1(K, M)$.

Since $\mathrm{Gal}(L_l/K)$ is a finite cyclic group, we have $H^1(\mathrm{Gal}(L_l/K), M) \cong \hat{H}^1(\mathrm{Gal}(L_l/K), M)$, the Tate-cohomology group. Since M is finite, we have $\#\hat{H}^1(\mathrm{Gal}(L_l/K), M) = \#\hat{H}^0(\mathrm{Gal}(L_l/K), M)$. Let σ generate $\mathrm{Gal}(L_l/K)$

and N_σ be the norm from L_l to K. The group $\hat{H}^0(\mathrm{Gal}(L_l/K), M)$ is isomorphic to the quotient of the kernel of $\sigma - 1$ on M by the image of N_σ on M. Since N_σ on M is the same as the composition of the norm from L to K and the multiplication by l map, we see that the image of N_σ on M is trivial. So $\hat{H}^0(\mathrm{Gal}(L_l/K), M) \cong M(K)$. □

Let C be a curve defined over the field K. When we write C we mean the set of points $C(\overline{K})$. A divisor on C is a formal finite sum of points on C; it could be denoted $\sum_{P \in C} n_P P$, with $P \in C$, $n_P \in \mathbb{Z}$, and where all but finitely many n_P are 0. The set of P's for which $n_P \neq 0$ is the support of the divisor. The degree of $\sum n_P P$ is $\sum n_P$. There is a natural action of $\mathrm{Gal}(\overline{K}/K)$ on divisors of C. A divisor fixed by $\mathrm{Gal}(\overline{K}/K)$ will be called a divisor over K. For example, the divisor $(\sqrt{3}, 0) + (-\sqrt{3}, 0) - 2(1, 1)$ would be a divisor over $\mathbb{Q}$ of degree 0 on a curve containing those three points.

The function field of C is denoted $\overline{K}(C)$ and its $\mathrm{Gal}(\overline{K}/K)$-invariants are denoted $K(C)$. If $f \in K(C)$ and $\sum n_P P$ is a divisor over K then we define $f(\sum n_P P) = \prod f(P)^{n_P}$. A priori, this is an element of $\overline{K}$; however, given the Galois-actions on both f and the divisor, we see that $\prod f(P)^{n_P} \in K$. If $f \in \overline{K}(C)$ then $\mathrm{div}(f) = \sum_{P \in C} \mathrm{ord}_P(f) P$ is its divisor. The divisor of a function is called a principal divisor. If $f \in K(C)$ then $\mathrm{div}(f)$ is a divisor over K. If $f, g \in \overline{K}(C)$ and their divisors have disjoint supports, then $f(\mathrm{div}(g)) = g(\mathrm{div}(f))$; this is called Weil reciprocity.

The degree 0 divisors of C form a group and the principal divisors form a subgroup. The quotient group is the Jacobian of C, which we will denote J. Here $J = J(\overline{K})$. We can give J the structure of an algebraic variety, but we will not need to do so in this article.

3. Description of the first pairing

Let us recall the pairing of Frey and Rück. Let p be a prime, $q = p^s$ for some positive integer s and $\mathbb{F}_q$ denote the finite field with q elements. Let C be a curve of genus greater than or equal to 1 over $\mathbb{F}_q$. (If C has genus 1, then J is just an elliptic curve.) Assume $C(\mathbb{F}_q)$ is non-empty. This is a safe assumption whenever the genus of C is small and q is sufficiently large, as is true in cryptographic applications. Let J be the Jacobian of C. If D is a degree 0 divisor, then we represent its divisor class in J by $[D]$. Since $C(\mathbb{F}_q)$ is non-empty, every divisor class in $J(\mathbb{F}_q)$ is represented by a degree 0 divisor over $\mathbb{F}_q$. Let $l \neq p$ be a prime dividing $\#J(\mathbb{F}_q)$. Here and in Section 4 we will make the assumption that the l-th roots of unity are contained in $\mathbb{F}_q$. In other words, we are assuming that $l | q - 1$.

There is a pairing from $J(\mathbb{F}_q)/lJ(\mathbb{F}_q) \times J[l](\mathbb{F}_q) \to \mathbb{F}_q^*/\mathbb{F}_q^{*l}$. Let $R \in J(\mathbb{F}_q)$ and $S \in J[l](\mathbb{F}_q)$. Choose degree 0 divisors D_1 and D_2 over $\mathbb{F}_q$, with disjoint supports, such that $R = [D_1]$ and $S = [D_2]$. Choose $f_{D_2} \in \mathbb{F}_q(C)$ where $\mathrm{div}(f_{D_2}) = lD_2$. Then $\langle R, S\rangle = f_{D_2}(D_1)$. We will see in Section 4 that this pairing is bilinear, non-degenerate and does not depend on the choices of D_1, D_2 or f_{D_2}.

4. Proof of non-degeneracy of the first pairing

From Section 2 we have an injection $J(\mathbb{F}_q)/lJ(\mathbb{F}_q) \overset{\delta_l}{\hookrightarrow} H^1(\mathbb{F}_q, J[l])$.

Lemma 2: *The map δ_l induces an isomorphism of $J(\mathbb{F}_q)/lJ(\mathbb{F}_q)$ and $H^1(\mathbb{F}_q, J[l])$.*

Proof: Since $J(\mathbb{F}_q)$ is a finite abelian group, we have $\#J(\mathbb{F}_q)/lJ(\mathbb{F}_q) = \#J[l](\mathbb{F}_q)$. From Proposition 1, we have $\#J[l](\mathbb{F}_q) = \#H^1(\mathbb{F}_q, J[l])$. □

Let $\mathcal{B} = \{[D_1], \ldots, [D_b]\}$ be a basis of $J[l](\mathbb{F}_q)$. Let $w_l : J[l] \to (\mu_l)^b$ by $[D] \mapsto (e_l([D],[D_1]), \ldots, e_l([D],[D_b]))$, where e_l is the l-Weil pairing. Let $\hat{w}_l$ be the induced map from $H^1(\mathbb{F}_q, J[l])$ to $H^1(\mathbb{F}_q, (\mu_l)^b)$.

Lemma 3: *The map $\hat{w}_l$ induces an isomorphism of $H^1(\mathbb{F}_q, J[l])$ and $H^1(\mathbb{F}_q, (\mu_l)^b)$.*

Proof: The following is a short exact sequence of $\mathrm{Gal}(\overline{\mathbb{F}_q}/\mathbb{F}_q)$-modules

$$0 \to \ker(w_l) \to J[l] \overset{w_l}{\to} (\mu_l)^b \to 0.$$

That w_l maps surjectively follows from the non-degeneracy of the Weil-pairing. Taking $\mathrm{Gal}(\overline{\mathbb{F}_q}/\mathbb{F}_q)$-invariants gives

$$0 \to \ker(w_l)(\mathbb{F}_q) \to J[l](\mathbb{F}_q) \overset{w_l}{\to} (\mu_l)^b \to H^1(\mathbb{F}_q, \ker(w_l)) \to H^1(\mathbb{F}_q, J[l]) \overset{\hat{w}_l}{\to} H^1(\mathbb{F}_q, (\mu_l)^b).$$

Let $a = \dim_{\mathbb{F}_l} \ker(w_l)(\mathbb{F}_q)$. We know $b = \dim_{\mathbb{F}_l} J[l](\mathbb{F}_q) = \dim_{\mathbb{F}_l} (\mu_l)^b$. From Proposition 1, we see $\dim_{\mathbb{F}_l} H^1(\mathbb{F}_q, \ker(w_l)) = a$ and $\dim_{\mathbb{F}_l} H^1(\mathbb{F}_q, J[l]) = \dim_{\mathbb{F}_l} H^1(\mathbb{F}_q, (\mu_l)^b) = b$. The dimensions of the seven vector spaces in the long exact sequence are $0, a, b, b, a, b, b$, respectively; so the last two are isomorphic. □

Let $\iota : H^1(\mathbb{F}_q, (\mu_l)^b) \to H^1(\mathbb{F}_q, \mu_l)^b$ be the canonical isomorphism. We have the inverse Kummer isomorphism $H^1(\mathbb{F}_q, \mu_l) \cong \mathbb{F}_q^*/\mathbb{F}_q^{*l}$. Let k denote the composition of ι with the product of b inverse Kummer isomorphisms. So $k : H^1(\mathbb{F}_q, (\mu_l)^b) \to (\mathbb{F}_q^*/\mathbb{F}_q^{*l})^b$ is an isomorphism.

For each $[D_i] \in \mathcal{B}$, choose $f_{D_i} \in \mathbb{F}_q(C)$ with the property that $\mathrm{div}(f_{D_i}) = lD_i$. We call a divisor D of C a *good divisor* if D has degree 0, is defined over $\mathbb{F}_q$ and its support does not intersect the supports of any of the D_i's. Let $D = \sum n_P P$ be a good divisor. We note that f_{D_i} induces an evaluation homomorphism from the group of good divisors to $\mathbb{F}_q^*$.

Lemma 4: *The map f_{D_i} induces a well-defined homomorphism from $J(\mathbb{F}_q)/lJ(\mathbb{F}_q)$ to $\mathbb{F}_q^*/\mathbb{F}_q^{*l}$.*

Proof: Since $C(\mathbb{F}_q)$ is non-empty, every element of $J(\mathbb{F}_q)$ contains a good divisor (see [22, Lemma 3, p. 166]. Let T_1, T_2 be two linearly equivalent good divisors. Then $T_1 - T_2 = \mathrm{div}(h)$ for some $h \in \mathbb{F}_q(C)$. We have

$$f_{D_i}(T_1 - T_2) = f_{D_i}(\mathrm{div}(h)) = h(\mathrm{div}(f_{D_i})) = h(lD_i) = h(D_i)^l \in \mathbb{F}_q^{*l}. \quad \square$$

We have a map $(f_{D_1}, \ldots, f_{D_b})$ from $J(\mathbb{F}_q)/lJ(\mathbb{F}_q)$ to $(\mathbb{F}_q^*/\mathbb{F}_q^{*l})^b$.

Theorem 5: The maps $(f_{D_1}, \ldots, f_{D_b})$ and $k \circ \hat{w}_l \circ \delta_l$ are the same as maps from $J(\mathbb{F}_q)/lJ(\mathbb{F}_q)$ to $(\mathbb{F}_q^*/\mathbb{F}_q^{*l})^b$.

Proof: Let $w_{l,i}$ be the map from $J[l] \rightarrow \mu_l$ given by $[U] \mapsto e_l([U], [D_i])$ and $\hat{w}_{l,i}$ be the map it induces on H^1's. Let k_i be the inverse Kummer isomorphism from the i-th component of $H^1(\mathbb{F}_q, \mu_l)^b$ to the i-th component of $(\mathbb{F}_q^*/\mathbb{F}_q^{*l})^b$. It suffices to show that $f_{D_i} = k_i \circ \hat{w}_{l,i} \circ \delta_l$ on $J(\mathbb{F}_q)/lJ(\mathbb{F}_q)$. Let $[D] \in J(\mathbb{F}_q)$ where D is a good divisor. From Lemma 4, the choice of D is unimportant. From [22, Lemma 3, p. 166], we can choose a degree 0 divisor T over $\overline{K}$, whose support does not intersect the supports of the D_i's, such that $[lT] = [D]$. The class of cocycles $\delta_l([D])$ includes the cocycle $(\sigma \mapsto [^\sigma T - T])$ with $\sigma \in \mathrm{Gal}(\overline{\mathbb{F}_q}/\mathbb{F}_q)$. So $\hat{w}_{l,i} \circ \delta_l([D])$ is the class of cocycles that includes $(\sigma \mapsto e_l([^\sigma T - T], [D_i]))$. The e_l-Weil pairing can be defined as follows. If h_1 and h_2 are functions on C with divisors lU_1 and lU_2 respectively, with disjoint supports, then $e_l([U_1], [U_2]) = h_2(U_1)/h_1(U_2)$. Let $\mathrm{div}(g) = lT - D$ with g defined over the field of definition of the divisor T. We have $\mathrm{div}(^\sigma g) = l^\sigma T - D$. So $\mathrm{div}(^\sigma g/g) = l(^\sigma T - T)$. Recall $\mathrm{div}(f_{D_i}) = lD_i$. We have

$$e_l([^\sigma T - T], [D_i]) = \frac{f_{D_i}(^\sigma T - T)}{(^\sigma g/g)(D_i)} = \frac{^\sigma \beta}{\beta}$$

where $\beta = f_{D_i}(T)/g(D_i)$. So we have

$$k_i \circ \hat{w}_{l,i} \circ \delta_l([D]) \equiv \beta^l \equiv \frac{f_{D_i}(lT)}{g(lD_i)} \equiv \frac{f_{D_i}(lT)}{f_{D_i}(lT - D)} \equiv f_{D_i}(D) \bmod \mathbb{F}_q^{*l}. \quad \square$$

This proof is essentially identical to that of [20, Thm 2.3].

Corollary 6: *The map* $(f_{D_1}, \ldots, f_{D_b})$ *from* $J(\mathbb{F}_q)/lJ(\mathbb{F}_q)$ *to* $(\mathbb{F}_q^*/\mathbb{F}_q^{*l})^b$ *is an isomorphism.*

Proof: This follows from Lemmas 2, 3, the fact that k is an isomorphism and Theorem 5. □

Corollary 7: *The pairing from* $J(\mathbb{F}_q)/lJ(\mathbb{F}_q) \times J[l](\mathbb{F}_q) \to \mathbb{F}_q^*/\mathbb{F}_q^{*l}$ *given by* $\langle [D], [D'] \rangle = f_{D'}(D)$ *is bilinear and non-degenerate.*

Proof: Non-degeneracy follows from Corollary 6. We have linearity on the left from Corollary 6. Taking the divisor is a homomorphism from $\mathbb{F}_{q^m}(C)^*$ to $\mathrm{Div}^0(C)(\mathbb{F}_{q^m})$; so we can extend this pairing to $J[l](\mathbb{F}_q)$ from its basis $\mathcal{B}$. This gives linearity on the right. □

5. Motivation for the pairing using isogenies over the base field

We start with the same set-up as before, except that we no longer assume that $l|q-1$. We will assume that $l^2 \nmid \#J(\mathbb{F}_q)$. For cryptographic purposes, l is usually taken to be a prime, such that $\frac{1}{l}\#J(\mathbb{F}_q)$ is very small compared to l; so this is a reasonable assumption.

Let m be the order of q in $\mathbb{F}_l^*$. We note that $\mathbb{F}_{q^m} = \mathbb{F}_q(\mu_l)$. The q-Frobenius map raises elements of $\overline{\mathbb{F}_q}$ to the q-th power. This induces a map on points of C, and hence on divisors on C and then on elements of J. Let π_q denote the map induced on J by the q-Frobenius map. Let us first show the existence of a subgroup of $J[l]$ on which π_q acts as multiplication by q. If λ is an eigenvalue of π_q on $J[l]$ then we let $J[l]^{(\lambda)}$ denote the λ-eigenspace. Since $\dim_{\mathbb{F}_l} J[l](\mathbb{F}_q) = 1$, we see that $1 (\mathrm{mod}\, l)$ is an eigenvalue of π_q on $J[l]$ and $J[l]^{(1)} = J[l](\mathbb{F}_q)$.

Proposition 8: *There is a subgroup of* $J[l]$ *of order* l*, which is isomorphic as a* $\mathrm{Gal}(\overline{\mathbb{F}_q}/\mathbb{F}_q)$*-module to* μ_l*.*

I thank Joseph Wetherell for suggesting the following proof.

Proof: The zeros of the characteristic polynomial of π_q come in complex conjugate pairs α, q/α. Since $1 (\mathrm{mod}\, l)$ is an eigenvalue with a 1-dimensional eigenspace in $J[l]$, we see that $q (\mathrm{mod}\, l)$ is an eigenvalue (an exercise shows that $l^2 \not| \#J(\mathbb{F}_q)$ implies this eigenspace has dimension 1). So π_q acts on $J[l]^{(q)}$ by multiplying by q, which is the same as the action of π_q on μ_l. □

As a consequence of this proposition, we note that $J[l]^{(q)}$ is a subgroup of $J(\mathbb{F}_q(\mu_l)) = J(\mathbb{F}_{q^m})$.

We will describe a pairing γ from $J[l]^{(1)} \times J[l]^{(q)}$ to $\mathbb{F}_{q^m}^*/\mathbb{F}_{q^m}^{*l}$. Let D and D_1 be degree 0 divisors over $\mathbb{F}_q$ and $\mathbb{F}_{q^m}$, respectively, with disjoint supports such that $[D] \in J[l]^{(1)}$ and $[D_1] \in J[l]^{(q)}$. Choose $f_{D_1} \in \mathbb{F}_{q^m}(C)$ such that $\mathrm{div}(f_{D_1}) = lD_1$.

Let $\gamma([D],[D_1]) = f_{D_1}(D) \in \mathbb{F}_{q^m}^*/\mathbb{F}_{q^m}^{*l}$.

Theorem 9: The pairing γ is bilinear and non-degenerate.

Note that γ pairs two 1-dimensional $\mathbb{F}_l$-vector spaces, which is elegant.

In order to prove this theorem, we first present some lemmas. Let $\hat{A}$ be the abelian variety $J/J[l]^{(q)}$ and let $\hat{\phi} : J \to \hat{A}$ be the natural quotient map (thus $J[\hat{\phi}]$, the kernel of $\hat{\phi}$ on $J(\overline{\mathbb{F}_q})$, is $J[l]^{(q)}$). We note $\hat{\phi}$ is an isogeny of degree l defined over $\mathbb{F}_q$. Let $\phi : A \to \hat{J}$ be the dual isogeny. Let $\lambda : J \to \hat{J}$ be the canonical principal polarization on $J/\mathbb{F}_q$ coming from J being the Jacobian of $C/\mathbb{F}_q$. By abuse of notation, we denote the composition $\lambda^{-1} \circ \phi$ by $\phi : A \to J$.

Lemma 10: *We have* $\#J(\mathbb{F}_q)/\phi A(\mathbb{F}_q) = l$.

Proof: From Cartier duality, since $J[\hat{\phi}]$ is isomorphic to μ_l as a $\mathrm{Gal}(\overline{\mathbb{F}_q}/\mathbb{F}_q)$-module, we have that $A[\phi]$ is isomorphic to $\mathbb{Z}/l\mathbb{Z}$ as a $\mathrm{Gal}(\overline{\mathbb{F}_q}/\mathbb{F}_q)$-module. Thus $\#A[\phi](\mathbb{F}_q) = l$. The following sequence is exact

$$0 \to A[\phi](\mathbb{F}_q) \to A(\mathbb{F}_q) \xrightarrow{\phi} J(\mathbb{F}_q) \to J(\mathbb{F}_q)/\phi A(\mathbb{F}_q) \to 0.$$

Since A and J are isogenous over $\mathbb{F}_q$, we have $\#A(\mathbb{F}_q) = \#J(\mathbb{F}_q)$. Thus $\#A[\phi](\mathbb{F}_q)$ and $\#J(\mathbb{F}_q)/\phi A(\mathbb{F}_q)$ are the same. □

From this lemma we see that the quotient map $J[l]^{(1)} = J[l](\mathbb{F}_q) \to J(\mathbb{F}_q)/\phi A(\mathbb{F}_q)$ is an isomorphism. As we will see, γ should really be thought of as a pairing on $J(\mathbb{F}_q)/\phi A(\mathbb{F}_q) \times J[l]^{(q)}$.

The following is an exact sequence of $\mathrm{Gal}(\overline{\mathbb{F}_q}/\mathbb{F}_q)$-modules

$$0 \to A[\phi] \to A(\overline{\mathbb{F}_q}) \xrightarrow{\phi} J(\overline{\mathbb{F}_q}) \to 0.$$

Taking $\mathrm{Gal}(\overline{\mathbb{F}_q}/\mathbb{F}_q)$-invariants gives us $J(\mathbb{F}_q)/\phi A(\mathbb{F}_q) \overset{\delta_\phi}{\hookrightarrow} H^1(\mathbb{F}_q, A[\phi])$.

Lemma 11: *The restriction map induces an isomorphism of* $H^1(\mathbb{F}_q, A[\phi])$ *and* $H^1(\mathbb{F}_{q^m}, A[\phi])$.

We let r_ϕ denote this restriction map.

Proof: We have the following inflation/restriction exact sequence

$$\begin{aligned} 0 \to H^1(\mathrm{Gal}(\mathbb{F}_{q^m}/\mathbb{F}_q), A[\phi](\mathbb{F}_{q^m})) &\overset{\inf}{\to} H^1(\mathbb{F}_q, A[\phi]) \overset{r_\phi}{\to} \\ \to H^1(\mathbb{F}_{q^m}, A[\phi])^{\mathrm{Gal}(\mathbb{F}_{q^m}/\mathbb{F}_q)} &\to H^2(\mathrm{Gal}(\mathbb{F}_{q^m}/\mathbb{F}_q), A[\phi]). \end{aligned} \tag{2}$$

Since m is the order of q in the group $\mathbb{F}_l^*$, we see $m \mid l-1$. So $\gcd(m,l)=1$. Since m and l are the exponents of the groups $\mathrm{Gal}(\mathbb{F}_{q^m}/\mathbb{F}_q)$ and $A[\phi]$, respectively, we have $H^i(\mathrm{Gal}(\mathbb{F}_{q^m}/\mathbb{F}_q), A[\phi]) = 0$ for all i. In addition, since $\mathrm{Gal}(\mathbb{F}_{q^m}/\mathbb{F}_q)$ acts trivially on $A[\phi]$ we have $H^1(\mathbb{F}_{q^m}, A[\phi])^{\mathrm{Gal}(\mathbb{F}_{q^m}/\mathbb{F}_q)} = H^1(\mathbb{F}_{q^m}, A[\phi])$. □

We now create an isomorphism of $H^1(\mathbb{F}_{q^m}, A[\phi])$ with another cohomology group. Let D_1 be a degree 0 divisor on C over $\mathbb{F}_{q^m}$, with the property that the divisor class $[D_1]$ generates $J[l]^{(q)} = J[\hat{\phi}]$.

Define the map $w_\phi : A[\phi] \to \mu_l$ by $w_\phi(R) = e_\phi(R, [D_1])$, where e_ϕ denotes the ϕ-Weil pairing. The map w_ϕ is an isomorphism of $\mathrm{Gal}(\overline{\mathbb{F}_{q^m}}/\mathbb{F}_{q^m})$-modules. Let $\hat{w}_\phi$ be the map induced on cohomology.

Lemma 12: *The map* $\hat{w}_\phi$ *is an isomorphism from* $H^1(\mathbb{F}_{q^m}, A[\phi])$ *to* $H^1(\mathbb{F}_{q^m}, \mu_l)$.

Proof: Isomorphisms of $\mathrm{Gal}(\overline{\mathbb{F}_{q^m}}/\mathbb{F}_{q^m})$-modules induce isomorphisms on cohomology. □

Let k denote the inverse Kummer isomorphism from $H^1(\mathbb{F}_{q^m}, \mu_l)$ to $\mathbb{F}_{q^m}^*/\mathbb{F}_{q^m}^{*l}$. Choose $f_{D_1} \in \mathbb{F}_{q^m}(C)$ such that $\mathrm{div}(f_{D_1}) = lD_1$.

Lemma 13: *The map* f_{D_1} *induces a well-defined homomorphism from* $J(\mathbb{F}_q)/lJ(\mathbb{F}_q)$ *to* $\mathbb{F}_{q^m}^*/\mathbb{F}_{q^m}^{*l}$.

The proof of this lemma is essentially identical to that of Lemma 4.

Lemma 14: *The maps* f_{D_1} *and* $k \circ \hat{w}_\phi \circ r_\phi \circ \delta_\phi$ *are the same from* $J(\mathbb{F}_q)/\phi A(\mathbb{F}_q)$ *to* $\mathbb{F}_{q^m}^*/\mathbb{F}_{q^m}^{*l}$.

Proof: Let r_l be the restriction map from $H^1(\mathbb{F}_q, J[l])$ to $H^1(\mathbb{F}_{q^m}, J[l])$. From Theorem 5 we know that f_{D_1} and $k \circ \hat{w}_{l,1} \circ r_l \circ \delta_l$ are the same from

$J(\mathbb{F}_q)/lJ(\mathbb{F}_q)$ to $\mathbb{F}_{q^m}^*/\mathbb{F}_{q^m}^{*l}$. There is an isogeny $\tau : J \rightarrow A$ with $\phi \circ \tau = l$. From the commutative diagram

$$\begin{array}{ccccccc} 0 \rightarrow & J[l] & \rightarrow & J(\overline{\mathbb{F}_q}) & \xrightarrow{l} & J(\overline{\mathbb{F}_q}) & \rightarrow 0 \\ & \downarrow \tau & & \downarrow \tau & & \downarrow 1 & \\ 0 \rightarrow & A[\phi] & \rightarrow & A(\overline{\mathbb{F}_q}) & \xrightarrow{\phi} & J(\overline{\mathbb{F}_q}) & \rightarrow 0 \end{array}$$

we get the following commutative diagram by taking $\mathrm{Gal}(\overline{\mathbb{F}_q}/\mathbb{F}_q)$-invariants

$$\begin{array}{ccc} J(\mathbb{F}_q)/lJ(\mathbb{F}_q) & \xrightarrow{\delta_l} & H^1(\mathbb{F}_q, J[l]) \\ \downarrow & & \downarrow \tau \\ J(\mathbb{F}_q)/\phi A(\mathbb{F}_q) & \xrightarrow{\delta_\phi} & H^1(\mathbb{F}_q, A[\phi]). \end{array}$$

From the compatibility of Weil pairings we have $e_l([D],[D_1]) = e_\phi(\tau([D]),[D_1])$. Thus the triangle of the following diagram commutes and so the whole diagram commutes

$$\begin{array}{ccccccc} J(\mathbb{F}_q)/lJ(\mathbb{F}_q) & \xrightarrow{\delta_l} & H^1(\mathbb{F}_q, J[l]) & \xrightarrow{r_l} & H^1(\mathbb{F}_{q^m}, J[l]) \searrow \hat{w}_{l,1} & & \\ \downarrow & & \downarrow & & \downarrow \tau & H^1(\mathbb{F}_{q^m}, \mu_q) & \xrightarrow{k} \mathbb{F}_{q^m}^*/\mathbb{F}_{q^m}^{*l}. \\ J(\mathbb{F}_q)/\phi A(\mathbb{F}_q) & \xrightarrow{\delta_\phi} & H^1(\mathbb{F}_q, A[\phi]) & \xrightarrow{r_\phi} & H^1(\mathbb{F}_{q^m}, A[\phi]) \nearrow \hat{w}_\phi & & \end{array}$$

From commutativity, f_{D_1} must factor through $\phi A(\mathbb{F}_q)$ and f_{D_1} and $k \circ \hat{w}_\phi \circ r_\phi \circ \delta_\phi$ are the same as maps from $J(\mathbb{F}_q)/\phi A(\mathbb{F}_q)$ to $\mathbb{F}_{q^m}^*/\mathbb{F}_{q^m}^{*l}$. □

This proof is essentially identical to that of [20, Thm 2.3].

Lemma 15: *The map f_{D_1} from $J(\mathbb{F}_q)/\phi A(\mathbb{F}_q)$ to $\mathbb{F}_{q^m}^*/\mathbb{F}_{q^m}^{*l}$ is an isomorphism.*

Proof: The map δ_ϕ is an injection. From Lemmas 11 and 12 and the fact that k is an isomorphism, we see that the composition $k \circ \hat{w}_\phi \circ r_\phi \circ \delta_\phi$ is an injective homorphism. From Lemma 10 we have $\#J(\mathbb{F}_q)/\phi A(\mathbb{F}_q) = l$ and since $\mathbb{F}_{q^m}^*$ is cyclic we have $\#\mathbb{F}_{q^m}^*/\mathbb{F}_{q^m}^{*l} = l$. Thus this injection must be an isomorphism. From Lemma 14, the composition is the same as f_{D_1}. □

We can now finish the proof of Theorem 9.

Proof: We have linearity on the left from Lemma 15. Taking the divisor is a homomorphism from $\mathbb{F}_{q^m}(C)^*$ to $\mathrm{Div}^0(C)(\mathbb{F}_{q^m})$. This gives linearity on the right. Non-degeneracy also follows from Lemma 15. □

6. Applications to elliptic curves

An interesting corollary of Proposition 8 is the following.

Corollary 16: *Let $E/\mathbb{F}_q$ be an elliptic curve. Let $l \nmid q$ be prime with $l | \#E(\mathbb{F}_q)$ and $l^2 \nmid \#E(\mathbb{F}_q)$. Assume $\mu_l \not\subseteq \mathbb{F}_q$ (i.e. $l \nmid q-1$). Then $E[l] \subseteq E(\mathbb{F}_q(\mu_l))$.*

Now let us explain what seems to be a surprising fact, which applies, for example, to trace 2 elliptic curves (i.e. where $\#E(\mathbb{F}_q) = \#\mathbb{F}_q^*$). Assume E is defined over $\mathbb{F}_q$, $l | \#E(\mathbb{F}_q)$, $l^2 \nmid \#E(\mathbb{F}_q)$ and $l | q-1$ (so $\mu_l \subseteq \mathbb{F}_q$). Let D be a degree 0 divisor of E over $\mathbb{F}_q$ such that $[D]$ generates $E[l](\mathbb{F}_q)$. If $f_D \in \mathbb{F}_q(E)$ has divisor lD then f_D induces an isomorphism of $E[l](\mathbb{F}_q)$ (which is generated by $[D]$) and $\mathbb{F}_q^*/\mathbb{F}_q^{*l}$. At first glance, this might seem to contradict the fact that the l-Weil pairing of an element with itself is trivial. One might imagine that one needs to use $f_{D'}$ for some $[D'] \in E[l] \setminus E[l](\mathbb{F}_q)$. There is no contradiction, however. What is happening is better explained by considering isogenies.

Let $A = E/\big(E[l](\mathbb{F}_q)\big)$ and $\hat{\phi}$ be the induced isogeny from E to A (note A is an elliptic curve also). So $E[\hat{\phi}] = E[l](\mathbb{F}_q)$. Let $\phi : A \rightarrow E$ be the dual isogeny (here we identify the elliptic curves with their dual abelian varieties as in [6]). Though $[D]$ is an element of $E[l](\mathbb{F}_q)$, it should be thought of as representing an element of $E(\mathbb{F}_q)/\phi A(\mathbb{F}_q)$. Let us show that $\#E(\mathbb{F}_q)/\phi A(\mathbb{F}_q) = l$. As $\mathrm{Gal}(\overline{\mathbb{F}_q}/\mathbb{F}_q)$-modules, we have $E[\hat{\phi}] \cong \mathbb{Z}/l\mathbb{Z}$; so by Cartier duality we have $A[\phi] \cong \mu_l$. But since $\mu_l \subseteq \mathbb{F}_q$ we have $\mu_l \cong \mathbb{Z}/l\mathbb{Z}$ as $\mathrm{Gal}(\overline{\mathbb{F}_q}/\mathbb{F}_q)$-modules. Since $A[\phi] \cong \mathbb{Z}/l\mathbb{Z}$ we see that $E(\mathbb{F}_q)/\phi A(\mathbb{F}_q)$ has size l. Since $l^2 \nmid \#E(\mathbb{F}_q)$, we see that $[D]$ generates $E(\mathbb{F}_q)/\phi A(\mathbb{F}_q)$.

The image of $[D] \in E(\mathbb{F}_q)/\phi A(\mathbb{F}_q)$ under δ_ϕ lands in $H^1(\mathbb{F}_q, A[\phi])$ and is a class of cocycles including, say, ξ. The image of ξ in $A[\phi]$ is then ϕ-Weil paired with $[D]$, a chosen generator of $E[\hat{\phi}]$. Since $\dim_{\mathbb{F}_l} A[\phi] = \dim_{\mathbb{F}_l} E[\hat{\phi}] = 1$, the ϕ-Weil pairing of any two non-trivial elements is non-trivial. From Lemma 15, this gives the isomorphism of $E[l](\mathbb{F}_q)$ and $\mathbb{F}_q^*/\mathbb{F}_q^{*l}$ by f_D.

References

1. Brown, K.S. *Cohomology of Groups*, Springer-Verlag, New York, 1982
2. Frey, G. and Rück, H.-G., *A remark concerning m-divisibility and the discrete logarithm in the divisor class group of curves*, Math. Comp. **62**, (1994), 865–874.
3. Lang, S. *Abelian Varieties*, Interscience Publishers, New York, 1959.
4. Schaefer, E.F., *Computing a Selmer group of a Jacobian using functions on the curve*, Math. Ann. **310**, (1998), 447–471.

5. Serre, J.-P., *Cohomologie Galoisienne*, Springer-Verlag, New York, 1973.
6. Silverman, J.H., *The Arithmetic of Elliptic Curves*, Springer-Verlag, New York, 1986.

Elliptic curve torsion points and division polynomials

Iftikhar A. Burhanuddin and Ming-Deh A. Huang

Department of Computer Science,
University of Southern California,
Los Angeles, CA 90089, USA.
burhanud,huang@usc.edu

We present two algorithms - p-adic and l-adic - to determine $E(\mathbb{Q})_{tors}$ the group of rational torsion points on an elliptic curve. Another algorithm we introduce is one which decides whether an elliptic curve over $\mathbb{Q}_p$ has a non-trivial p-torsion part and this comes into play in the p-adic torsion computation procedure. We also make some remarks about the discriminant of the m-division polynomial of an elliptic curve and the information it reveals about torsion points.

1. Introduction

The Mordell-Weil theorem says that given an elliptic curve E over a number field K, the group of K-rational points $E(K)$ is finitely generated. This implies that the group of K-rational torsion $E(K)_{tors}$ is finite. A theorem of Mazur states the groups which can appear as $E(K)_{tors}$, when $K = \mathbb{Q}$. The purpose of this paper is to introduce methods which efficiently compute elliptic curve rational torsion.

We begin by briefly recalling the current approaches to determine $E(\mathbb{Q})_{tors}$. Firstly, one can compute torsion in a brute force fashion using the Nagell-Lutz theorem, which states that torsion points are integral and bounded in magnitude, but this technique can be computationally expensive. This naive method was superseded by Doud's complex analytic cubic time algorithm [5], where the input length is the size of the coefficients of the elliptic curve.

Garcia-Selfa et al [7] proposed a softly quadratic time algorithm ("softly" refers to the fact that sub-linear factors are ignored), where they compute with the Tate Normal Form of an elliptic curve. Their procedure uses Loos' root-finding algorithm as a blackbox routine and does not use any information about how the discriminants of F_m (polynomials which arise in their algorithm) are related to the discriminant of the elliptic curve. And hence a different prime is selected to compute the roots of F_m for each m.

In §4 we devise a polynomial-time algorithm (polynomial in $\log p$ and the size of the discriminant of the curve) that decides whether a given elliptic curve over $\mathbb{Q}_p$ has a non-trivial p-torsion part. The algorithm has two subroutines, the first procedure computes $\#E_0(\mathbb{Q}_p)[p]$ and the second determines $\#E(\mathbb{Q}_p)[p]$ when E has split multiplicative reduction. The triviality of this group would imply the triviality of $E(\mathbb{Q})[p]$ and therefore this decisional procedure finds its way into our rational torsion computation algorithm.

The roots of division polynomials correspond to torsion points of the elliptic curve and §2 introduces these well-studied polynomials. Our algorithms essentially perform root finding on these polynomials. We introduce in §6 an algorithm to compute the p-torsion part using a p-adic approach (except when $p = 2$, which is discussed in §5). This algorithm has a worst case (determinstic or expected) time complexity which is softly quadratic in the size of the discriminant of the elliptic curve. An l-adic algorithm is devised in §8 to compute $E(\mathbb{Q})_{tors}$ with a worst case softly quadratic running time. A randomized avatar of this method runs expectedly in softly linear number of bit operations.

The basic idea of our algorithms is given an elliptic curve E over $\mathbb{Q}$ we view it as a curve over $\mathbb{Q}_l$ and use Hensel lifting (whenever it is efficient) to compute $E(\mathbb{Q}_l)[m]$, the $\mathbb{Q}_l$-rational m-torsion points, to desired precision ($m = p = l$ in the p-adic approach). The values of m we investigate are dictated by Mazur's result and the sufficient precision to work with is supplied by the Nagell-Lutz theorem. We then check to see if these points are in $E(\mathbb{Q})[m]$, the group of m-torsion rational points on E. We discuss time complexity analysis of the above torsion computation procedures in §9.

In the l-adic algorithm the choice of the prime l rests on the fact that the prime support of the m-division polynomial equals the prime support of m and the prime support of the discriminant of the elliptic curve, which we prove in §7.1. This relationship between the discriminants enables us to use a single "good" prime to compute the m-torsion for all m. In order to relate $\Delta(f_m)$ the discriminant of f_m the m-division polynomial to Δ the discriminant of the elliptic curve, we symbolically computed the discriminants of these polynomials using Magma for small values of m. This led us to discover a formula for $\Delta(f_m)$. In §7.2 we establish the equivalence of this formula when m is odd to a lemma of Stark [19]. Finally, we discuss current and future directions of research in §10.

We would like to thank S. Kamienny and W. Raskind for insightful discussions about the arithmetic of elliptic curves. The authors are grateful to P. Gaudry for giving us pointers to fast p-adic computation procedures, to H. Stark for providing us with a reference to his result and to W. Stein for spurring us to think about elliptic curves with split multiplicative reduction in terms of the Tate curve. We were supported in part by the following NSF grants CCR-9820778 and CCR-0306393.

2. Division Polynomials

Let K be a number field and $\overline{K}$ be an algebraic closure of K. Let E be an elliptic curve over K given by a Weierstrass equation of the form $y^2 = x^3 + ax + b$, where $a, b \in R$, where R is the ring of integers of K.

We begin by presenting definitions and theorems concerning torsion points and polynomials which characterize them. Define division polynomials Ψ_m recursively as follows:

$$\begin{aligned}
&\Psi_1 = 1,\ \Psi_2 = 2y,\ \Psi_3 = 3x^4 + 6ax^2 + 12bx - a^2,\\
&\Psi_4 = (2x^6 + 10ax^4 + 40bx^3 - 10a^2x^2 - 8bax - 2a^3 - 16b^2)\Psi_2,\\
&\Psi_{2k+1} = \Psi_{k+2}\Psi_k^3 - \Psi_{k-1}\Psi_{k+1}^3, k \geq 2\\
&\Psi_{2k} = (\Psi_{k+2}\Psi_{k-1}^2 - \Psi_{k-2}\Psi_{k+1}^2)\Psi_k/\Psi_2, k \geq 2.
\end{aligned}$$

Define for $m > 2$, $f_m = \Psi_m$, when m is odd and $f_m = \Psi_m/\Psi_2$, when m is even. Observe that f_m (also referred to as division polynomials) are univariate. Let $d = \deg f_m$, which is equal to $\frac{m^2-1}{2}$, m is odd and $\frac{m^2-4}{2}$ otherwise. The leading coefficient of f_m is m, when m is odd and $m/2$ when m is even.

The x-coordinates of the m-torsion points of E correspond to the roots of f_m in the following way [1, Corollary III.7]: Let $P \in E(\overline{K})$, such that P is not a 2-torsion point then $P \in E(\overline{K})[m] \Leftrightarrow f_m(x(P)) = 0$. Recall that $E(\overline{K})[m] = (\mathbb{Z}/m\mathbb{Z}) \times (\mathbb{Z}/m\mathbb{Z})$ [17, Corollary 6.4b].

We will now define the discriminant of a polynomial and related notions [3, §3.3.2]. Let S be an integral domain with quotient field L and $\overline{L}$ be an algebraic closure of L. Let $g \in S[X]$ with $n = \deg(g)$, $lc(g)$ be its leading coefficient and α_i be the roots of g in $\overline{L}$. Define the discriminant of g to be

$$\Delta(g) = lc(g)^{n-1+deg(g')} \prod_{1 \leq i < j \leq n} (\alpha_i - \alpha_j)^2.$$

Let $f = x^3 + ax + b$ and hence $\Delta(f) = -(4a^3 + 27b^2)$. The discriminant of the elliptic curve is defined to be $\Delta(E) = -16(4a^3 + 27b^2)$ (also denoted as Δ). From here on we will assume that E is an elliptic curve over $\mathbb{Q}$ given by $y^2 = x^3 + ax + b$, with $a, b \in \mathbb{Z}$. This implies that the division polynomials have integral coefficients and hence their discriminants $\Delta(f_m) \in \mathbb{Z}$, which is clear from the definition of the discriminant in terms of the Sylvester matrix [3, Lemma 3.3.4].

3. $E(\mathbb{Q})_{tors}$ and Hensel's lemma

The purpose of this section is to present some background material [17, Chapter VIII] before we introduce our elliptic curve rational torsion algorithm. A corollary of the Mordell-Weil Theorem states that $E(\mathbb{Q})_{tors}$ is a finite group. To determine this group the methods which are currently in use are guided by the following theorems.

Nagell-Lutz Theorem: Let E be an elliptic curve over $\mathbb{Q}$ given by

$$y^2 = x^3 + ax + b, \ \ a, b \in \mathbb{Z}.$$

Suppose $O \neq P \in E(\mathbb{Q})_{tors}$ then $x(P), y(P) \in \mathbb{Z}$ and either $y(P) = 0$ or $y(P)^2 | (4a^3 + 27b^2)$.

Mazur's Theorem:

$$E(\mathbb{Q})_{tors} \cong \begin{cases} \mathbb{Z}/n\mathbb{Z}, & 1 \leq n \leq 10, 12 \quad \text{case (i)}, \\ \mathbb{Z}/2\mathbb{Z} \times \mathbb{Z}/2n\mathbb{Z}, & 1 \leq n \leq 4 \quad \text{case (ii)}. \end{cases}$$

Observe that an elliptic curve over $\mathbb{Q}$ can be transformed into the form which appears in the Nagell-Lutz theorem using a change of coordinates [17, Pages 46-50].

Suppose $P \in E(\mathbb{Q})_{tors} \setminus E(\mathbb{Q})[2]$ then $x(P)$ will be a root of $x^3 + ax + b - y(P)^2$. The Nagell-Lutz theorem tells us that $y(P)^2 | (4a^3 + 27b^2)$ which implies $x(P) | (b - (4a^3 + 27b^2)/k)$ for some $k \in \mathbb{Z}$. If $P \in E(\mathbb{Q})[2]$ and non-trivial then reasoning similar to the above leads to $x(P) | b$ since $y(P) = 0$. Hence the coordinates of the torsion points are $O(C)$ in magnitude, where $C = \max(|a|^3, |b|^2)$.

The brute-force approach to compute torsion is to try out all the possible values for $y(P)^2$ such that it divides $4a^3 + 27b^2$. In the worst case this is computationally expensive as it involves factoring and also $4a^3 + 27b^2$ might have many square divisors giving rise to many possibilities [5].

Instead our algorithms, l-adic and p-adic perform root finding on division polynomials using the following variant of the Hensel's lemma [6, Lemma 2.1]: Let $u \in \mathbb{Z}_p$ and $h \in \mathbb{Z}_p[x]$. Let k be such that $p^k || h'(u)$ and assume $p^{n+k}|h(u)$ for some $n > k$. Let

$$\delta = \frac{p^{-k}h(u)}{p^{-k}h'(u)}$$

and $v = u - \delta$. Then $v \equiv u \bmod p^n$, $p^{2n}|h(v)$ and $p^k||h'(v)$.

Hensel's lemma leads to an efficient (softly linear time) method to lift points provided the roots of the polynomial separate early enough (k is bounded by a constant or log of the required p-adic precision), which would ensure that the initialization procedure (where the root is computed modulo p^k) does not take more than softly-linear time. When the roots of the polynomial fail to separate quickly, we resort to a p-adic polynomial factorization algorithm [15]. This factorization procedure takes as input a monic polynomial and hence we will have to use the standard trick of multiplying f_m by m^{d-1} and make a change of variables replacing $m \cdot x$ by x to render it monic.

In the torsion computation algorithms both l-adic and p-adic which follow, before the output is returned we need to check whether a $E(\mathbb{Q}_l)$ torsion point calculated to appropriate precision is in $E(\mathbb{Q})_{tors}$. This is possible because a priori we know the magnitude of the rational torsion points. Also negative integers (which are an infinite power series) are detected by noticing a recurring $l-1$ in the truncated l-adic expansion. They can be recovered as follows: $\sum_{i=0}^{\infty} a_i l^i = -(l - a_0 + \sum_{i=1}^{\log_l C}(l-1-a_i)l^i)$, where the left hand side represents the negative integer as an l-adic number.

4. Deciding whether $E(\mathbb{Q}_p)[p]$ is non-trivial

We know that $E(\overline{\mathbb{Q}}_l)[p] = \mathbb{Z}/p\mathbb{Z} \times \mathbb{Z}/p\mathbb{Z}$. Suppose $p > 2$ and $\#E(\mathbb{Q}_l)[p] = p^2$ then the Weil Pairing [17, Corollary III.8.1.1] would imply $\mu_p \subset \mathbb{Q}_l^*$, which would hold if $p|l-1$, a contradiction. Therefore in particular $\#E(\mathbb{Q}_p)[p] = 1, p$.

The question of efficiently determining whether $\#E(\mathbb{Q}_p)[p] = p$ is one of independent interest. Our motivation to look at this problem is by viewing an elliptic curve E over $\mathbb{Q}$ as an elliptic curve over $\mathbb{Q}_p$, $E(\mathbb{Q}_p)[p]$ being trivial would imply $E(\mathbb{Q})[p]$ is trivial (since the latter injects into the former). And this is useful information in $E(\mathbb{Q})_{tors}$ computing procedures as we shall see in the coming sections.

The algorithms presented in this section work with an elliptic curve over $\mathbb{Q}_p$. We will assume that we are presented with an elliptic curve E over $\mathbb{Q}$ and a prime $p > 2$. We make this choice to simplify the time complexity and p-adic precision analysis of the algorithms (otherwise in the worst case

- split multiplicative reduction and $p|v_p(\Delta)$ - we will require as input the coefficients of the curve over $\mathbb{Q}_p$ to p-digits of p-adic accuracy, where v_p is the p-adic valuation of $\mathbb{Q}_p$). Given an elliptic curve over the rationals we will use Tate's algorithm [18, Chapter IV.9] to compute the minimal Weierstrass equation of E at p. And by abuse of notation we will denote by E both the original elliptic curve and its minimal Weierstrass equation at p. Also we will refer to the associated discriminants of the former and latter as Δ and hopefully what we mean will be clear from the context.

It is straightforward to determine the type of reduction modulo p of the elliptic curve using Δ the minimal discriminant at p. Let $y^2 = x^3 + ax + b$ be the minimal Weierstrass equation of E an elliptic curve over $\mathbb{Q}_p$, where $p > 3$ then E is said to have good (bad) reduction at p if $v_p(\Delta) = 0$ ($v_p(\Delta) > 0$). On the other hand, the type of bad reduction – multiplicative or additive – can be determined as follows: E has multiplicative reduction if and only if $v_p(\Delta) \geq 1$ and $v_p(ab) = 0$ and it has additive reduction if and only if $v_p(a), v_p(b) \geq 1$ [17, Exercise VII.7.1(b)]. Also E has split multiplicative reduction if and only if $-2\overline{ab}$ is a square in $\mathbb{F}_p$ (otherwise E has non-split multiplicative reduction) [13, Page 27]. Other local information (for example $\#E(\mathbb{Q}_p)/E_0(\mathbb{Q}_p)$) is obtained using Tate's algorithm.

The proof of following theorem will keep us occupied for the remainder of this section:

Theorem 17: There exists an algorithm which takes as input an elliptic curve over $\mathbb{Q}$ and a prime $p > 2$ and decides whether $\#E(\mathbb{Q}_p)[p] = p$. It has a worst case time complexity which is polynomial in the size of p and the size of the discriminant of the elliptic curve.

4.1. *Computing* $\#E_0(\mathbb{Q}_p)[p]$

This following algorithm determines $\#E_0(\mathbb{Q}_p)[p]$, in other words computes $\#E(\mathbb{Q}_p)[p]$ when there are no p-torsion points which reduce to a singular point.

Algorithm 1: Let E be an elliptic curve over $\mathbb{Q}_p$ given by a minimal Weierstrass equation, where $p > 2$.

Input. We are given the coefficients of E, modulo p^2 and the type of reduction.

Output. TRUE if $\#E_0(\mathbb{Q}_p)[p] = p$ and FALSE if $\#E_0(\mathbb{Q}_p)[p] = 1$.

(1) $n \leftarrow \#\overline{E}_{ns}(\mathbb{F}_p)$.
(2) If $p \nmid n$ return FALSE.

(3) Pick a non-trivial point $\overline{P} \in \overline{E}_{ns}(\mathbb{F}_p)[p]$.
(4) Lift $\overline{P}$ to $P \in E_0(\mathbb{Q}_p) \setminus E_1(\mathbb{Q}_p)$ using Hensel's lemma such that $P \equiv \overline{P} \bmod p$. We only need to determine $x(P)$ modulo p^2.
(5) Compute $x([p-1]P) \bmod p^2$ using the repeated squaring trick [10, Page 23] and the elliptic curve group law formulae [17, Algorithm 2.3].
(6) If $x([p-1]P) \equiv x(P) \bmod p^2$ return TRUE. Otherwise return FALSE.

Lemma 18: *The above algorithm works as desired.*

Proof: First we recall a fact about the structure of $\overline{E}_{ns}(\mathbb{F}_p)$ in the case of bad reduction [17, Exercise III.3.5]: $\overline{E}_{ns}(\mathbb{F}_p) \cong \mathbb{F}_p^+$, $\mathbb{F}_p^*$ or $\{t \in L^* \mid N_{L/\mathbb{F}_p}(t) = 1\}$ where $L = \mathbb{F}_p(\alpha_1, \alpha_2)$ and α_1, α_2 are the slopes of tangent lines in the non-split multiplicative reduction case.

Note that $E_1(\mathbb{Q}_p) \cong \hat{E}(p\mathbb{Z}_p) \cong \hat{G}_a(p\mathbb{Z}_p)$ is torsion-free. Now the short exact sequence $0 \to E_1(\mathbb{Q}_p) \to E_0(\mathbb{Q}_p) \to \overline{E}_{ns}(\mathbb{F}_p) \to 0$ [17, Proposition VII.2.1] gives rise to the following long exact sequence via the extended snake lemma [12, Lemma II.4.1]:

$$0 \to E_0(\mathbb{Q}_p)[p] \to \overline{E}_{ns}(\mathbb{F}_p)[p] \xrightarrow{\phi} \hat{G}_a(p\mathbb{Z}_p)/p\hat{G}_a(p\mathbb{Z}_p) \to E_0(\mathbb{Q}_p)/pE_0(\mathbb{Q}_p) \to \overline{E}_{ns}(\mathbb{F}_p)/p\overline{E}_{ns}(\mathbb{F}_p) \to 0$$

If $\gcd(n,p) = 1$ then $\overline{E}_{ns}(\mathbb{F}_p)[p] = 0$ which implies that $E_0(\mathbb{Q}_p)[p] = 0$. This case takes care of split multiplicative reduction as we have $\#\overline{E}_{ns}(\mathbb{F}_p) = \#\mathbb{F}_p^* = p-1$ and of the non-split case as we have $\#\overline{E}_{ns}(\mathbb{F}_p) = 1, p-1, p+1, p^2-1$.

If $\gcd(n,p) \neq 1$ (due to good reduction or additive reduction) and we pick a point $P \neq O$ in $\overline{E}_{ns}(\mathbb{F}_p)[p]$, then appealing to the lemma below tells us that $E_0(\mathbb{Q}_p)[p]$ being non-trivial is equivalent to $x([p-1]P) \equiv x(P) \bmod p^2$. Now we observe that when we compute $x([p-1]P)$ by the squaring trick, the denominators are p-adic units (part (2) of the lemma) and the group law formulae hold modulo p^2. This suggests that only the coefficients of elliptic curve E and of the coordinates of the point P modulo p^2 contribute towards the computation. This completes the proof of the theorem. □

Lemma 19: *Let E be an elliptic curve over $\mathbb{Q}_p$ given by a minimal Weierstrass equation. If $Q \in E_0(\mathbb{Q}_p) \setminus E_1(\mathbb{Q}_p)$ and $\overline{Q} \in \overline{E}_{ns}(\mathbb{F}_p)[p]$ then*

(1) $[i]Q \in E_0(\mathbb{Q}_p) \setminus E_1(\mathbb{Q}_p)$, $i = 1, \ldots, p-1$,
(2) $x([i]Q) \not\equiv x([j]Q) \bmod p$, $0 < j < i < p$ *and* $i + j < p$,
(3) $x([p-k]Q) \equiv x([k]Q) \bmod p$ *and* $y([p-k]Q) \equiv -y([k]Q) \bmod p$ *(in particular* $y([p-k]Q) \not\equiv y([k]Q) \bmod p$*),* $0 < k < p$,

(4) $[p]Q \in E_i(\mathbb{Q}_p) \setminus E_{i+1}(\mathbb{Q}_p) \Leftrightarrow v_p(x([p]Q)) = -2i \Leftrightarrow v_p(x([p-1]Q) - x(Q)) = i,\ i \geq 1$,

(5) $x([p-1]Q) - x(Q) \equiv 0 \bmod p^2 \Leftrightarrow \phi = 0$, *where* $\phi : \overline{E}_{ns}(\mathbb{F}_p)[p] \to \hat{G}_a(p\mathbb{Z}_p)/p\hat{G}_a(p\mathbb{Z}_p)$.

Proof:

(1) Suppose $[i]Q \in E_1(\mathbb{Q}_p)$ then $[i]\overline{Q} = O$ which is a contradiction since $\gcd(i,p) = 1$.

(2) Suppose $x([i]Q) \equiv x([j]Q) \bmod p$. This assumption combined with the fact that $[i]\overline{Q}, [j]\overline{Q} \neq O$ implies that $[i]\overline{Q} = \pm([j]\overline{Q})$ and hence $[i \pm j]\overline{Q} = O$. This is a contradiction as $\gcd(i \pm j, p) = 1$.

(3) Let $R := [p-k]Q$. Therefore $\overline{R} = [p-k]\overline{Q} = -[k]\overline{Q}$. Hence $x(\overline{R}) = x([k]\overline{Q})$ and $y(\overline{R}) = -y([k]\overline{Q})$.

(4) From part (3) we know that $x([p-k]Q) \equiv x([k]Q) \bmod p$. Say $v_p(x([p-k]Q) - x([k]Q)) = i$. We also know that $y([p-k]Q) \not\equiv y([k]Q) \bmod p$. From the group law formulae to calculate $[p]Q$ (say using $[k]Q$ and $[p-k]Q$), it follows that $v_p(x([p]Q)) = -2i$ which is equivalent to $v_p(y([p]Q)) = -3i$ [13, Proof of Theorem 7.1(c)]. And therefore $[p]Q \in E_i(\mathbb{Q}_p) \setminus E_{i+1}(\mathbb{Q}_p)$.

(5) $x([p-1]Q) - x(Q) \equiv 0 \bmod p^2$ implies $[p]Q \in E_2(\mathbb{Q}_p)$ by part (4). This implies that $\phi(\overline{Q}) = (\log_E \circ \lambda \circ [p])(Q) = 0 \in \hat{G}_a(p\mathbb{Z}_p)/p\hat{G}_a(p\mathbb{Z}_p)$. Here $\lambda(R) = -x(R)/y(R)$, where $R \in E_1(\mathbb{Q})$ and $\log_E(z) = z + O(z^2)$, where $z \in \hat{E}(p\mathbb{Z}_p)$.
(On the other hand $x([p-1]Q) - x(Q) \not\equiv 0 \bmod p^2$ implies $[p]Q \in E_1(\mathbb{Q}_p) \setminus E_2(\mathbb{Q}_p)$ by part (4). This implies that $\phi(\overline{Q}) = (\log_E \circ \lambda \circ [p])(Q) \neq 0$). □

In step 3 we merely need to pick $x_0 \in \mathbb{F}_p$ such that $(\frac{x_0^3 + ax_0 + b}{p}) = 1$ and computing the Legendre symbol can be done in $O(\log^2 p)$ [3, Algorithm 1.4.12]. Hensel lifting can be performed in almost linear time. Step 5 would consume $\mathcal{O}(\log^2 p)$ bit operations. And therefore in the good reduction case the overall time complexity of the algorithm is dominated by the point counting routine which takes time $\mathcal{O}(\log^4 p)$ [1, Page 117].

In the case of good reduction if $p > 5$ we have two cases $\gcd(n,p) = 1$ and $n = p$. If $p = 3, 5$ we have a third case — $\gcd(n,p) = p$ and $n \neq p$ — the only instances of which are $n = 2p$ by the Hasse bound. And this is the reason we pick $\overline{P} \in \overline{E}(\mathbb{F}_p)[p]$ in the algorithm.

We will now use the output of the above algorithm to decide whether $E(\mathbb{Q}_p)[p]$ is non-trivial with the help of the following fact [18, Corollary

IV.9.2]: Let E be an elliptic curve over $\mathbb{Q}_p$. Then we have the following exact sequence:

$$0 \to E_0(\mathbb{Q}_p) \to E(\mathbb{Q}_p) \to G \to 0$$

where if E has split multiplicative reduction over $\mathbb{Q}_p$, then G is a cyclic group of order $v(\Delta) = -v(j)$, in the additive scenario the group order is at most 4 and in the non-split multiplicative instance it is either 1 or 2.

In order to weed out the spurious cases we impose some conditions.

Lemma 20: *Algorithm 1 correctly computes* $\#E(\mathbb{Q}_p)[p]$ *provided either*

- *E has good reduction or*
- *E has additive reduction and $p > 3$, or*
- *E has additive reduction, $p = 3$ and $\gcd(\#G, 3) = 1$, or*
- *E has non-split multiplicative reduction or*
- $\#E_0(\mathbb{Q}_p)[p] = p$.

Proof: In the case of good reduction $E_0 = E$ and $\overline{E}_{ns} = \overline{E}$ and the lemma follows. If E has bad reduction we have the following long exact sequence: $0 \to E_0(\mathbb{Q}_p)[p] \to E(\mathbb{Q}_p)[p] \to G[p] \to E_0(\mathbb{Q}_p)/pE_0(\mathbb{Q}_p) \to E(\mathbb{Q}_p)/pE(\mathbb{Q}_p) \to G/pG \to 0$

Under the first 4 assumptions of the lemma we have $G[p] = 0$ and assuming the fifth case holds then $\#E(\mathbb{Q}_p)[p] = p$, and hence in all the scenarios $E_0(\mathbb{Q}_p)[p] \cong E(\mathbb{Q}_p)[p]$. □

4.2. *The algorithm when E has split multiplicative reduction at p*

To deal with the split multiplicative case we use the theory of the Tate curve [18, Sections 3-5].

Algorithm 2: An elliptic curve E over $\mathbb{Q}_p$ with split multiplicative reduction given by a minimal Weierstrass equation, where $p > 2$.

Input. $j(E)$, the j-invariant of E up to two significant p-adic digits and $v_p(j(E))$.

Output. TRUE if $\#E(\mathbb{Q}_p)[p] = p$ and FALSE if $\#E(\mathbb{Q}_p)[p] = 1$.

(1) $g \leftarrow -v_p(j(E))$.
(2) If $p \nmid g$ then return FALSE.
(3) $s_0 + s_1 p \leftarrow p^g \cdot j(E) \bmod p^2$.
(4) $u_1 \leftarrow -(\frac{1}{p} \cdot \frac{s_0^{p-1}-1}{(p-1)s_0^{p-2}}) \bmod p$.

(5) $t_1 \leftarrow \frac{s_1 - u_1}{s_0} \bmod p$.
(6) If $t_1 = 0$ return TRUE else return FALSE.

Lemma 21: *The algorithm works as claimed.*

Proof: Now let us view E as the Tate curve E_q, with $q \in \mathbb{Q}_p^*$ (by abuse of notation we will refer to both of them as E). $G = E(\mathbb{Q}_p)/E_0(\mathbb{Q}_p)$ which is a cyclic group of order $\#G = v_p(\Delta) = v_p(q) = -v_p(j(E))$. Furthermore $0 < v_p(\Delta) < \infty$ and hence step (1) is well-defined.

Recall that $E(\mathbb{Q}_p) = \mathbb{Q}_p^*/q^{\mathbb{Z}}$ [18, Theorem V.3.1(d)], $E_0(\mathbb{Q}_p) \cong \mathbb{Z}_p^*$ [18, Page 432]. We have $\mathbb{Z}_p^*[p] = 1$ (since $x^p - 1$ has only the trivial root of unity in $\mathbb{Q}_p$) and hence $\#E_0(\mathbb{Q}_p)[p] = 1$.

Now by the snake lemma, the short exact sequence

$$0 \to \mathbb{Z}_p^* \to \mathbb{Q}_p^*/q^Z \xrightarrow{v_p} G \to 0$$

gives us the following long exact sequence

$$0 \to \mathbb{Q}_p^*/q^Z[p] \xrightarrow{v_p} G[p] \xrightarrow{\delta} \mathbb{Z}_p^*/\mathbb{Z}_p^{*p} \to (\mathbb{Q}_p^*/q^Z)/(\mathbb{Q}_p^*/q^Z)^p \to G/pG \to 0$$

$G = \mathbb{Z}/v_p(q)\mathbb{Z}$ and $G[p]$ is generated by $\frac{v_p(q)}{p}$. Now $\mathbb{Z}_p^* = \mu_{p-1} \cdot (1 + p\mathbb{Z}_p)$, where μ_{p-1} are the $p - 1^{\text{st}}$ roots of unity in $\mathbb{Z}_p^*$. This tells us that $\mathbb{Z}_p^*/\mathbb{Z}_p^{*p} \cong (1 + p\mathbb{Z}_p)/(1 + p^2\mathbb{Z}_p) \cong \mathbb{Z}/p\mathbb{Z}$.

Observe that $\delta = 0 \Leftrightarrow \mathbb{Q}_p^*/q^Z[p] \cong G[p]$, therefore the question is how do we determine whether $\delta = 0$. In the case that $G[p] = 0$, that is, when $p \nmid v_p(q)$ then $\#E(\mathbb{Q}_p)[p] = 1$ and the correctness of step (2) of the algorithm follows. Now let us consider the case when $p | v_p(q)$. By the definition of the connecting homomorphism δ, we have $\delta(\frac{v_p(q)}{p}) = p^{\frac{v_p(q)}{p}p}/q^{\mathbb{Z}} + \mathbb{Z}_p^{*p}$, where $p^{v_p(q)}/q^{\mathbb{Z}} \in \mathbb{Z}_p^*$. If $p^{v_p(q)}/q \in \mathbb{Z}_p^{*p}$ then $\delta = 0$ which would imply $\#\mathbb{Q}_p^*/q^Z[p] = p$, otherwise $G[p] \cong \mathbb{Z}_p^*/\mathbb{Z}_p^{*p}$ and $\#\mathbb{Q}_p^*/q^Z[p] = 1$.

To check whether $p^{v_p(q)}/q \in \mathbb{Z}_p^{*p}$, firstly we will need $v_p(q)$ which is equal to $v_p(\Delta)$. The q parameter is obtained by working with the $j(E)$, the j-invariant of E [18, Lemma V.5.1]. Specifically $q \equiv j(E)^{-1} \bmod p^{2 \cdot v_p(\Delta)}$ and hence $p^{v_p(q)}/q \equiv j(E) \bmod p^{2 \cdot v_p(\Delta)}$. We want to ascertain whether the unit $p^{v_p(q)}/q \in \mathbb{Z}_p^*$ is actually in $\mathbb{Z}_p^{*p} \cong (1+p^2\mathbb{Z}_p)$ and therefore it follows that we need to compute the unit to 2 digits of p-adic precision since $v_p(q) > p > 2$.

To decide if $s \in \mathbb{Z}_p^*$ is in fact an element of $\mathbb{Z}_p^{*p}$ we do the following: Suppose $s = s_0 + s_1 p + \ldots$, then we can express it as a product of a $p - 1^{\text{st}}$-root of unity (say $u = u_0 + u_1 p + \ldots$, which is obtained by Hensel lifting s_0 to a root of $x^{p-1} - 1$ in $\mathbb{Z}_p^*$) and a 1-unit ($t = t_0 + t_1 p + \ldots$). Now working modulo p^2, we can decide whether an element of $\mathbb{Z}_p^*$ is in $\mathbb{Z}_p^{*p} = 1 + p^2\mathbb{Z}_p$ ($\Leftrightarrow t_1 = 0$). □

Given Δ, the time to compute $v_p(\Delta)$ is $\mathcal{O}(\log \Delta \log p)$ [20, Theorem 9.17]. Step (4) of the algorithm where we compute the $p-1^{\text{st}}$-root of unity using Hensel lifting to 2 p-adic digits will cost $\mathcal{O}(\log^2 p)$ (where we work modulo p^2 to compute $s_0^{p-1} - 1$ since p divides it).

4.3. *The complete algorithm*

Algorithm 3: An elliptic curve E over $\mathbb{Q}$ given by a Weierstrass equation $y^2 = x^3 + ax + b$, with $a, b \in \mathbb{Z}$.

Input. We are given the coefficients of E and a prime $p > 2$.

Output. TRUE if $\#E(\mathbb{Q}_p)[p] = p$ and FALSE if $\#E(\mathbb{Q}_p)[p] = 1$.

(1) Compute the minimal Weierstrass equation of E at p using Tate's algorithm.
(2) If E has additive reduction, $p = 3$ and $\#G[3] = 3$ then using f_3 and either some initialization and Hensel lifting of the singular point or p-adic factorization algorithm [15] determine whether $\#E(Q_3)[3] = 1$ or 3 and return FALSE, TRUE respectively.
(3) If E has split multiplicative reduction then return output of algorithm 2.
(4) Return output of algorithm 1.

The time complexity of Step 1 - Tate's algorithm - is analyzed in §9. If Hensel lifting will cost more than softly quadratic time, we resort to a general purpose p-adic factorization algorithm whose expected number of bit operations is $\mathcal{O}(\log^2 \Delta)$, fixing $p = 3$. It follows from our discussion that the time complexity above algorithm is polynomial in $\log p$ and $\log \Delta$ and this completes the proof of theorem 17.

Computing $E(\mathbb{Q}_p)[p]$ efficiently seems to be an interesting problem. In certain scenarios we can use Hensel lifting with f_p and the appropriate initialization, which is dictated by the discriminant of f_p, otherwise we can resort to p-adic polynomial factorization algorithms. When p is small and fixed this method can be used to compute $E(\mathbb{Q}_p)[p]$ in softly quadratic time, a subroutine employed in our p-adic torsion rational computation procedure which is described over the following two sections.

We would like to remark that working with the Tate curve to explicitly and efficiently compute torsion points on an elliptic curve with split multiplicative reduction seems to be an interesting problem due to the power series used to express the coefficients and coordinates of the Tate curve [18, Theorem V.3.1].

5. Computing $E(\mathbb{Q})[2]$

In this section we will look at determining the 2-torsion part of $E(\mathbb{Q})$. This procedure will come into play when we compute $E(\mathbb{Q})_{tors}$ in the following section. Also similar techniques can be pursued to compute $E(\mathbb{Q})[2^i]$ working with f_{2^i} division polynomials, where $i = 2, 3$.

Given an elliptic curve over $\mathbb{Q}$ and a prime $l > 2$, we use Tate's algorithm to obtain a minimal Weierstrass equation at l. We begin by listing conditions to avoid singular torsion contribution instances referred to as the S2T (Singular 2-Torsion) scenarios.

Lemma 22: $E_0(\mathbb{Q}_l)[2] = E(\mathbb{Q}_l)[2]$ *provided either*

(1) E has good reduction at l or
(2) E has split multiplicative reduction and $\gcd(v_l(\Delta), 2) = 1$ *or*
(3) E has additive reduction and $G[2] = 0$ *or*
(4) E has non-split multiplicative reduction $G[2] = 0$ *or*
(5) $\#E_0(\mathbb{Q}_l)[2] = 4$.

Proof: The proof follows for case (i) since $E = E_0$ and $\overline{E}_{ns} = \overline{E}$. Applying the extended snake lemma gives us

$$0 \to E_0(\mathbb{Q}_l)[2] \to E(\mathbb{Q}_l)[2] \to G[2] \to E_0(\mathbb{Q}_l)/2E_0(\mathbb{Q}_l) \to E(\mathbb{Q}_l)/2E(\mathbb{Q}_l) \to G/2G \to 0$$

Either of the cases $(ii) - (iv)$ of the lemma gives us $G[2] = 0$ and therefore $E_0(\mathbb{Q}_l)[2] \cong E(\mathbb{Q}_l)[2]$ For case (5), it follows from the fact that $\#E(\overline{\mathbb{Q}}_l)[2] = 4$. □

Algorithm 4: Let E be an elliptic curve over $\mathbb{Q}$ given by a Weierstrass equation and a prime $l > 2$. Suppose we want to compute the $E(Q_l)[2]$ to $O(\log \Delta)$ digits of l-adic accuracy.

Input. We are given the coefficients of E and l.

Output. $< T_i >$, where $T \in E(\mathbb{Q}_p)[2]$ and $i = 1, 2$.

(1) Compute the minimal Weierstrass equation of E at l using Tate's algorithm.
(2) $n \leftarrow \#\overline{E}_{ns}(\mathbb{F}_l)$.

(3) If $\gcd(n, 2) = 1$ and $!S2T$ then Return $< O >$.
(4) Generate non-trivial elements in $\overline{E}_{ns}(\mathbb{F}_l)[2]$, say $\overline{P_i}$.
(5) Lift $\overline{P_i}$ to $P_i \in E_0(\mathbb{Q}_l)[2]$ using $f = x^3 + ax + b$.
(6) If !S2T then Return $< P_i >$ and $< O >$.
(7) If S2T then we lift $\overline{Q}$, the singular point on $\overline{E}(\mathbb{F}_l)$ to $Q_j \in E(\mathbb{Q}_l)[2] \setminus E_0(\mathbb{Q}_l)[2]$ using $x^3 + ax + b$. Return $< P_i >, < Q_j >$ and $< O >$.

Lemma 23: *The above algorithm works as desired.*

Proof: We use the fact that the 2-torsion points have 0 as their y-coordinates. Since $E_1(\mathbb{Q}_l)[2] = 0$ and $\mathbb{Z}_l/2\mathbb{Z}_l = 0$, the appropriate application of the snake lemma sequence tells us that $E_0(\mathbb{Q}_l)[2] \cong \overline{E}_{ns}(\mathbb{F}_l)[2]$.

If we are not in the S2T case, by lemma 22 we have $E_0(\mathbb{Q}_l)[2] \cong E(\mathbb{Q}_l)[2]$. Hence if $\gcd(n, 2) = 1$, there are no non-trivial $E(\mathbb{Q}_l)[2]$ points. On the other hand if $\gcd(n, 2) \neq 1$, we might obtain 1 or 3 non-trivial $E(\mathbb{Q}_l)[2]$ points by lifting $\overline{E}_{ns}(\mathbb{F}_l)[2]$ points.

Now if we are in the S2T case, there might be a contribution of $0, 1$ or 2, $E(\mathbb{Q}_l)[2]$ points from the singular point.

The above algorithm is essentially a root-finding algorithm for the polynomial f. We use Hensel lifting when we can lift points efficiently provided $v_l(\Delta)$ is bounded (see §9). Otherwise we use an l-adic polynomial factorization algorithm [15]. Fixing the prime l the time complexity in the former case is $\mathcal{O}(\log \Delta)$ deterministic time and the latter approach takes $\mathcal{O}(\log^2 \Delta)$ expected time. Therefore a clever choice of the prime makes the algorithm compute faster. □

6. The p-adic algorithm

Given an elliptic curve E over $\mathbb{Q}$ this procedure computes $E(\mathbb{Q})_{tors}$ p-adically. We first compute the 2-torsion part l-adically, where $l = 3, 5$ or 7 as described in the previous section. In order to compute m-torsion, where m is the power of a prime $p > 2$, we view E as a curve over $\mathbb{Q}_p$ and using the appropriate factor of the m-division polynomial compute $E(\mathbb{Q}_p)[m]$ points to desired precision and then check to see if they are rational. The integers m we have to consider are implied by Mazur's theorem.

If p is the prime of good reduction and $p^i || m$, then we can lift a $\mathbb{F}_p$-rational root of f_{p^i} using Hensel's lemma with $k = i$ because of the following lemma due to Satoh [16, Lemma 2.7]: Let E be an ordinary elliptic curve over $\mathbb{Q}_p$ given by a minimal Weierstrass equation $y^2 = x^3 + ax + b$ that has good reduction at p, where $p > 2$ and $ab \neq 0$. Suppose $E(\mathbb{Q}_p^{ur})[p^i] \neq O$. If

$P \in E(\mathbb{Q}_p^{ur})[p^i]$ is a non-trivial point, then $v_p(\partial\Psi_{p^i}(x(P))) = i$.

The above lemma will play a role in the algorithm when p happens to be 3, 5 or 7 and we compute 3, 9, 5 and 7 torsion points respectively.

The lemma deals with ordinary elliptic curves and hence before we apply the lemma we should rule out the supersingular case. The following is a test for supersingularity [1, Page 46]: Let E be an elliptic curve over $\mathbb{Q}_p$ that had good reduction at p. E is supersingular if and only if

- $p \geq 5$ and $\#\overline{E}(\mathbb{F}_p) = p + 1$.
- $p = 2, 3$ and $\#\overline{E}(\mathbb{F}_p) = 1$, $p + 1$ or $2p + 1$.

Also if E is supersingular then $E(\mathbb{Q}_p)[p] = O$ and hence $E(\mathbb{Q})[p] = O$.

In the following algorithm we compute a non-zero element of $E(\mathbb{Q}_p)[p]$ by solving the p-division polynomial to $O(\log_p C)$ p-adic precision. We compute the p-adic expansions of the points using either Hensel lifting or p-adic polynomial factorization algorithms (see §9).

Algorithm 5:

Input. An elliptic curve E in the form $y^2 = x^3 + ax + b$, with $a, b \in \mathbb{Z}$.

Output. $< T, t >_i$, where $T \in E(\mathbb{Q})[t]$ and $i = 1, 2$ and these points are the generators of $E(\mathbb{Q})_{tors}$.

(1) Compute $E(\mathbb{Q})[2]$. Use a small prime ($l = 3, 5, 7$) to compute $E(\mathbb{Q}_l)[2]$. $r \leftarrow \#E(\mathbb{Q})[2] - 1$. Let $R_1, \ldots, R_r$ be the non-trivial 2-torsion points.

(2) If $r = 0$ then
  - (a) For $p = 3, 5, 7$ do the following:
    - i. Compute $\#E(\mathbb{Q}_p)[p]$.
    - ii. If $\#E(\mathbb{Q}_p)[p]$ == 1 then goto start of the loop and iterate with next prime.
    - iii. Compute a point Q using f_p.
    - iv. If $p = 5, 7$ then Return $< Q, p >$.
    - v. If $p = 3$, then try to compute S, a non-trivial 9-torsion point using f_9. If successful then
      - Return $< S, 9 >$.
      - else Return $< Q, 3 >$.
  - (b) Return $< O, 1 >$.

(3) If $r = 1$ then
  - (a) For $p = 3, 5$ do the following:
    - i. Compute $\#E(\mathbb{Q}_p)[p]$.

ii. If $\#E(\mathbb{Q}_p)[p] == 1$ then goto start of the loop and iterate with next prime.
iii. Compute a point Q using f_p.
iv. $U \leftarrow R_1 + Q$.
v. If $p = 5$ then Return $< U, 10 >$.
vi. If $p = 3$ then try and compute V, a non-trivial 4-point using f_4. If successful then
   - Return $< Q + V, 12 >$.
   - else Return $< U, 6 >$.

(b) Try and compute a non-trivial W, a non-trivial 4-torsion point using f_4. If successful then
   - try and compute Z, a non-trivial 8-torsion point using f_8. If successful then
     - Return $< Z, 8 >$.
     - else Return $< W, 4 >$.

(c) Return $< R_1, 2 >$.

(4) If $r = 3$ then

(a) For $p = 3$ do the following:
   i. Compute $\#E(\mathbb{Q}_p)[p]$.
   ii. If $\#E(\mathbb{Q}_p)[p] == 1$ then exit loop.
   iii. Compute a point Q using f_p.
      A. $U \leftarrow R_1 + Q$.
      B. Return $< U, 6 >$ and $< R_2, 2 >$.

(b) For R_1, R_2, R_3 do the following:
   i. Try and compute W, a non-trivial 4-torsion point using f_4. If successful
      - then try and compute Z, a non-trivial 8-torsion point using f_8. If successful then
        - Return $< Z, 8 >$ and $< R_1, 2 >$.
        - else Return $< W, 4 >$ and $< R_1, 2 >$.

(c) Return $< R_1, 2 >$ and $< R_2, 2 >$.

Theorem 24: The above algorithm works as desired.

Proof: Observe that if $\#E(\mathbb{Q})[2] = 4$ then we are in case (ii) of Mazur's classification. Conversely suppose case (ii) holds then assuming $\#E(\mathbb{Q})[2] =$

1, 2 leads us to contradictions. Therefore we first compute points in $E(\mathbb{Q}_l)[2]$ and check to see if we get 1, 2, 4 points in $E(\mathbb{Q})[2]$ (equivalently x^3+ax+b has 0, 1, 3 roots in $\mathbb{Q}$).

The algorithm works since the map $E(\mathbb{Q})[p] \to E(\mathbb{Q}_p)[p]$ is injective and for $p > 2$, $\#E(\mathbb{Q}_p)[p] = 1, p$. The rest of the algorithm proceeds in a case-by-case fashion aided by implications of Mazur's result. □

7. Discriminant of the division polynomial

In this section we prove a few facts about the discriminant of f_m, in particular Lemma 27, which makes l-adic torsion computation efficient.

7.1. *Prime Support of* $\Delta(f_m)$

Lemma 25: *Let* $m = 2k+1 > 1$ *be an integer.*

(1) $f || f'_m$
(2) $2^2 | f'_m$
(3) $m | f'_m$
(4) $m^{d-1} | \Delta(f_m)$

Proof:

(1) We will prove this by induction on m. $f'_3 = 12f$ and the base case holds. Suppose $f|f'_i$ for all odd $i < m$. Let us assume k is odd (in the even case a similar argument applies). We have $\Psi_{2k+1} = f_{2k+1} = f_{k+2}f_k^3 - (f_{k-1}\Psi_2)(f_{k+1}\Psi_2)^3 = f_{k+2}f_k^3 - f_{k-1}f_{k+1}^3\Psi_2^4 = f_{k+2}f_k^3 - 2^4 \cdot f_{k-1}f_{k+1}^3 f^2$. Now $f'_{2k+1} = f'_{k+2}f_k^3 + 3 \cdot f_{k+2}f_k^2 f'_k - 2^4 \cdot (f_{k-1}f_{k+1}^3)' f^2 - 2^5 \cdot f_{k-1}f_{k+1}^3 f f'$. f divides f'_{k+2} and f'_k by the inductive assumption and hence f divides each of the terms in f'_{2k+1}. In particular f exactly divides f'_{2k+1}, otherwise f_{2k+1} would have repeated roots.
(2) The argument is similar to the one above for part 1.
(3) By [2, Theorem 1 and Corollary 1] we have $m|(\Psi_m^2)'$ for any m and $p \nmid \Psi_m^2$ for any odd prime p. Suppose $m = \prod_i p_i$, where p_i are odd primes. From the above, $m|\Psi_m\Psi'_m$ which implies $p_i|\Psi_m\Psi'_m$. Also $p_i \nmid \Psi_m$. Therefore $p_i|\Psi'_m$ and therefore $m|\Psi'_m$.
(4) The statement follows from part 3, $lc(f_m) = m$ and the matrix definition of the discriminant, where the coefficients of f'_m are repeated on $\frac{m^2-1}{2}$ rows. □

Lemma 26: *Let $m = 2k > 2$ be an integer.*

(1) $k|\Delta(f_m)$
(2) $2^2|f'_m$
(3) $m|\Delta(f_m)$

Proof:

(1) Recall that $lc(f_m) = k$ and the coefficient of x^{d-1} in f_m is 0. Consider the matrix associated to $R(f_m, f'_m)$. The first column of this matrix has k at entry $(1,1)$ and $k\frac{m^2-4}{2}$ at $(\frac{m^2-4}{2}, 1)$ and 0 elsewhere. The second column of this matrix has k at entry $(2,2)$ and $k\frac{m^2-4}{2}$ at $(\frac{m^2-4}{2}+1, 2)$ and 0 elsewhere. Hence $k^2|R(f_m, f'_m)$ and $k|\Delta(f_m)$ by the definition of discriminant.

(2) $2^2|f'_4$ and the base case holds. Suppose $2^2|f'_i$ for all even $i < m$. Now $f'_{2k} = (f_{k+2}f^2_{k-1} - f_{k-2}f^2_{k+1})f'_k + (f'_{k+2}f^2_{k-1} + f_{k+2}\cdot 2\cdot f_{k-1}f'_{k-1} - f'_{k-2}f^2_{k+1} - f_{k-2}\cdot 2\cdot f_{k+1}f'_{k+1})f_k$. Let us assume that k is odd (similar analysis for the even case). From the previous lemma we know that 2^2 divides f'_k, f'_{k+2}, f'_{k-2}. By the inductive hypothesis 2^2 also divides f'_{k-1}, f'_{k+1}. Hence $2^2|f'_m$.

(3) Follows from part 1 and 2. □

Lemma 27: *Prime support of $\Delta(f_m)$ = prime support of m ∪ prime support of Δ, where $m > 2$ is an integer.*

Proof: Lemmas 25 and 26 state that 2 and m divide $\Delta(f_m)$. Hence it suffices to consider the primes which are relatively prime to m and prove that the prime support of $\Delta(f_m)$ equals the set of primes where E has bad reduction over $\mathbb{Q}$.

Consider E to be an elliptic curve over $\mathbb{Q}_p$. We will take a minimal Weierstrass equation denoted again by E and let its discriminant be Δ. Let L be a finite extension of $\mathbb{Q}_p$. Let $\mathfrak{p} = (\pi)$ be the prime over p in the ring of integers of L, $\mathbb{F}_\mathfrak{p}$ the residue field and e the ramification index.

Let $x = u^2x' + r, y = u^3y' + su^2x' + t$ be a change of coordinates giving a minimal Weierstrass equation for E/L denoted by E'. The discriminant Δ' for E' satisfies $\Delta' = u^{-12}\Delta$ and hence $v_\mathfrak{p}(\Delta') = -12v_\mathfrak{p}(u) + v_\mathfrak{p}(\Delta)$.

Let x_i and x'_i, $1 \le i \le d$ be the roots of the m-division polynomial associated to E and E' respectively. For $i \neq j$, we have $v_\mathfrak{p}(x'_i - x'_j) = v_\mathfrak{p}(\frac{x_i - r}{u^2} - \frac{x_j - r}{u^2}) = v_\mathfrak{p}(\frac{x_i - x_j}{u^2})$ and hence $v_\mathfrak{p}(x_i - x_j) = 2v_\mathfrak{p}(u) + v_\mathfrak{p}(x'_i - x'_j)$.

Now let us consider the valuation of the discriminant of the m-division polynomial associated to E over $\mathbb{Q}_p$:

$$v_{\mathfrak{p}}(\Delta(f_m)) = (2d-2)v_{\mathfrak{p}}(lc(f_m)) + 2\sum_{1\le i<j\le d}(2v_{\mathfrak{p}}(u) + v_{\mathfrak{p}}(x'_i - x'_j))$$

Suppose $(m,p) = 1$. Observe that $v_{\mathfrak{p}}(\cdot) = ev_p(\cdot)$. Since $v_p(m) = 0$, we have $v_{\mathfrak{p}}(lc(f_m)) = v_{\mathfrak{p}}(m) = 0$ when m is odd and when m is even, we have $v_{\mathfrak{p}}(lc(f_m)) = v_{\mathfrak{p}}(m/2) = 0$.

Case 1. Let E be an elliptic curve with potential good reduction over $\mathbb{Q}_p$. Let $L = \mathbb{Q}_p(E(\overline{\mathbb{Q}}_p)[m])$ be a finite extension of $\mathbb{Q}_p$ over which E has good reduction [18, Proposition IV.10.3] which means $v_{\mathfrak{p}}(\Delta') = 0$ and therefore $v_{\mathfrak{p}}(u) = v_{\mathfrak{p}}(\Delta)/12$.

Moreover the reduction modulo $\mathfrak{p}$ map $E(L)[m] \to \overline{E}(\mathbb{F}_{\mathfrak{p}})[m]$ is injective [17, Proposition VII.3.1 b] and hence $v_{\mathfrak{p}}(x'_i - x'_j) = 0$ for all $i \neq j$ and hence $v_{\mathfrak{p}}(\Delta(f_m)) = d(d-1)/6 \cdot v_{\mathfrak{p}}(\Delta)$ which implies that

$$v_p(\Delta(f_m)) = \frac{d(d-1)}{6}\cdot v_p(\Delta)$$

Hence if p is a prime of good reduction for $E/\mathbb{Q}$ then $p \nmid \Delta(f_m)$ and if p is a prime of bad (additive) reduction then $p|\Delta(f_m)$.

Case 2. Let E be an elliptic curve with potential multiplicative reduction over $\mathbb{Q}_p$. Let $L \supset \mathbb{Q}_p(E(\overline{\mathbb{Q}}_p)[m])$ be a finite extension of $\mathbb{Q}_p$ over which E has (split) multiplicative reduction which means $v_{\mathfrak{p}}(\Delta') > 0$ and $v_{\mathfrak{p}}(c_4) = 0$. We know that $v_{\mathfrak{p}}(c'_4) = v_{\mathfrak{p}}(u^{-4}c_4)$ therefore $v_{\mathfrak{p}}(u) = v_{\mathfrak{p}}(c_4)/4$. Also $j = c_4^3/\Delta$ and this implies $v_{\mathfrak{p}}(c_4) = \frac{1}{3}\cdot(v_{\mathfrak{p}}(\Delta) + v_{\mathfrak{p}}(j))$.

$$v_{\mathfrak{p}}(\Delta(f_m)) = \frac{d(d-1)}{6}\cdot(v_{\mathfrak{p}}(\Delta) + v_{\mathfrak{p}}(j)) + \sum_{1\le i<j\le d} 2v_{\mathfrak{p}}(x'_i - x'_j)$$

Now $v_{\mathfrak{p}}(\Delta) + v_{\mathfrak{p}}(j) = 0$ or > 0 depending on whether E has multiplicative or additive reduction over $\mathbb{Q}_p$. Also in either case there exist i, j such that $v_{\mathfrak{p}}(x'_i - x'_j) > 0$ (x-coordinates of points which reduce to the singular point). Hence $v_{\mathfrak{p}}(\Delta(f_m)) > 0$ which implies $v_p(\Delta(f_m)) > 0$.

Therefore if p is a prime of bad (additive or multiplicative) reduction for E over $\mathbb{Q}$ then $p|\Delta(f_m)$. □

7.2. *Discriminant formula*

While investigating the discriminant of the m-division polynomials we stumbled upon a precise formula which expresses $\Delta(f_m)$ in terms of m

and the discriminant of the elliptic curve E. Based on symbolically computing discriminants of f_m, the m-division polynomials for $3 \leq m \leq 12$, we arrived at the following:

$$\Delta(f_m) = \begin{cases} (-1)^{\frac{m-1}{2}} m^{d-1} \Delta^{\frac{d(d-1)}{6}} & m \text{ odd}, \\ 2^4 m^{d-4} \Delta^{\frac{d(d-1)}{6}} & m \text{ even}. \end{cases}$$

The above formula in the odd case turns out to be equivalent to a lemma of Stark [19], which he proved using a complex-analytic approach, in particular Kronecker's second limit formula. Stark's result deals with an elliptic curve E given in Weierstrass normal form by $y^2 = 4x^3 - g_2 x - g_3$, where g_2 and g_3 are rational. This equation can be parameterized by the Weierstrass $\wp$-function, $x = \wp(w), y = \wp'(w)$ whose period lattice is written in the form $\Omega = \omega_0[1, z]$, where z is in $\mathfrak{h}$, the upper half plane. Suppose N is odd and v_1 and v_2 run through a set of representatives of $N^{-1}\Omega \bmod \Omega$. It is shown that

$$\prod_{v_1 \not\equiv 0, v_2 \not\equiv 0, v_1 \pm v_2 \not\equiv 0} [\wp(v_1) - \wp(v_2)] = \pm N^{-2(N^2-3)} \Delta(E)^{(N^2-1)(N^2-3)/6}. \quad (1)$$

Note that the above equation is a product over of differences of torsion points, which are distinct and not inverses of each other.

Now we will establish the equivalence between our formula and that of Stark. Let m be an odd number and the roots of f_m be x_i, $1 \leq i \leq d$, then by the definition of the discriminant of a polynomial as stated in the §2,

$$\Delta(f_m) = lc(f_m)^{2d-2} \prod_{1 \leq i < j \leq d} (x_i - x_j)^2$$

where $lc(f_m) = m$ and $d = deg\ f_m = \frac{m^2-1}{2}$. Using our formula modulo sign, we have

$$\prod_{i \neq j} (x_i - x_j) = \prod_{1 \leq i < j \leq d} (x_i - x_j)^2 = \frac{\Delta(f_m)}{m^{2d-2}} = m^{-\frac{m^2-3}{2}} \Delta^{\frac{(m^2-1)(m^2-3)}{24}}. \quad (2)$$

The above equation is a product of differences of distinct x-coordinates of the torsion points. Now comparing Eq. 4 with Eq. 5 and taking $m = N$ we see that they are equivalent up to fourth root, which is explained by the difference in the products.

8. The l-adic algorithm

As stated in the introduction our algorithm works as follows: given an elliptic curve E over $\mathbb{Q}$ we view it as a curve over $\mathbb{Q}_l$ and using the appropriate factor of the m-division polynomial we compute the $\mathbb{Q}_l$-rational m-torsion points to desired precision and then check to see if they are in fact in $E(\mathbb{Q})$. The prime l is chosen such that E over $\mathbb{Q}_l$ has good reduction. The integers m we have to consider are dictated by Mazur's theorem.

Algorithm 6: Input. An elliptic curve E in the form $y^2 = x^3 + ax + b$, with $a, b \in \mathbb{Z}$.

Output. $< T, t >_i$, where $T \in E(\mathbb{Q})[t]$ and $i = 1, 2$ and these points are the generators of $E(\mathbb{Q})_{tors}$.

(1) Pick a prime $l > 7$ such that $l \nmid \Delta$.
(2) Compute $E(\mathbb{Q}_l)[2]$ using f.
(3) $r \leftarrow \#E(\mathbb{Q})[2] - 1$. Let $R_1, \ldots, R_r$ be the non-trivial 2-torsion points.
(4) If $r = 0$ then
   (a) For $p = 3, 5, 7$ do the following:
      i. If $p \nmid \#\overline{E}(\mathbb{F}_l)$ then goto start of the loop and iterate with next prime.
      ii. Compute $Q \in E(\mathbb{Q}_l)[p] \setminus \{O\}$ using f_p.
      iii. If $Q \notin E(\mathbb{Q})$ then goto start of the loop and iterate with next prime.
      iv. If $p = 5, 7$ then Return $< Q, p >$.
      v. If $3^2 | \#\overline{E}(\mathbb{F}_l)$ then
         - Compute $S \in E(\mathbb{Q}_l)[9] \setminus E(\mathbb{Q}_l)[3]$ using f_9. If $S \notin E(\mathbb{Q})$ then Return $< Q, 3 >$ else Return $< S, 9 >$.
         - else Return $< Q, 3 >$.
   (b) Return $< O, 1 >$.
(5) If $r = 1$ then
   (a) For $p = 3, 5$ do the following:
      i. If $p \nmid \#\overline{E}(\mathbb{F}_l)$ then goto start of the loop and iterate with next prime.
      ii. Compute $Q \in E(\mathbb{Q}_l)[p] \setminus \{O\}$ using f_p.
      iii. If $Q \notin E(\mathbb{Q})$ then goto start of the loop and iterate with next prime.
      iv. $U \leftarrow R_1 + Q$.
      v. If $p = 5$ then Return $< U, 10 >$.
      vi. If $4 \nmid \#\overline{E}(\mathbb{F}_l)$ then Return $< U, 6 >$.
      vii. Compute $V \in E(\mathbb{Q}_l)[4] \setminus E(\mathbb{Q}_l)[2]$ using f_4.
      viii. If $V \in E(\mathbb{Q})$ then

- Return $< V + Q, 12 >$.
- else Return $< U, 6 >$.

(b) If $4 \nmid \#\overline{E}(\mathbb{F}_l)$ then Return $< R_1, 2 >$.
(c) Compute $W \in E(\mathbb{Q}_l)[4] \setminus E(\mathbb{Q}_l)[2]$ using f_4.
(d) If $W \in E(\mathbb{Q})$ then

- If $8 \nmid \#\overline{E}(\mathbb{F}_l)$ then Return $< W, 4 >$.
- Compute $Z \in E(\mathbb{Q}_l)[8] \setminus E(\mathbb{Q}_l)[4]$ using f_8.
- If $Z \in E(\mathbb{Q})$ then
  - Return $< Z, 8 >$.
  - else Return $< W, 4 >$.

(e) Return $< R_1, 2 >$.

(6) If $r = 3$ then

(a) Do the following:

i. If $3 \nmid \#\overline{E}(\mathbb{F}_l)$ then exit loop.
ii. Compute a point $Q \in E(\mathbb{Q}_l)[3] \setminus \{O\}$ using f_3.
iii. If $Q \in E(\mathbb{Q})$ then

A. $U \leftarrow R_1 + Q$.
B. Return $< U, 6 >$ and $< R_2, 2 >$.

(b) Do the following:

i. If $8 \nmid \#\overline{E}(\mathbb{F}_l)$ then Return $< R_1, 2 >$ and $< R_2, 2 >$.
ii. Compute $W \in E(\mathbb{Q}_l)[4] \setminus E(\mathbb{Q}_l)[2]$ using f_4.
iii. If $W \in E(\mathbb{Q})$ then

- If $16 \nmid \#\overline{E}(\mathbb{F}_l)$ then Return $< W, 4 >$ and $< R_2, 2 >$.
- Compute $Z \in E(\mathbb{Q}_l)[8] \setminus E(\mathbb{Q}_l)[4]$ using f_8.
- If $Z \in E(\mathbb{Q})$ then
  - Return $< Z, 8 >$ and $< R_2, 2 >$.
  - else Return $< W, 4 >$ and $< R_2, 2 >$.

(c) Return $< R_1, 2 >$ and $< R_2, 2 >$.

Theorem 28: The algorithm works as desired.

Proof: Our choice of prime $l > 7$ implies that E has good reduction at l and hence the reduction map $E(\mathbb{Q}_l)[m] \to \overline{E}(\mathbb{F}_l)[m]$ is injective for all m we consider ($m = 2, 3, 4, 5, 7, 8, 9$) since $(m, l) = 1$. Suppose we want to compute the roots of the m-division polynomial. If $m | \#\overline{E}(\mathbb{F}_l)$ then the roots of f_m are distinct modulo l and therefore we can lift an $\mathbb{F}_l$-rational root of f_m to $\mathbb{Q}_l$ using Hensel's lemma with $k = 0$. Finally since we know that the magnitude of the rational torsion points, we check to see if the $\mathbb{Q}_l$-rational p-torsion computed to appropriate precision are in fact rational torsion. □

9. Time complexity analysis of the algorithms

Let $M(N)$ denote the bit operations required to multiply two numbers of size N. We will assume that a fast integer multiplication algorithm like Schőnhage-Strassen [20, Theorem 8.24] is used in which case $M(N) = O(N \log N \log\log N) = \mathcal{O}(N)$ (in soft-Oh notation). An integer of size N can be expressed p-adically using a recursive procedure called radix conversion [20, Theorem 9.17] in $O(M(N \log p) \log N)$ time. Due to the quadratic convergence of the Hensel lifting, if the desired precision is N then we can perform lifting in $O(M(N \log p))$ bit operations [20, Theorem 9.26] (assuming the degree of the polynomial is constant).

Given an elliptic curve with integral coefficients and discriminant Δ, we use Tate's algorithm [4, Chapter 3.2] to compute the minimal Weierstrass equation at a prime p, the type of reduction at p and the local index $\#G = E(Q_p)/E_0(Q_p)$ (data which is used in p-adic algorithm). Let $\gamma = \max\{\Delta, p\}$. Then time complexity of Tate's algorithm is $\mathcal{O}(\log \gamma)$ (plus the time compute the number of roots of certain quadratic and cubic congruences modulo p).

The primes we work with are very small in size compared to $\log \Delta$ - constant length in the p-adic algorithm and $\mathcal{O}(\log\log \Delta)$ in the l-adic procedure (see below) - hence in our context the running time of Tate's procedure is $\mathcal{O}(\log \Delta)$.

9.1. *The p-adic algorithm*

We recall that the discriminant of a polynomial gives us an upper bound on when the roots of the polynomial separate and in practice Hensel lifting might be efficiently usable at an "earlier" stage. The discriminant of the division polynomial depends on the discriminant of the elliptic curve and we shall see that p-adic information regarding the latter will to lead to faster algorithms. We note that in certain scenarios, $v_p(\Delta)$ can be read from a table [18, Page 365].

Since the primes p we work with are small $p = 3, 5, 7$ the time to compute $\#\overline{E}(\mathbb{F}_p)$ (when p is a prime of good reduction) is negligible. Once we find an approximate root we use Hensel's lifting to compute a $\mathbb{Q}_p$-rational root up to $O(\log_p C)$ accuracy and the time complexity of this operation is $O(M((\log_p C)\log p)) = \mathcal{O}(\log C) = \mathcal{O}(\log \Delta)$ bit operations. Hence the idea is to find an approximate root of the division polynomial in $\mathcal{O}(\log \Delta)$ such that the overall time complexity of lifting is still softly linear. In our context, if for $p = 3, 5$ and 7, $v_p(\Delta) =$ a constant or $O(\log\log \Delta)$ then $v_p(\Delta(f_p)) =$ constant or $O(\log\log \Delta)$ and we end up with the lifting operation being softly linear. In particular from the formula for the discriminant we know that $v_p(\Delta(f_p)) = \frac{p^2-3}{2} + \frac{(p^2-3)(p^2-1)}{24} v_p(\Delta)$. Hence if p is a prime of good reduction the roots of f_p separate at a level which is independent of Δ. Moreover in this situation computation to find an approximate root is reduced drastically due to Satoh's lemma. When $m = 2, 4, 8$, we compute l-adically with f, f_4, f_8 respectively where l is either $3, 5$ or 7. Reasoning similar to the above tells us that if $v_l(\Delta) =$ a constant or $O(\log\log \Delta)$ then lifting costs softly linear time.

On the other hand without the above bounds on the valuation of the discriminant, Hensel lifting may be expensive and in this case we resort to factoring f_m using p-adic and l-adic (when $m = 2, 4, 8$) polynomial factorization algorithms [15] whose time complexity after fixing the prime and the degrees of the polynomials is expected softly quadratic in $\log \Delta$. In this scenario instead of naive lifting techniques we choose polynomial factorization algorithms because the former could take exponential time in $\log \Delta$.

Therefore in the worst case the overall time complexity of computing $E(\mathbb{Q})_{tors}$ is $\mathcal{O}(\log^2 \Delta)$ deterministic or expected depending on the flow of control through the algorithm.

9.2. *The l-adic algorithm*

To deterministically find a prime $l > 7$ which does not divide Δ takes $\mathcal{O}(\log^2 \Delta)$ [7]. We compute with f and m-division polynomials, where $m = 3, 4, 5, 7, 8, 9$. and the size of the coefficients of these polynomials are bounded by $O(\log C) = O(\log \Delta)$ [7]. Hence by resorting to a randomized algorithm [20, Corollary 18.12 (ii)], the expected running time to pick a good prime is $\mathcal{O}(\log \Delta)$. In either case the magnitude of this prime is $\mathcal{O}(\log \Delta)$ and hence the time to compute $\#\overline{E}(\mathbb{F}_l)$ is negligible and to naively find an $\mathbb{F}_l$-rational root of the division polynomial is $O(l)$ (or expected $\mathcal{O}(\log l)$ [20, Corollary 14.16)]) operations in $\mathbb{F}_l$. Once we find an approximate root we use Hensel lifting (with $k = 0$) to a compute $\mathbb{Q}_l$-rational root up to $O(\log_l C)$ accuracy and the time complexity of this operation is $O(M((\log_l C)\log l)) = \mathcal{O}(\log C) = \mathcal{O}(\log \Delta)$ bit operations.

Therefore the finding the good prime routine dictates the overall running time of the l-adic algorithm, which is $\mathcal{O}(\log^2 \Delta)$ deterministic time or an expected running time of $\mathcal{O}(\log \Delta)$. The idea is that if we can find a good prime which is small in magnitude quickly in $\mathcal{O}(\log \Delta)$ time then the complexity of computing $E(\mathbb{Q})_{tors}$ is softly linear.

10. Conclusions

We have presented two algorithms to compute elliptic curve rational torsion both of which run in worst case softly quadratic time (deterministic or expected in the p-adic approach depending on the branch of execution and deterministic in the l-adic algorithm). We believe the l-adic algorithm is better because it is more elegant and is faster in theory as the randomized version of the l-adic algorithm runs in expected softly linear time. Our choice of using a p-adic polynomial factorization algorithm in the p-adic algorithm was a philosophical one as we wanted to compute p-adically as much as possible. Therefore in practice a hybrid algorithm, using a combination of p-adic and l-adic methods might work faster.

To extend the ideas in this paper to devise efficient algorithms to compute elliptic curve torsion over number fields, analogues to the theorems of Nagell-Lutz and Mazur might prove useful. [17, Exercise VIII.8.11] generalizes the former, while structural results à la Mazur's result exist for number field extensions of degrees $2, 3$ [9], [8] respectively and for higher extension degrees there are the bounds of Osterlé and Parent [14]. We note that these results are not essential in our context as we can instead obtain an upper bound for the size of the torsion group of the given elliptic curve by computing the size of group of the elliptic curve over the residue field for a few primes. This endeavor of computing $E(K)_{tors}$ will require working with extensions of $\mathbb{Q}$ and $\mathbb{Q}_p$, their associated rings of integers and residue fields. Hence arithmetic in these objects will have to be analyzed to develop efficient algorithms.

References

1. Blake I.F., et al. *Elliptic curves in cryptography.* London Mathematical Society Lecture Note Series, 265. Cambridge University Press, Cambridge, 1999.
2. Cassels, J. W. S. *A note on the division values of* $\wp(u)$. Proc. Cambridge Philos. Soc., 45 (1949), 167–172.
3. Cohen, H. *A course in computational algebraic number theory.* Graduate Texts in Mathematics, 138. Springer-Verlag, Berlin, 2000.

4. Cremona, J. *Algorithms for Modular Elliptic Curves*. Available at http://www.maths.nott.ac.uk/personal/jec/book/fulltext/index.html
5. Doud D. *A procedure to calculate torsion of elliptic curves over* $\mathbb{Q}$. Manuscripta Math. 95 (1998), no. 4, 463–469.
6. Fouquet, M., et al. *An extension of Satoh's algorithm and its implementation*, J. Ramanujan Math. Soc., 15 (2000), no. 4, 281–318.
7. Garcia-Selfa, I. et al. *Computing the rational torsion of an elliptic curve using Tate normal form*. J. Number Theory 96 (2002), no. 1, 76–88.
8. Jeon, D., et al. *On the torsion of elliptic curves over cubic number fields*. Acta Arith., 113 (2004), no .3, 291–301.
9. Kamienny, S., Mazur B. *Rational torsion of prime order in elliptic curves over number fields*. With an appendix by A. Granville. Columbia University Number Theory Seminar (New York, 1992). Astérisque No. 228 (1995), 3, 81–100.
10. Koblitz, N. *A course in number theory and cryptography*. Graduate Texts in Mathematics, 114. Springer-Verlag, New York, 1994.
11. Mazur, B. *Modular curves and the Eisenstein ideal*. Inst. Hautes Études Sci. Publ. Math. 47 (1977), 33–186.
12. Milne, J. S. *Class Field Theory*. Available at http://jmilne.org/math/
13. Milne, J. S. *Elliptic Curves*. Available at http://jmilne.org/math/
14. Parent, P. *Bornes effectives pour la torsion des courbes elliptiques sur les corps de nombres*. J. Reine Angew. Math. 506 (1999), 85–116.
15. Pauli, S. *Factoring polynomials over local fields*. J. Symbolic Comput. 32 (2001), no. 5, 533–547.
16. Satoh, T. *The canonical lift of an ordinary elliptic curve over a finite field and its point counting*. J. Ramanujan Math. Soc. 15 (2000), no. 4, 247–270.
17. Silverman, J. H. *The arithmetic of elliptic curves*. Graduate Texts in Mathematics, 106. Springer-Verlag, New York, 1986.
18. Silverman, J. H. *Advanced topics in the arithmetic of elliptic curves*. Graduate Texts in Mathematics, 151. Springer-Verlag, New York, 1994.
19. Stark, H. M. *The Theorem of Coates-Wiles revisited*. Number theory related to Fermat's last theorem (Cambridge, Mass., 1981), pp. 349–362, Progr. Math., 26, Birkhäuser, Boston, Mass., 1982.
20. von zur Gathen, J., Gerhard J. *Modern Computer Algebra*. Cambridge University Press, Cambridge, 2003.

Detecting complex multiplication

Jeffrey D. Achter

Department of Mathematics
Colorado State University
Fort Collins, CO 80523-1874
j.achter@colostate.edu

We give an efficient, deterministic algorithm to decide if two abelian varieties over a number field are isogenous. From this, we derive an algorithm to compute the endomorphism ring of an elliptic curve over a number field.

1. Introduction

In this paper, we answer two fundamental decision problems about elliptic curves over number fields. Specifically, we explain how to detect whether two elliptic curves over a number field are isogenous, and how to decide whether an elliptic curve has complex multiplication. These algorithms rely on Lemma 2, which actually applies to abelian varieties of any dimension, and Proposition 4, respectively.

In each case, we answer a question about a variety over a number field by examining its reduction at finitely many primes. At this level of generality, such a strategy is common in algorithmic number theory. For example, a common method for computing modular polynomials – that is, bivariate polynomials whose roots are j-invariants of elliptic curves related by an isogeny of fixed degree – is to perform the analogous computation over various finite fields, and then to lift the result using the Chinese remainder theorem. In contrast, we will see that to answer the decision problems posed here, one need not ever lift an object to characteristic zero.

The engine driving the machines presented here is Faltings's paper on the Mordell conjecture. Milne observed in *Mathematical Reviews* that Faltings "seems to give an algorithm for deciding when two abelian varieties over a number field are isogenous." In this paper, we further refine the proof of [7, Theorem 5] to the point where it literally yields an efficient algorithm for the isogeny decision problem.

At a crucial stage in that argument, Faltings shows that the isogeny class of X is determined by the action of $\mathrm{Gal}(L/K)$ on $X[\ell](\bar{K})$, where $[L:K]$ has effectively bounded degree and ramification but is difficult to compute directly. He therefore works with $\widetilde{L}$, the compositum of all possible such extensions of K, a large but still finite extension of K. An appeal to the Chebotarev density theorem guarantees that there is a finite set of primes T of K such that $\{\left(\frac{\widetilde{L}/K}{\mathfrak{p}}\right) : \mathfrak{p} \in T\} = \mathrm{Gal}(\widetilde{L}/K)$. Therefore, X and Y are isogenous if and only if the reductions $X_{\mathfrak{p}}$ and $Y_{\mathfrak{p}}$ are isogenous for each $\mathfrak{p} \in T$.

We derive an algorithm for detecting isogeny by showing that it suffices to use a set of primes $\mathfrak{p}$ with absolute norm smaller than some constant B. Effective Chebotarev-type theorems [3, 8] let us calculate a suitable B solely in terms of the degree and ramification data of L, without requiring recourse to the compositum $\widetilde{L}$.

Subsequently, we show how to use this result to test the hypothesis that an elliptic curve E has complex multiplication by a field F. Briefly, after a finite extension of the base field, there exists an elliptic curve E' with complex multiplication by F. Even without computing E' explicitly, we can use Lemma 2 to detect whether E and E' are geometrically isogenous, and thus check whether E has complex multiplication by F.

We start by reviewing literature concerning effective Chebotarev density theorems, and explain how to use ℓ-adic representations to detect isogeny between abelian varieties. The reader may wish to skip Section 2.1 on first reading, and turn directly to Section 2.2.

In the next section, we use these considerations to design algorithms for elliptic curves over number fields. In Section 3.1, we describe an algorithm to determine whether two elliptic curves are isogenous. In Section 3.2, we combine the results of the previous section with new results on complex multiplication to give an algorithm which decides whether a given elliptic curve has complex multiplication.

Several improvements are available to improve the efficiency of these methods. In the interest of streamlining the exposition, these suggestions are gathered as a series of remarks in the final section.

2. Background

As discussed above, the method of this paper is to apply effective Chebotarev bounds to Faltings's proof of the Tate conjecture in order to construct efficient algorithms; we review these results in Sections 2.1 and 2.2, respectively.

2.1. *Effective Chebotarev density theorems*

The Galois group of a finite extension of number fields L/K is generated by the Frobenius elements of primes of L lying over primes $\mathfrak{p}$ of K. We collect here various results from the literature which place upper bounds on the size of the primes necessary in order for their Artin symbols to generate $\mathrm{Gal}(L/K)$. Throughout, we will use (GRH) to highlight bounds which rely on the generalized Riemann hypothesis, and (U) to denote bounds which hold unconditionally.

For an extension of fields L/K, we let $\Delta_{L/K}$ denote the discriminant and $\mathbf{N}_{L/K}$ the norm map. For a prime ideal $\mathfrak{p}$ of K, let $\kappa(\mathfrak{p})$ be the residue field $\mathcal{O}_K/\mathfrak{p}$ and let $p_\mathfrak{p}$ be the characteristic of that field.

Let S be a finite set of places of K and N a nonnegative integer. We will express our Chebotarev-type bounds in terms of the following quantities:

$$\begin{aligned}
\Delta^*(K,S,N) &:= \left|\Delta_{K/\mathbb{Q}}\right|^N \Big(N \cdot \prod_{\mathfrak{p}\in S} p_\mathfrak{p}^{1-1/N}\Big)^{N\cdot[K:\mathbb{Q}]} \\
B_{(\mathrm{LO})}(K,S,N) &:= 70 \cdot (\log \Delta^*(K,S,N))^2 \\
B_{(\mathrm{BS})}(K,S,N) &:= (4\log \Delta^*(K,S,N) + 2.5N \cdot [K:\mathbb{Q}] + 5)^2 \\
B_{(\mathrm{GRH})}(K,S,N) &:= \min\{B_{(\mathrm{LO})}(K,S,N), B_{(\mathrm{BS})}(K,S,N)\}.
\end{aligned}$$

Let $c_{(\mathrm{U})}$ be the effective constant A_1 of [8], and let

$$B_{(\mathrm{U})}(K,S,N) := \begin{cases} \Delta^*(K,S,N)^{c_{(\mathrm{U})}} & K \supsetneq \mathbb{Q} \\ 2\Delta^*(K,S,N)^{c_{(\mathrm{U})}} & K = \mathbb{Q} \end{cases}.$$

Finally, let

$$\begin{aligned}
\mathcal{T}_{(\mathrm{GRH})}(K,S,N) &:= \{\mathfrak{p} \subset K : \mathbf{N}_{K/\mathbb{Q}}\mathfrak{p} \le B_{(\mathrm{GRH})}(K,S,N) \text{ and } \mathfrak{p} \notin S\} \\
\mathcal{T}_{(\mathrm{U})}(K,S,N) &:= \{\mathfrak{p} \subset K : \mathbf{N}_{K/\mathbb{Q}}\mathfrak{p} \le B_{(\mathrm{U})}(K,S,N) \text{ and } \mathfrak{p} \notin S\}.
\end{aligned}$$

Lemma 1: *Let K be a finite extension of $\mathbb{Q}$, and let $S \subset K$ be a finite set of prime ideals. Let L/K be a Galois extension with $[L:K] \le N$ unramified outside S. For any $\sigma \in Gal(L/K)$, there exists $\mathfrak{p} \in \mathcal{T}_{(\mathrm{U})}(K,S,N)$ and a* *prime $\mathfrak{q}$ of L dividing $\mathfrak{p}$ such that $Fr_{\mathfrak{q}} = \sigma$. If the generalized Riemann hypothesis holds, then $\mathfrak{p}$ may be taken in $\mathcal{T}_{(\mathrm{GRH})}(K, S, N)$.*

Proof: The statement combines several different effective Chebotarev density theorems. For a conjugacy class $C \subset \mathrm{Gal}(L/K)$, each gives an effective upper bound for the norm of the smallest prime $\mathfrak{p}$ such that $\left(\frac{L/K}{\mathfrak{p}}\right) = C$, computed in terms of the absolute discriminant of L. By [15, Proposition 5], $|\Delta_{L/\mathbb{Q}}| \leq \Delta^*(K, S, N)$; thus, in the sequel, we may replace each occurrence of $|\Delta_{L/\mathbb{Q}}|$ in [3, 8, 9] with $\Delta^*(K, S, N)$.

By [8, Theorem 1.1], any conjugacy class $C \subset \mathrm{Gal}(L/K)$ occurs as $\left(\frac{L/K}{\mathfrak{p}}\right)$ for some $\mathfrak{p} \in \mathcal{T}_{(\mathrm{U})}(K, S, N)$. Now suppose that the generalized Riemann hypothesis holds. Lagarias and Odlyzko prove [9] that a bound of the form $B_{(\mathrm{LO})}$ suffices, and Oesterle shows [15, 2.5] that the constant is at most 70. The bound $B_{(\mathrm{BS})}$ is obtained by Bach and Sorenson in [3, Theorem 5.1], again under the assumption of the generalized Riemann hypothesis.

Since the Frobenius elements $\mathrm{Fr}_{\mathfrak{q}}$ of all primes lying over a prime $\mathfrak{p}$ of K form the conjugacy class $\left(\frac{L/K}{\mathfrak{p}}\right)$, the result follows. □

2.2. *Abelian varieties and Galois modules*

Let X/K be an abelian variety, and let ℓ be a rational prime such that X has good reduction at all primes of K lying over ℓ. The ℓ-adic Tate module of X is $T_\ell(X) := \lim_{\overleftarrow{n}} X[\ell^n](\bar{K})$; let $V_\ell(X) := T_\ell(X) \otimes_{\mathbb{Z}} \mathbb{Q}$ be the rational Tate module. Then $T_\ell(X)$ is a $\mathbb{Z}_\ell$-representation of $\mathrm{Gal}(\bar{K}/K)$, while $V_\ell(X)$ is a $\mathbb{Q}_\ell$ representation of $\mathrm{Gal}(\bar{K}/K)$. It has long been known that these representations encode detailed arithmetic information about X.

In fact, Faltings proves the Tate conjecture; the canonical map $\mathrm{End}(X) \otimes_{\mathbb{Z}} \mathbb{Z}_\ell \to \mathrm{End}(T_\ell X)^{\mathrm{Gal}(K)}$ is an isomorphism. Consequently [7, Corollary 2] two abelian varieties are X and Y are isogenous if and only if $V_\ell X$ and $V_\ell Y$ are isomorphic as $\mathrm{Gal}(\bar{K}/K)$-modules.

We denote the reduction of an abelian variety X/K at a prime of good reduction $\mathfrak{p}$ by $X_{\mathfrak{p}}$; it is an abelian variety over $\kappa(\mathfrak{p})$.

The following result was proved by Serre [15, 8.3] in the special case where $\dim X = \dim Y = 1$ and $K = \mathbb{Q}$, but the the absolute constant given there is ineffective.

Lemma 2: *Let X and Y be g-dimensional abelian varieties over a number field K. Let S be a set of places of K containing all primes of bad reduction of X and Y, and let ℓ be a rational prime which is relatively prime to* *each place of S. Let* $\nu_g(\ell) = |GL_{2g}(\mathbb{Z}/\ell)|$*. Then X and Y are isogenous if and only if* $X_\mathfrak{p}$ *and* $Y_\mathfrak{p}$ *are isogenous for all* $\mathfrak{p} \in \mathcal{T}_{(\mathrm{U})}(K, S, \nu_g(\ell)^2)$*. If the generalized Riemann hypothesis is true, then X and Y are isogenous if and only if* $X_\mathfrak{p}$ *and* $Y_\mathfrak{p}$ *are isogenous for all* $\mathfrak{p} \in \mathcal{T}_{(\mathrm{GRH})}(K, S, \nu_g(\ell)^2)$*.*

Proof: Our proof is closely modelled on that of [7, Theorem 5] and [12, Theorem 23.7]. Let $\mathcal{T} = \mathcal{T}_{(\mathrm{GRH})}(K, S, \nu_g(\ell)^2)$ if the generalized Riemann hypothesis is to be assumed, and let $\mathcal{T} = \mathcal{T}_{(\mathrm{U})}(K, S, \nu_g(\ell)^2)$ otherwise. The key point is that the isogeny class of an abelian variety over a number field is determined by the Galois representation on its rational Tate module. Lemma 1 lets us detect the isomorphism class of a Galois representation using only the Frobenius elements over the finite set of primes $\mathcal{T}$.

Let $\rho : \mathrm{Gal}(\bar{K}/K) \to \mathrm{Aut}(T_\ell X) \times \mathrm{Aut}(T_\ell Y)$ be the product representation. Since X and Y both have good reduction outside S, $\mathrm{Gal}(\bar{K}/K)$ acts on $T_\ell X \times T_\ell Y$ via some quotient $\mathrm{Gal}(E/K)$ with E unramified outside S. Let R be the subring of $\mathrm{End}(T_\ell X) \times \mathrm{End}(T_\ell Y)$ generated over $\mathbb{Z}_\ell$ by $\{\rho(\sigma) : \sigma \in \mathrm{Gal}(E/K)\}$. We will show that R is in fact generated, again over $\mathbb{Z}_\ell$, by the actions of $\mathrm{Fr}_\mathfrak{q}$ for primes $\mathfrak{q}$ of E lying over $\mathfrak{p} \in \mathcal{T}$.

By Nakayama's Lemma, it suffices to prove that these Frobenius elements, acting on $(T_\ell X/\ell) \times (T_\ell Y/\ell) = X[\ell](\bar{K}) \times Y[\ell](\bar{K})$, generate $(R/\ell)^\times$. Now, the action of $\mathrm{Gal}(E/K)$ on $X[\ell](\bar{K}) \times Y[\ell](\bar{K})$ factors through $\mathrm{Gal}(L/K)$, where $[L : K]$ is a finite Galois extension of degree at most $|\mathrm{Aut}(X[\ell](\bar{K})) \times \mathrm{Aut}(Y[\ell](\bar{K}))| = \nu_g(\ell)^2$. By Lemma 1, $\{\mathrm{Fr}_\mathfrak{q} : \mathfrak{q}|\mathfrak{p} \in \mathcal{T}\} = \mathrm{Gal}(L/K)$. Therefore, $\{\rho(\mathrm{Fr}_\mathfrak{q}) : \mathfrak{q}|\mathfrak{p} \in \mathcal{T}\}$ generates R/ℓ over $\mathbb{Z}/\ell$, and this same set generates R over $\mathbb{Z}_\ell$.

If $X_\mathfrak{p}$ and $Y_\mathfrak{p}$ are isogenous for some prime of good reduction $\mathfrak{p}$, then [20, Theorem 1] $V_\ell X$ and $V_\ell Y$ are isomorphic as $\mathrm{Gal}(\kappa(\mathfrak{p}))$-modules. The hypothesis that $X_\mathfrak{p}$ and $Y_\mathfrak{p}$ are isogenous for $\mathfrak{p} \in \mathcal{T}$ implies that, for each $\mathrm{Fr}_\mathfrak{q}$ with $\mathfrak{q}|\mathfrak{p} \in \mathcal{T}$, $\mathrm{tr}(\mathrm{Fr}_\mathfrak{q}|T_\ell X) = \mathrm{tr}(\mathrm{Fr}_\mathfrak{q}|T_\ell Y)$. Extending $\mathbb{Z}_\ell$-linearly, we have $\mathrm{tr}(\sigma|T_\ell X) = \mathrm{tr}(\sigma|T_\ell Y)$ for each $\sigma \in \mathrm{Gal}(\bar{K}/K)$, so that [4, §12.1, Proposition 3] $V_\ell X$ and $V_\ell Y$ are isomorphic as $\mathrm{Gal}(\bar{K}/K)$-modules. By the Tate conjecture [7, Corollary 2] X and Y are isogenous. □

3. Algorithms for elliptic curves

3.1. *Detecting isogenous elliptic curves*

The isogeny class of an elliptic curve E over a finite field κ is uniquely determined by $|E(\kappa)|$. Indeed, by [20, Theorem 1] the isogeny class of E is determined by its characteristic polynomial of Frobenius, which has the form $T^2 - aT + |\kappa|$. Since the number of points on an elliptic curve with such a characteristic polynomial is $|\kappa|+1-a$, we see that two elliptic curves over κ are isogenous if and only if they have the same number of points over κ.

Any efficient algorithm for counting points on elliptic curves over finite fields, such as Schoof's method [14] which requires $O(\log^9 |\kappa|)$ bit operations, therefore yields an efficient method for deciding if two elliptic curves are isogenous.

More generally, any efficient algorithm for computing the action of Frobenius on $T_\ell X$ for a class of abelian varieties X, such as Jacobians of hyperelliptic curves, can decide if two such abelian varieties are isogenous. (Note that, in dimension greater than one, the action of Frobenius is *not* uniquely determined by its trace. Data such as the characteristic polynomial of Frobenius, rather than just the trace of Frobenius, is required to detect the isogeny class of an abelian variety over κ.)

We now turn our attention to number fields. In principle, Faltings's theorem affords us a choice of methods for determining whether two elliptic curves E_1 and E_2 over a given number field are isogenous. For instance, Masser and Wüstholz [11] use transcendence theory to give an explicit upper bound on the minimal degree of an isogeny between two elliptic curves. One could then try to enumerate all curves related to E_1 by an isogeny of given degree [22], and check if E_2 is isomorphic to any of them. One could also simply try to see if E_1 and E_2 satisfy a modular equation of suitable degree. Each of these operations carries a nontrivial computational cost [1]. Moreover, the best known *constant* appearing in such degree bounds [13, Théorème 1] is larger than 10^{61}; such a method remains of theoretical, rather than practical, interest.

Alternatively, an efficient algorithm follows from Lemma 2. Given two elliptic curves E_1 and E_2 over a common number field K, compute the discriminant Δ_i of E_i, and thence the set of primes S_i for which E_i has bad reduction. If $S_1 \neq S_2$, then E_1 and E_2 are not isogenous [17, Corollary 2]. Otherwise, choose a rational prime ℓ relatively prime to each element of $S := S_1 = S_2$. By Lemma 2, E_1 and E_2 are isogenous if and only if $E_{1,\mathfrak{p}}$ and $E_{2,\mathfrak{p}}$ are isogenous for for each $\mathfrak{p} \in \mathcal{T}(K, S, (\ell^2-1)^2(\ell^2-\ell)^2)$; this last condition may be checked using point-counting for each $E_{i,\mathfrak{p}}$. (Again, if a method is available for computing the characteristic polynomial of Frobenius, then the same method works for detecting isogeny of abelian varieties of dimension g; one simply computes at all primes with norm less than $B(K, S, \nu_g(\ell)^2)$.)

We remark that exhibiting infinitely many primes $\mathfrak{p}$ for which $E_{1,\mathfrak{p}}$ and

$E_{2,\mathfrak{p}}$ are isogenous does not prove that E_1 and E_2 are isogenous. Indeed, suppose that E_1 and E_2 have complex multiplication by distinct fields F_1 and F_2, respectively. On one hand, E_1 and E_2 are not isogenous, since the rational ring of endomorphisms is an isogeny invariant. On the other hand, we will see below that $E_{1,\mathfrak{p}}$ and $E_{2,\mathfrak{p}}$ are both supersingular, and thus isogenous, for all primes $\mathfrak{p}$ of K for which $p_{\mathfrak{p}}$ is inert in each extension $F_i/\mathbb{Q}$.

3.2. *Detecting complex multiplication*

Let E be an elliptic curve over a number field K. The endomorphism ring $\mathrm{End}(E)$ of E is isomorphic either to $\mathbb{Z}$ or to an order $\mathcal{O}$ in a quadratic imaginary field, F. In the latter case, we say that E has complex multiplication by F. (More generally, we will say that an elliptic curve over an arbitrary field has complex multiplication by F if its endomorphism ring contains an order in F.)

Elliptic curves with complex multiplication are prominent in primality testing and cryptography [2] and other aspects of algorithmic number theory [6]. Motivated by this, one might seek an algorithm for determining whether a given elliptic curve E over a number field K has complex multiplication. In [5], the author describes two methods. The first is a probabilistic algorithm which runs in polynomial time in the inputs; the second runs in deterministic polynomial time, but the constants appearing in the analysis of the running time are ineffective. In this section, we use Lemmas 1 and 2 to give an efficient, effective algorithm to determine whether an elliptic curve has complex multiplication. We start by collecting a body of facts about elliptic curves with complex multiplication. The subsequent algorithm follows naturally from these observations.

Deuring investigated the relationship between the arithmetic of F and the reductions $E_{\mathfrak{p}}$ at primes of K. (For the moment, we ignore primes of bad reduction.) He proved (see [21, Exemple b]) that $E_{\mathfrak{p}}$ is ordinary if and only if $p_{\mathfrak{p}}$, the rational prime lying under $\mathfrak{p}$, splits in F. Invoking the Chebotarev density theorem for F, we see that E has ordinary reduction at half the primes of K, and supersingular reduction at the others.

Conversely, if $\mathrm{End}(E) \cong \mathbb{Z}$, so that E *does not* have complex multiplication, then supersingular primes have density zero [16, IV-13, Exercise 1]. This basic observation leads to a probabilistic method, detailed in [5], for checking whether an elliptic curve has complex multiplication. Broadly speaking, finding may primes of supersingular reduction provides evidence for the hypothesis that E has complex multiplication.

In the sequel, we will use Lemma 2 (and the accompanying discussion at the end of Section 3.1) to describe a deterministic algorithm to test whether an elliptic curve E has complex multiplication. By this, we mean that the algorithm is guaranteed to terminate after a finite, explicitly computable number of operations, and that the output is a verifiable proof that E does (or does not) have complex multiplication.

It is convenient to assume that the elliptic curve of interest has no automorphisms other than $\{\pm 1\}$, and that it has good reduction everywhere. The former condition is equivalent to the assertion that E *does not* have complex multiplication by $\mathbb{Q}(\sqrt{-1})$ or $\mathbb{Q}(\sqrt{-3})$, which is easily verified by checking that $j(E) \notin \{0, 1728\}$.

The latter condition holds, possibly after a finite extension of the base field, for an elliptic curve with complex multiplication. (This assertion is equivalent to the result of Weber [18, C.11.2.a] that an elliptic curve with complex multiplication has integral j-invariant.) Concretely, let $\mathcal{N}$ be the product of all primes of bad reduction of E, and let $K_1 = K(\sqrt{\mathcal{N}})$. Suppose that E has complex multiplication by a field whose only roots of unity are 1 and -1. A special case of [17, Theorem 7] shows that E_{K_1} has good reduction at all places of K_1.

Henceforth, we will assume that E/K has everywhere good reduction.

We now show that it is easy to find a prime of ordinary reduction for E, and thereby find a candidate ring of endomorphisms for E.

Lemma 3: *Let E/K be an elliptic curve over a number field with complex multiplication and good reduction everywhere. Let S be the set of rational primes ramified in the extension $K/\mathbb{Q}$. Then there exists a prime $\mathfrak{p}$ of K lying over a rational prime p with $p \leq B_{(\mathrm{U})}(\mathbb{Q}, S, 2)$ such that E has good, ordinary reduction at $\mathfrak{p}$. If the generalized Riemann hypothesis is true, then p may be taken less than or equal to $B_{(\mathrm{GRH})}(\mathbb{Q}, S_{\mathbb{Q}}, 2)$.*

Proof: If $\mathrm{End}(E) \otimes \mathbb{Q}$ is isomorphic to a quadratic imaginary field F, then K necessarily contains F [10, Theorem 3.1.1]. In particular, the support of the discriminant of F over $\mathbb{Q}$ is contained in the support of the discriminant of K over $\mathbb{Q}$, so that F is unramified outside S. Moreover, E has ordinary reduction at a prime $\mathfrak{p}$ over the rational prime p if and only if p splits in F. The Chebotarev density theorem (Lemma 1) guarantees the existence of such a p with $p \leq B_{(\mathrm{U})}(\mathbb{Q}, S_{\mathbb{Q}}, 2)$. (If the generalized Riemann hypothesis is true, then p may be taken to be at most $B_{(\mathrm{GRH})}(\mathbb{Q}, S_{\mathbb{Q}}, 2)$.) □

Since supersingular primes have density zero for an elliptic curve *with-*

out complex multiplication, it seems unlikely that one would encounter an E/K without a small (in the sense of Lemma 3) ordinary prime. Still, if this were to happen, one could then conclude that the elliptic curve had endomorphism ring equal to $\mathbb{Z}$.

Let $\mathfrak{p}$ be a prime of ordinary reduction of E. Then the ring $\mathbb{Z}[\mathrm{Fr}_{\mathfrak{p}}] \subseteq \mathrm{End}(E_{\mathfrak{p}})$ is isomorphic to an order in a quadratic imaginary field F. (To see this, use the result of Deuring [18, Theorem V.3.1], paralleling the result in characteristic zero, that the endomorphism ring of an ordinary elliptic curve is either $\mathbb{Z}$ or an order in a quadratic imaginary field. Moreover, the Frobenius endomorphism cannot have a real conjugate [21, Exemple a], so it must actually generate a quadratic imaginary field.) Moreover, given the number of points on $E_{\mathfrak{p}}$, one can determine the field $F = \mathrm{Frac}\,\mathbb{Z}[\mathrm{Fr}_{\mathfrak{p}}]$. This is a candidate field of (rational) endomorphisms of E, and we show in Proposition 4 how to test the hypothesis that E truly does have complex multiplication by F. (At this stage of the calculation, one knows that E has complex multiplication by *some* quadratic imaginary field if and only if it has complex multiplication by F.)

It is known [18, Corollary C.11.1] that there is a finite extension K' of K and an elliptic curve E'/K' with complex multiplication by F. At this point, one could simply compute the j-invariant of E' and check whether $j(E)$ and $j(E')$ are conjugate under $\mathrm{Gal}(\mathbb{Q})$. However, to compute the polynomial over $\mathbb{Q}$ which $j(E')$ satisfies takes time $O(|\Delta_{F/\mathbb{Q}}|^2(\log|\Delta_{F/\mathbb{Q}}|)^2)$ [2] (see also [6, 7.6]).

Now, if one could construct E' efficiently, one could use Lemma 2 to test whether E and E' are isogenous, since isogenous elliptic curves have commensurable rings of endomorphisms. Even without knowing E' explicitly, however, we can efficiently test whether the two curves are (geometrically) isogenous.

Proposition 4: *Suppose that E/K has good reduction everywhere. Let F be a quadratic imaginary subfield of K whose only roots of unity are -1 and 1, and let $h^*(F) = 2\sqrt{\Delta_{F/\mathbb{Q}}}/\pi$. Then E has complex multiplication by F if and only if for each prime $\mathfrak{p} \in K$ lying over a rational prime p with* $\mathbf{N}\mathfrak{p} \le B_{(\mathrm{U})}(K, \emptyset, h^*(F)\nu_2(2)^2)$, *either:*

- *$E_{\mathfrak{p}}$ is supersingular and F is inert or ramified at p, or*
- *End$(E_{\mathfrak{p}}) \otimes \mathbb{Q} \cong F$, and F is split at p.*

If the generalized Riemann hypothesis is true, then it suffices to consider those primes with norm at most $B_{(\mathrm{GRH})}(K, \emptyset, h^*(F)\nu_2(2)^2)$.

Proof: If the generalized Riemann hypothesis is to be assumed, we write B for $B_{(\mathrm{GRH})}$ and $\mathcal{T}$ for $\mathcal{T}_{(\mathrm{GRH})}$; otherwise, these symbols denote $B_{(\mathrm{U})}$ and $\mathcal{T}_{(\mathrm{U})}$, respectively. Note that the statement is equivalent to the assertion that E has complex multiplication by F if and only if the same is true of $E_{\mathfrak{p}}$ for each prime $\mathfrak{p} \in B(K, \emptyset, h^*(F)\nu_2(2)^2)$. Since there is a natural inclusion $\mathrm{End}(E) \hookrightarrow \mathrm{End}(E_{\mathfrak{p}})$ for each prime $\mathfrak{p}$ [10, Theorem 2.3.2], if E has complex multiplication by F then the same is true of $E_{\mathfrak{p}}$ for each prime $\mathfrak{p}$, and in particular for those in $B(K, \emptyset, h^*(F)\nu_2(2)^2)$.

Having secured this, we focus on the converse. There exists an elliptic curve over a field K' with complex multiplication by F if and only if K' contains the Hilbert class field of F [18, Theorem C.11.2]. Moreover, since the only roots of unity in F are $\{\pm 1\}$, we may assume that E' has good reduction everywhere [17, Theorem 9].

Therefore, let K' be the compositum of K and the Hilbert class field of F, and let E'/K' be an elliptic curve with everywhere good reduction and complex multiplication by F. The original elliptic curve E has complex multiplication by F if and only if $E_{K'}$ and E' are isogenous over some finite extension of K'. (An analytic construction, as in [18, C.11], shows that E and E' are isogenous over $\mathbb{C}$; this isogeny must then descend to some finite K'/K [19, Theorem II.2.2].) Equivalently, E has complex multiplication by F if and only if $E_{K'}$ is isogenous to some twist of E'.

Let $N = \nu_2(2)^2$ and suppose that, for all primes $\mathfrak{q} \in \mathcal{T}(K', \emptyset, N)$, $E_{\mathfrak{q}}$ has complex multiplication by F. Then $E_{\mathfrak{q}}$ and $E'_{\mathfrak{q}}$ are isogenous up to a quadratic twist, and there exists a twist E'' of E' such that $E_{\mathfrak{q}}$ and $E''_{\mathfrak{q}}$ are isogenous for all $\mathfrak{q} \in \mathcal{T}(K', \emptyset, N)$. By Lemma 2, E and E'' are isogenous, and thus E has complex multiplication by F.

If the prime $\mathfrak{q}$ of K' lies over the prime $\mathfrak{p}$ of K, then $\mathbf{N}_{K'/\mathbb{Q}}(\mathfrak{q}) \leq \mathbf{N}_{K/\mathbb{Q}}(\mathfrak{p})$. In particular, each prime $\mathfrak{q} \in \mathcal{T}(K', \emptyset, N)$ lies over a prime $\mathfrak{p}$ with $\mathbf{N}_{K/\mathbb{Q}}\mathfrak{p} \leq B(K', \emptyset, N)$. Moreover, $E_{\mathfrak{q}}$ is the base change $E_{\mathfrak{p}} \times \kappa(\mathfrak{q})$, and thus $E_{\mathfrak{q}}$ has complex multiplication by F if and only if $E_{\mathfrak{p}}$ does. We have thus shown that E has complex multiplication by F if and only if the same is true for each reduction $E_{\mathfrak{p}}$ with $\mathbf{N}_{K/\mathbb{Q}}\mathfrak{p} \leq B(K', \emptyset, N)$. Now, K' is an unramified extension of K of degree at most $h^*(F)$. Therefore,

$$\begin{aligned}\Delta^*(K', \emptyset, N) &= \left|\Delta_{K'/\mathbb{Q}}\right|^N N^{N\cdot[K':\mathbb{Q}]} = \left|\Delta_{K/\mathbb{Q}}^{[K':K]}\right|^N \cdot N^{N\cdot[K':K]\cdot[K:\mathbb{Q}]} \\ &= \left|\Delta_{K/\mathbb{Q}}\right|^{N\cdot[K':K]} (N^{N\cdot[K':K]})^{[K:\mathbb{Q}]} \leq \Delta^*(K, \emptyset, h^*(F)N),\end{aligned}$$

and the result follows. □

Taken together, the results of this subsection suggest the following algorithm for determining whether an elliptic curve E has complex multiplication. First, check whether $j(E) \in \{0, 1728\}$; if so, the answer is yes; if not, one continues. Second, construct $K(\sqrt{\mathcal{N}})$, and verify that $E/K(\sqrt{\mathcal{N}})$ has good reduction everywhere; if this fails, then [17, Theorem 7] E does not have complex multiplication. Otherwise, replace K with $K(\sqrt{\mathcal{N}})$, and use Lemma 3 to find a candidate field F of endomorphisms. Finally, Proposition 4 allows us to test if E/K has complex multiplication by F.

4. Algorithmic considerations

We close with some remarks which may allow more efficient implementation of these algorithms.

4.1. *Chebotarev density theorem*

It is sometimes possible to improve the bounds given in Lemma 1. In special cases where $[K : \mathbb{Q}]$ and N are both small, [3, Table 1] provides even tighter bounds for the norm of the smallest prime ideal with given Artin symbol.

Moreover, if either the bound $B_{(\mathrm{BS})}$ or $B_{(\mathrm{U})}$ is used, then it suffices to consider those primes $\mathfrak{p}$ of K with norm a rational prime.

4.2. *Bounds in Lemma 2*

The term $\nu(g) = |\mathrm{GL}_{2g}(\mathbb{Z}/\ell)|$ arises in the proof of Lemma 2 as the size of the automorphism group of $X[\ell]$. If X further comes equipped with a polarization over K of degree prime to ℓ, then the action of $\mathrm{Gal}(\bar{K}/K)$ on $T_\ell X$ commutes with the induced symplectic pairing. Therefore, if one further makes the assumption in Lemma 2 that X and Y admit polarizations over K of degree relatively prime to ℓ, then $\nu_g(\ell)$ may be replaced by $|\mathrm{GSp}_{2g}(\mathbb{Z}/\ell)|$.

4.3. *Candidate fields of complex multiplication*

In Section 3.2, we suggested using Lemma 3 to find a candidate ring of endomorphisms of E. Alternatively, if $\mathrm{End}_K(E)$ is an order in a quadratic imaginary field then $\mathrm{End}_K(E) \subseteq K$, so that $\mathrm{End}_K(E) \otimes \mathbb{Q}$ is a quadratic imaginary subfield of K. Therefore, one can enumerate each such subfield F_i of K, and apply Proposition 4 to each; E has complex multiplication by some field if and only if it has complex multiplication by one of the F_i.

4.4. *Bounds in Proposition 4*

The proof of Proposition 4 shows that it suffices to consider those primes of K with norm at most, e.g., $70 \cdot (h^*(F) \log \Delta^*(K, \emptyset, \nu_2(2)^2))^2$ if the Lagarias and Odlyzko bound is to be used; the analogous improvement may be made in each of the other bounds, as well. Moreover, one can replace the (perhaps pessimistic) bound $h^*(F)$ with the actual class number of F; this class number can be computed in time $O(\left|\Delta_{F/\mathbb{Q}}\right|^{1/4+\epsilon})$, and even $O(\left|\Delta_{F/\mathbb{Q}}\right|^{1/5+\epsilon})$ if the generalized Riemann hypothesis is assumed [6, 5.4]. Finally, if one could verify that the Hilbert class field of F is already contained in K, then one would know (in the notation of the proof) that $K = K'$, and one could replace $h^*(F)$ with 1. However, it is not clear to the author how to verify this condition, short of actually computing the Hilbert class field.

4.5. *Detecting potential complex multiplication*

It is not hard to adopt the observations of Section 3.2 to test whether E *potentially* has complex multiplication, in the sense that $\mathrm{End}_{\bar{K}}(E) \otimes \mathbb{Q}$ is a quadratic imaginary field. One needs to replace Lemma 3 with an upper bound for the size of a prime of ordinary reduction; as noted there, we expect in practice that it is quite easy to find such a prime. This generates a candidate field F of rational endomorphisms for $E_{\bar{K}}$. One can then apply Proposition 4 to check whether $E_{K.F}$ has complex multiplication by F.

Acknowledgments

I thank Siman Wong for helpful discussions, and Tony Shaska for organizing this conference.

References

1. Amod Agashe, Kristin Lauter, and Ramarathnam Venkatesan. Constructing elliptic curves with a known number of points over a prime field. In *High primes and misdemeanours: lectures in honour of the 60th birthday of Hugh Cowie Williams*, volume 41 of *Fields Inst. Commun.*, pages 1–17. Amer. Math. Soc., Providence, RI, 2004.
2. A. O. L. Atkin and F. Morain. Elliptic curves and primality proving. *Math. Comp.*, 61(203):29–68, 1993.
3. Eric Bach and Jonathan Sorenson. Explicit bounds for primes in residue classes. *Math. Comp.*, 65(216):1717–1735, 1996.
4. N. Bourbaki. *Éléments de mathématique. 23. Première partie: Les structures fondamentales de l'analyse. Livre II: Algèbre. Chapitre 8: Modules et anneaux semi-simples.* Actualités Sci. Ind. no. 1261. Hermann, Paris, 1958.

5. Denis Charles. Complex multiplication tests for elliptic curves. 2004, math.NT/0409501.
6. Henri Cohen. *A course in computational algebraic number theory*, volume 138 of *Graduate Texts in Mathematics*. Springer-Verlag, Berlin, 1993.
7. G. Faltings. Endlichkeitssätze für abelsche Varietäten über Zahlkörpern. *Invent. Math.*, 73(3):349–366, 1983.
8. J. C. Lagarias, H. L. Montgomery, and A. M. Odlyzko. A bound for the least prime ideal in the Chebotarev density theorem. *Invent. Math.*, 54(3):271–296, 1979.
9. J. C. Lagarias and A. M. Odlyzko. Effective versions of the Chebotarev density theorem. In *Algebraic number fields: L-functions and Galois properties (Proc. Sympos., Univ. Durham, Durham, 1975)*, pages 409–464. Academic Press, London, 1977.
10. Serge Lang. *Complex multiplication*, volume 255 of *Grundlehren der Mathematischen Wissenschaften [Fundamental Principles of Mathematical Sciences]*. Springer-Verlag, New York, 1983.
11. D. W. Masser and G. Wüstholz. Estimating isogenies on elliptic curves. *Invent. Math.*, 100(1):1–24, 1990.
12. J.S. Milne. Abelian varieties. 1998, http://www.jmilne.org/math/.
13. Federico Pellarin. Sur une majoration explicite pour un degré d'isogénie liant deux courbes elliptiques. *Acta Arith.*, 100(3):203–243, 2001.
14. René Schoof. Counting points on elliptic curves over finite fields. *J. Théor. Nombres Bordeaux*, 7(1):219–254, 1995. Les Dix-huitièmes Journées Arithmétiques (Bordeaux, 1993).
15. Jean-Pierre Serre. Quelques applications du théorème de densité de Chebotarev. *Inst. Hautes Études Sci. Publ. Math.*, (54):323–401, 1981.
16. Jean-Pierre Serre. *Abelian l-adic representations and elliptic curves*. Advanced Book Classics. Addison-Wesley Publishing Company Advanced Book Program, Redwood City, CA, 1989.
17. Jean-Pierre Serre and John Tate. Good reduction of abelian varieties. *Ann. of Math. (2)*, 88:492–517, 1968.
18. Joseph H. Silverman. *The arithmetic of elliptic curves*, volume 106 of *Graduate Texts in Mathematics*. Springer-Verlag, New York, 1986.
19. Joseph H. Silverman. *Advanced topics in the arithmetic of elliptic curves*, volume 151 of *Graduate Texts in Mathematics*. Springer-Verlag, New York, 1994.
20. John Tate. Endomorphisms of abelian varieties over finite fields. *Invent. Math.*, 2:134–144, 1966.
21. John Tate. Classes d'isogénie des variétés abéliennes sur un corps fini (d'apres T. Honda). *Seminaire Bourbaki*, 352:95–110, 1968/1969.
22. Jacques Vélu. Isogénies entre courbes elliptiques. *C. R. Acad. Sci. Paris Sér. A-B*, 273:A238–A241, 1971.

Simple numerical uniformatization of elliptic curves

M. Seppälä

Department of Mathematics
Florida State University
Tallahassee, FL 32306
E-mail: Mika.Seppala@fsu.edu
and
Department of Mathematics and Statistics
PO Box 68
FI-00014 University of Helsinki
Finland
E-mail: Mika.Seppala@Helsinki.Fi

The numerical uniformization problem of algebraic curves, a part of Hilbert's 22nd problem about the uniformization of analytic equations [11], is to find algorithmically a discontinuous Möbius group uniformizing a given Riemann surface or a algebraic curve. In spite of intensive studies since 1890's this problem remains unsolved in its most general form [1, 3–5, 7–10, 12–17, 20, 22–24].

P.J. Myrberg found in [15] a numerical way to approximate a special Schottky uniformization of hyperelliptic curves. The results of Myrberg, with some generalizations and proofs in the language of modern mathematics, have been presented in [2, 18, 21].

When adapted to elliptic curves, the algorithm of Myrberg yields a simple and straightforward numerical uniformization which converges rapidly. The purpose of this paper is to explain Myrberg's uniformization in the case of elliptic curves and to show how the classical uniformization of elliptic curves can be derived from Myrberg's uniformization.

1. Myrberg's algorithm for elliptic curves

Consider an elliptic curve C defined by the affine equation

$$y^2 = (x - \lambda_1)(x - \lambda_2)(x - \lambda_3)(x - \lambda_4) \tag{1}$$

The algorithm is based on iteration of certain auxiliary functions which will be introduced next.

1.1. *Opening of slots*

Let a and b be two distinct points in the complex plane $\mathbb{C}$, and let I be the straight line segment from a to b. Let $m = \frac{a+b}{2}$ and $s = \frac{b-a}{2}$. Using a suitable choice of the branch of the square-root, the equation

$$\frac{z-(m+2s)}{z-(m-2s)} = \left(\frac{w-(m+s)}{w-(m-s)}\right)^2 \tag{2}$$

defines a mapping $z \mapsto w = \phi(z)$ which is defined in the complement of I and maps the complement of I on the outside of a disk with center at m and with radius $|s|$.

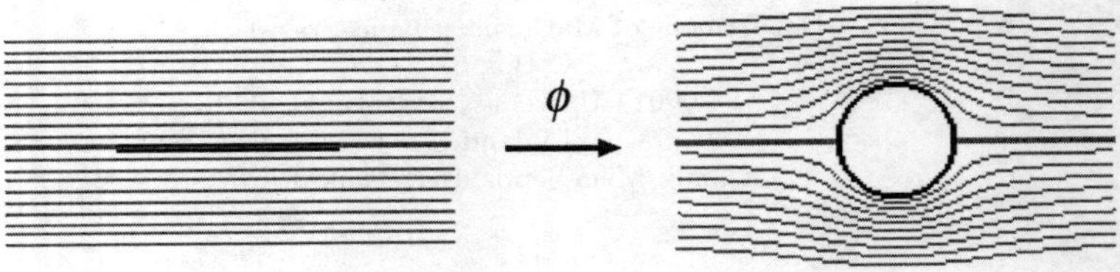

Fig. 3. The slot I is the black interval above left. Its complement gets mapped onto the complement of a disk by the mapping ϕ. The disk on the right has the same center as the slot on the left. The diameter of the disk is one half of the length of the slot.

Consider the elliptic rotation by angle π of the Riemann sphere $\hat{\mathbb{C}}$ with fixed-points $m+s$ and $m-s$. This mapping e is given by

$$e(z) = \frac{mz - m^2 + s^2}{z - m} \tag{4}$$

The mapping e maps the outside of the disk in Figure (3) onto the inside of the same disk. The composed mapping $e \circ \phi$ corresponds to the other choice of the branch of the square root when solving w in terms of z from equation (2).

The inverse of the opening mapping ϕ is the mapping

$$\Phi(w) = \frac{w^2 - wm + s^2}{w - m}. \tag{5}$$

1.2. *Parameterizations of elliptic curves*

Consider now a projective algebraic curve C defined by the affine equation (1). Observe that the mapping $(z, y) \mapsto z$ is a 2 to 1 branched covering of the curve C onto the Riemann sphere $\hat{\mathbb{C}}$.

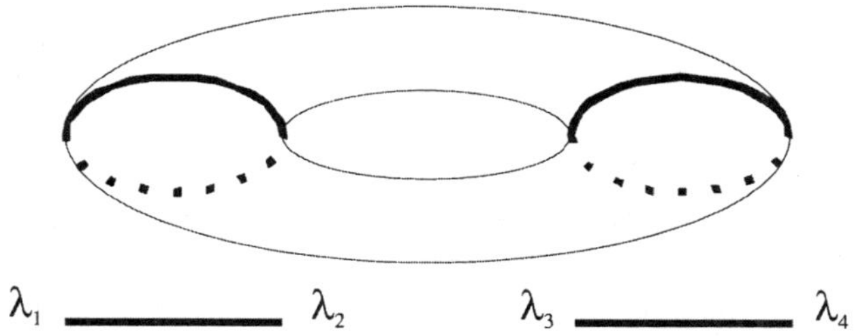

Fig. 6. The algebraic curve defined by (1) is a torus. It is a 2 to 1 covering of the Riemann sphere with four branch points. The slots below the torus both correspond to simple closed curves on the torus. Each of these curves passes through two branch points. The parameters λ_j of equation (1) are the end points of the slots. This is the classical way to view the curve C defined by (1).

Let I_j be a straight line segment with end points λ_{2j-1} and λ_{2j}, $j = 1, 2$. Assume that these line segments do not intersect. On the complement of $I_1 \cup I_2$

$$\pi(z) = \left(z, \sqrt{(z - \lambda_1)(z - \lambda_2)(z - \lambda_3)(z - \lambda_4)}\right) \tag{7}$$

is a well defined holomorphic function. It maps the complement of $I_1 \cup I_2$ onto a half of the torus in Figure (6). The other half of the torus corresponds to the other choice of the branch of the square root in (7).

Now one would like to extend the mapping (7) so that its image will cover the whole curve C. That is not possible, without some modifications, since the complement of the domain of definition of the mapping (7) is a union of two slots. For the extension we need an auxiliary mapping that opens slots to disks. To this end we use the mapping ϕ to open the first slots I_1 and I_2. In this process other slots will be created and they also have to be opened.

Apply the above opening mapping to open the first slot I_1 associated to the elliptic curve C defined by 1. Set

$$a = \lambda_1 \text{ and } b = \lambda_2.$$

Let ϕ be the corresponding opening mapping, Φ the inverse of ϕ, and let $D_1 = \phi(I_1)$ and $I_2^1 = \phi(I_2)$.

The composed mapping $\pi \circ \Phi$ is a mapping of the exterior of $D_1 \cup I_2^1$ onto a half of the curve C. Let $i = C \to C$, $i(z,y) = (z, -y)$. Extend the mapping

$$\pi \circ \Phi : \hat{\mathbb{C}} - (D_1 \cup I_2^1) \to C$$

to $D_1 - e(I_2^1)$ by

$$i \circ \pi \circ \Phi \circ e.$$

Extended in this way the mapping $\pi \circ \Phi$ covers now all of the curve C except a simple closed path going through the branch points $(\lambda_3, 0)$ and $(\lambda_4, 0)$. To extend the mapping $\pi \circ \Phi$ so that its image covers also that path one has to open the slot I_2^1.

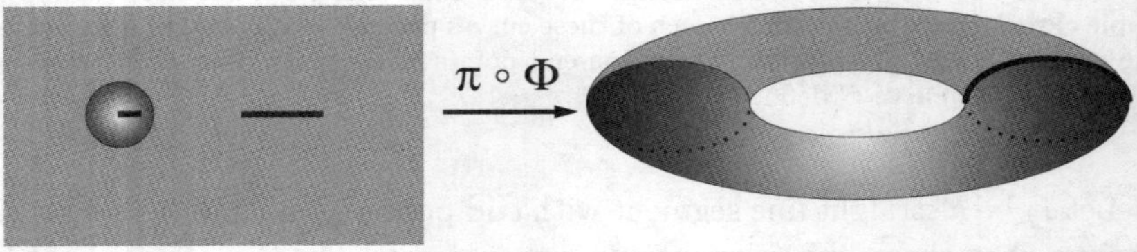

Fig. 7. The mapping $\pi \circ \Phi$ maps the outside of the disk and the slot I_2^1 (marked by the thick black line) on the left onto the upper half of the torus on the right. The elliptic rotation e maps the slot I_2^1 onto a much shorter slot $I_2^2 = e(I_2^1)$ inside the disk. Using the rotation e one extends the mapping $\pi \circ \Phi$ to the inside of the disk (excluding I_2^2). The elliptic rotation e corresponds to the mapping $(z, y) \mapsto (z, -y)$ of the algebraic curve C onto itself.

Each time when a slot is being opened a new slot will be created. The new slot is the image of the slot that has not been opened yet under the corresponding elliptic rotation that maps the outside of a disk onto its inside.

Iterate this process, and keep on opening the new slots that get created. Let $\phi_j, j = 1, 2, \ldots$ be the opening mappings that get formed when iterating the process described above. Form the sequences $(\lambda_j^k), j = 1, \ldots, 4$, setting

$$\lambda_j^1 = \phi_1(\lambda_j), \text{ and } \lambda_j^{k+1} = \phi_k(\lambda_j^k) \text{ for } j = 1, \ldots, 4. \quad (8)$$

A qualitative study regarding the behavior of the opening mappings shows that these sequences converge; see [21]. Convergence is fast.

Let now

$$\lambda_j^\infty = \lim_{k\to\infty} \lambda_j^k,$$

and let e_j be the elliptic rotation with rotation angle π and with fixed-points λ_{2j-1}^∞ and $\lambda_{2j}^\infty, j = 1, 2$.

Theorem 9: The group G generated by the loxodromic Möbius transformation $g = e_1 \circ e_2$ acts discontinuously in the complement Ω of the fixed-points of g and

$$C = \Omega/G$$

where C is the elliptic curve defined by (1).

Proof of this result follows from the above construction. The only point, which has not been discussed in the above, is the convergence of the sequences defined by (8). This has been shown, in a more general case, [21, Theorem 3], see also [15, 18].

Observe that by a conjugation by a Möbius transformation we may assume that the Möbius transformation g in Theorem (9) has 0 and ∞ as its repelling and attracting fixed points respectively. Then g is of the form $g(z) = \mu z$ with $|\mu| > 1$. The complex parameter μ parameterizes the moduli space of elliptic curves.

The usual classical uniformization of the elliptic curve C is by a lattice

$$L = \langle z \mapsto z + 1, z \mapsto z + \tau \rangle .$$

This maps to the uniformization constructed above by the mapping

$$z \mapsto e^{zi}$$

Thus Myrberg's construction gives a new way of finding also the classical uniformization of elliptic curves by lattices.

1.3. *Parameterization of hyperelliptic curves*

The above construction can be carried out in the case of general hyperelliptic curves without any changes. We consider projective hyperelliptic curves given by an affine equation of the type

$$y^2 = (z - \lambda_1)(z - \lambda_2) \cdots (z - \lambda_{2p+2}) \tag{10}$$

and assume that the branch points λ_j are distinct and ordered in such a way the slots $I_j, j = 1, \ldots, p + 1$, do not intersect. Here I_j is the straight line segment with end-points λ_{2j-1} and λ_{2j}. These assumptions do not limit the generality. The only difference to the elliptic curve case is that there are more slots.

By opening slots one creates, as before, the sequences (λ_j^k), $k = 1, \ldots, p+1$. Let

$$\lambda_j^\infty = \lim_{k\to\infty} \lambda_j^k, \quad j = 1, \ldots, p+1.$$

The above sequences converge and the limit points are distinct [21]. Let e_j be the elliptic rotation of angle π fixing λ_{2j-1}^∞ and $\lambda_{2j}^\infty, j = 1, \ldots, p+1$. The Möbius transformations $e_j \circ e_i, i \neq j$, are either loxodromic or hyperbolic. The Möbius group $G = \langle e_j \circ e_i | i, j = 1, \ldots, p+1\rangle$ acts discontinuously in the complement Ω of its limit set and the quotient Ω/G is the hyperelliptic curve defined by the equation (10). For details see [21].

2. Generalizations of Myrberg's algorithm

It appears possible to generalize the above to curves defined by affine equations of the type

$$y^n = (z - \lambda_1)(z - \lambda_2)\cdots(z - \lambda_{2p+2}). \tag{11}$$

In stead of the opening mapping defined by equation 2 we open the respective slots by mappings defined by the equation

$$\frac{z - (m + 2s)}{z - (m - 2s} = \left(\frac{w - (m + s)}{w - (m - s)}\right)^n \tag{12}$$

The elliptic rotations e of angle π will have to be replaced by elliptic rotations of angle $2\pi/n$.

It was pointed out by Myrberg in [15] that this will work. The details are, however, not contained in his paper and have to be worked out.

References

1. Burnside, W. (1893). Note on the equation $y^2 = x(x^4 - 1)$. *Proc. London Math. Soc.* **24**(1): 17-20.
2. Buser, P. and M. Seppälä (1996). Computing on Riemann surfaces. *Topology and Teichmüller spaces (Katinkulta, 1995)*: 5-30.
3. Buser, P. and R. Silhol (2001). Geodesics, periods, and equations of real hyperelliptic curves. *Duke Math. J.* **108**(2): 211-250.
4. Gianni, P., M. Seppälä, et al. (1998). Riemann surfaces, plane algebraic curves and their period matrices. *J. Symbolic Comput* **26**(6): 789–803.
5. Hejhal, D. A. (1974). Sur les paramétres accessoires pour l'uniformisation de Schottky. *C. R. Acad. Sci. Paris Sér. A* **279**: 713-716.

6. HEJHAL, D. A. (1974). Sur les paramétres accessoires pour l'uniformisation de Schottky. *C. R. Acad. Sci. Paris Sér. A* **279**: 695-697.
7. HEJHAL, D. A. (1976). Sur les paramétres accessoires pour l'uniformisation fuchsienne. *C. R. Acad. Sci. Paris Sér. A-B* **282**(8): Ai, A403-A406.
8. HEMPEL, J. A. (1988). On the uniformization of the n-punctured sphere. *Bull. London Math. Soc.* **20**(2): 97-115.
9. HEMPEL, J. A. and S. J. SMITH (1989). Uniformisation of the twice-punctured disc – problems of confluence. *Bull. Austral. Math. Soc.* **39**(3): 369-387.
10. HIDALGO, R. (2001). Numerical uniformization of hyperelliptic-M-symmetric Riemann surfaces. *Proyecciones* **20**(3): 351-365.
11. HILBERT, D. (1900). Uniformisirung analytischer Beziehungen mittelst automorpher Functionen. *Akad. Wiss. Göttingen*: 290-291.
12. KEEN, L. (1983). Accessory parameters and the uniformization of punctured tori. *Modular functions in analysis and number theory* **5**: 132-149.
13. KEEN, L., H. E. RAUCH, et al. (1979). Moduli of punctured tori and the accessory parameter of Lamé's equation. *Trans. Amer. Math. Soc.* **255**: 201-230.
14. KRA, I. Accessory parameters for punctured spheres. *Trans. Amer. Math. Soc.* **313**(2): 589-617, (1989).
15. MYRBERG, P. J. (1920). Über die Numerische Ausführung der Uniformisierung. *Acta Soc. Sci. Fenn.* **XLVIII**(7): 1-53.
16. RANKIN, R. A. (1958). Sir Edmund Whittaker's work on automorphic functions. *Proc. Edinburgh Math. Soc.* **11**: 25-30.
17. RODRIGUEZ, R. E. and V. GONZÁLEZ-AGUILERA (1997). Fermat's quartic curve, Klein's curve and the tetrahedron. *Extremal Riemann surfaces (San Francisco, CA, 1995)* **201**: 43-62.
18. SEMMLER, K. D. and M. SEPPÄLÄ (1995). *Numerical uniformization of hyperelliptic curves.* ISSAC 1995.
19. SEMMLER, K.-D. and M. SEPPÄLÄ (1995). *Numerical uniformization of hyperelliptic curves.* ISSAC, Montreal, Canada.
20. SEPPÄLÄ, M. (1996). Computational conformal geometry. *Algorithms in algebraic geometry and applications (Santander, 1994)* **143**: 365-372.
21. SEPPÄLÄ, M. (2004). Myrberg's numerical uniformization of hyperelliptic curves. *Ann. Acad. Sci. Fenn. Math.* **29**: 3-20.
22. SMITH, S. J. and J. A. HEMPEL (1989). The accessory parameter problem for the uniformization of the twice-punctured disc. *J. London Math. Soc. (2)* **40**(2): 269-279.
23. WAGNER, M. (2001). Numerische Uniformisierung hyperelliptischer Kurven, EPFL.
24. WHITTAKER, E. T. (1898). On the connexion of algebraic functions with automorphic functions. *Phil. Trans.* **192A**: 1-32.

On the moduli space of Klein four covers of the projective line

D. Glass & R. Pries *

Department of Mathematics
Columbia University
New York, NY 10027
glass@math.columbia.edu

101 Weber Building
Colorado State University
Fort Collins, CO 80523-1874
pries@math.colostate.edu

We study the sublocus $\mathcal{H}_{g,n}$ of $\mathcal{M}_g$ whose points correspond to curves which are $(\mathbb{Z}/2\mathbb{Z})^n$ covers of the projective line. In the case that $n = 2$, we describe a set of components of $\mathcal{H}_{g,2}$ and all of the intersections between these components. Unlike the case of the hyperelliptic locus $\mathcal{H}_{g,1}$ which is well-known to be connected, we show that $\mathcal{H}_{g,n}$ is not connected when $n = 2$ and $g \geq 4$ or when $n = 3$ and $g \geq 5$.

1. Introduction

In this paper, we study the locus of smooth curves of genus g which have multiple commuting involutions. More precisely, we study the sublocus $\mathcal{H}_{g,n}$ of $\mathcal{M}_g$ whose points correspond to curves which are $(\mathbb{Z}/2\mathbb{Z})^n$ covers of the projective line. This locus is a generalization of the hyperelliptic locus $\mathcal{H}_g$. In the case that $n = 2$, we describe a set of components of $\mathcal{H}_{g,2}$ and all of the intersections between these components, Theorem 9. Unlike the case of $\mathcal{H}_g$ $(n = 1)$ which is well-known to be connected, we show that $\mathcal{H}_{g,n}$ is not connected when $n = 2$ and $g \geq 4$ or when $n = 3$ and $g \geq 5$, Propositions 6 and 14.

*The second author was partially supported by NSF grant DMS-04-00461.

This paper can be viewed as part of a larger project to understand the sublocus S_g of the moduli space $\mathcal{M}_g$ of smooth curves of genus g whose points correspond to curves with automorphisms. For $g \geq 4$, S_g is the singular locus of $\mathcal{M}_g$, [8]. In [2], Cornalba describes the irreducible components of S_g. The generic point of one of these components corresponds to a curve whose automorphism group is cyclic of prime order.

It is not known in general how the irreducible components of S_g intersect. One approach is to describe all Hurwitz loci and the inclusions between them using algorithms that measure the action of the braid group, see for example [6], [7]. The approach in this paper is complementary in that we restrict to one choice of automorphism group and search for results which are true for all genera.

We would like to thank T. Shaska for inviting us to contribute to this volume.

2. Hurwitz Spaces

Let k be an algebraically closed field of characteristic p where $p = 0$ or $p > 2$. Let G be an elementary abelian 2-group of order 2^n. In Section 2.1, we recall results on the moduli space $H_{g,n}$ parametrizing G-Galois covers $f : X \to \mathbb{P}^1_k$ where X is a smooth projective k-curve of genus g. In Section 2.2, we describe the connected components of $H_{g,n}$.

2.1. *Background*

Let $F_{g,n}$ be the contravariant functor which associates to any k-scheme Ω the set of isomorphism classes of $(\mathbb{Z}/2)^n$-Galois covers $f_\Omega : X_\Omega \to \mathbb{P}^1_\Omega$ where X is a flat Ω-curve whose fibres are smooth projective curves of genus g and where the branch locus B of f_Ω is a simple horizontal divisor. In other words, the branch locus consists of Ω-points of $\mathbb{P}^1_\Omega$ which do not intersect. Since each inertia group is a cyclic group of order 2, the Riemann-Hurwitz formula [4, Corollary 2.4] implies $g = 2^{n-2}|B| - 2^n + 1$. We assume that $g \geq 1$.

It is well-known that there exists a coarse moduli space $H_{g,n}$ for the functor $F_{g,n}$ which is of finite type over $\mathbb{Z}[1/2]$. (For example, see [10, Chapter 10] when $k = \mathbb{C}$ and [11, Theorem 4] when $p > 2$.) There is a natural morphism $\tau : H_{g,n} \to \mathcal{M}_g$ whose fibres have dimension three. The morphism τ associates to any Ω-point of $H_{g,n}$ the isomorphism class of X_Ω, where $f_\Omega : X_\Omega \to \mathbb{P}^1_\Omega$ is the corresponding cover of Ω-curves. The fibres have dimension three since X_Ω is isotrivial if and only if after an étale base change from Ω to Ω' there is a projective linear transformation ρ such that $\rho f_{\Omega'}$ is constant.

We denote by $\mathcal{H}_{g,n}$ the image $\tau(H_{g,n})$ in $\mathcal{M}_g$. Given a smooth connected k-curve X, then X corresponds to a point of $\mathcal{H}_{g,n}$ if and only if $G \subset \mathrm{Aut}(X)$ with quotient $X/G \simeq \mathbb{P}^1_k$. Note that $\mathcal{H}_{g,1}$ is simply the locus $\mathcal{H}_g$ of hyperelliptic curves in $\mathcal{M}_g$.

There is also a natural morphism $\beta : H_{g,n} \to \mathbb{P}^{|B|}$ which is proper and étale over the image. The morphism β associates to any Ω-point of $H_{g,n}$ the Ω-point of $\mathbb{P}^{|B|}$ determined by the branch locus of the associated cover. More specifically, β associates to any cover $f_\Omega : X_\Omega \to \mathbb{P}^1_\Omega$ the Ω-point $[c_0 : \ldots : c_{|B|}]$ of $\mathbb{P}^{|B|}$ where c_i are the coefficients of the polynomial whose roots are the branch points of f_Ω. Note that the k-points of the image of β correspond to polynomials with no multiple roots.

It is often more useful to describe the branch locus of f_Ω directly as an Ω-point of $(\mathbb{P}^1)^{|B|}$. This can be done by considering an ordering of the branch points of f_Ω. The branch locus of a cover corresponding to a k-point of $H_{g,n}$ can be any k-point of $(\mathbb{P}^1)^{|B|} - \Delta$ where Δ is the weak diagonal consisting of points having at least two equal coordinates. In particular, for any Ω-point $(b_1, \ldots, b_{2g+2})$ of $(\mathbb{P}^1)^{2g+2} - \Delta$ there is a unique hyperelliptic cover $f_\Omega : X_\Omega \to \mathbb{P}^1_\Omega$ branched at $\{b_1, \ldots, b_{2g+2}\}$. Also the curve X_Ω has genus g.

In [3, Section 3.2], we proved several results about the points of $H_{g,n}$. First, by [3, Lemma 3.3], a cover $f : X \to \mathbb{P}^1$ corresponds to a point of $H_{g,n}$ if and only if X has genus g and $f : X \to \mathbb{P}^1$ is isomorphic to the normalized fibre product over $\mathbb{P}^1$ of n smooth hyperelliptic covers $C_i \to \mathbb{P}^1$ whose branch loci B_i satisfy a strong disjointedness condition. In the case $n = 2$, this condition merely says that $B_1 \neq B_2$. Second, suppose $f : X \to \mathbb{P}^1$ is the normalized fibre product over $\mathbb{P}^1$ of n smooth hyperelliptic covers $C_i \to \mathbb{P}^1$ with branch loci B_i. If $H \subset G$ has order 2^{n-1}, in [3, Lemma 3.2] we determined the branch locus of the hyperelliptic cover $X/H \to \mathbb{P}^1$ in terms of the branch loci B_i. For example, in the case $n = 2$, the quotient of X by the third involution of $\mathbb{Z}/2 \times \mathbb{Z}/2$ is a hyperelliptic cover of $\mathbb{P}^1$ with branch locus $(B_1 \cup B_2) \backslash (B_1 \cap B_2)$. Finally, for $n \geq 2$, by [3, Corollary 3.4], the locus $\mathcal{H}_{g,n}$ has dimension $(g + 2^n - 1)/2^{n-2} - 3$ if $g \equiv 1 \bmod 2^{n-2}$ and is empty otherwise. For example, $\mathcal{H}_{g,2}$ is non-empty of dimension g for all g and $\mathcal{H}_{g,3}$ has dimension $(g+1)/2$ if g is odd and is empty if g is even.

We will frequently refer to the following consequence of a theorem of Kani and Rosen. For example, it indicates that if $X \in \mathcal{H}_{g,2}$ then $g(X) =$

$g(X/\alpha_1) + g(X/\alpha_2) + g(X/\alpha_1\alpha_2)$ where α_1, α_2, and $\alpha_1\alpha_2$ denote the three involutions of $\mathbb{Z}/2 \times \mathbb{Z}/2$.

Theorem 1: Suppose X is a smooth projective curve of genus g and $G \subset \mathrm{Aut}(X)$. Suppose $H_i \subset G$ for $1 \leq i \leq r$ are subgroups of G for which $\cup_{i=1}^r H_i = G$ and $H_i \cap H_j = \{\mathrm{Id}\}$ for $i \neq j$. Then $(r-1)g(X) + |G|g(X/G) = \sum_{i=1}^r |H_i| g(X/H_i)$.

Proof: This follows immediately from [5], in which the authors prove that $(r-1)\,\mathrm{Jac}(X) + |G|\,\mathrm{Jac}(X/G)$ is isogenous to $\sum_{i=1}^r |H_i|\,\mathrm{Jac}(X/H_i)$. □

2.2. *Connected components of $H_{g,n}$*

This section contains some remarks on the components of $H_{g,n}$ and $\mathcal{H}_{g,n}$. Let $N = 2^n - 1$. We fix an ordering of the N subgroups $H_1, \ldots, H_N$ of G of order 2^{n-1} so that $G = \prod_{i=1}^n G/H_i$.

Proposition 2: *The Hurwitz space $H_{g,n}$ is the disjoint union of connected schemes $H_{g,n,\vec{g}}$ indexed by tuples $\vec{g} = (g_1, \ldots, g_N)$ with $g_i \in \mathbb{N}$ and $\Sigma_{i=1}^N g_i = g$.*

The dimension of each non-empty connected component $H_{g,n,\vec{g}}$ is the same, [3, Corollary 3.4]. However, some of the components $H_{g,n,\vec{g}}$ can be empty.

Proof: A point of $H_{g,n}$ corresponds to a cover $f : X \to \mathbb{P}^1$ along with an isomorphism between G and a subgroup of $\mathrm{Aut}(X)$. The cover f has N hyperelliptic quotients $C_1 \to \mathbb{P}^1, \ldots, C_N \to \mathbb{P}^1$ which are ordered by the fixed choice of ordering of the subgroups of index two of G. Let $\vec{g} = (g_1, \ldots, g_N)$ be the genera of $C_1, \ldots, C_N$. Using [3, Lemma 3.2], one can show that $\Sigma_{i=1}^N g_i = g$, see [3, Proposition 3.5]. For any tuple $\vec{g}$ with $g_i \in \mathbb{N}$ and $\Sigma_{i=1}^N g_i = g$, let $H_{g,n,\vec{g}}$ denote the locus of isomorphism classes of covers $f : X \to \mathbb{P}^1$ in $H_{g,n}$ whose hyperelliptic quotients have genera $\vec{g}$. The intersection of two of the strata $H_{g,n,\vec{g}}$ is empty since the tuple corresponding to the cover $f : X \to \mathbb{P}^1$ is well-defined.

To show $H_{g,n,\vec{g}}$ is connected, it is sufficient to show that its fibre over $\mathbb{C}$ is connected by [11, Theorem 4]. By GAGA [11], it is sufficient to work in the topological category. Consider two covers $f : X \to \mathbb{P}^1$ and $f' : X' \to \mathbb{P}^1$ corresponding to points in $H_{g,n,\vec{g}}$. Let B (resp. B') be the branch locus of f (resp. f'). The chosen ordering of the subgroups of index two of G yields a well-defined bijection between the hyperelliptic quotients $C_i \to \mathbb{P}^1$ of f and

$C'_i \to \mathbb{P}^1$ of f' for $1 \leq i \leq N$. Since f and f' are in the same locus $H_{g,n,\vec{g}}$, the genus of C_i and the genus of C'_i are equal. This implies that there is a bijection $\iota : B \to B'$ so that b is a branch point of $C_i \to \mathbb{P}^1$ if and only if $\iota(b)$ is a branch point of $C'_i \to \mathbb{P}^1$ for $1 \leq i \leq N$.

There is a morphism $\delta : [0,1] \to (\mathbb{P}^1_{\mathbb{C}})^{|B|} - \Delta$ so that: B is the set of coordinates of $\delta(0)$; B' is the set of the coordinates of $\delta(1)$; $b \in B$ is the jth coordinate of $\delta(0)$ if and only if $\iota(b)$ is the jth coordinate of $\delta(1)$; and the image of δ avoids Δ.

Consider the composition δ_1 of δ with the natural morphism $(\mathbb{P}^1)^{|B|} \to \mathbb{P}^{|B|}$. Since β is étale, δ_1 lifts to a unique morphism $\tilde{\delta} : [0,1] \to (H_{g,n})_{\mathbb{C}}$ with $\tilde{\delta}(0) = f$. In fact, the image of $\tilde{\delta}$ is in $(H_{g,n,\vec{g}})_{\mathbb{C}}$. To see this, note that the lifting $\tilde{\delta}$ yields a deformation of f which must preserve the size of the branch locus and the genus of each hyperelliptic quotient of f. The fact that $G = \prod_{i=1}^{n} G/H_i$ implies that f (resp. f') is the normalized fibre product of the first n hyperelliptic covers $C_i \to \mathbb{P}^1$ (resp. $C'_i \to \mathbb{P}^1$), see [3, Lemma 3.3]. It follows that $\tilde{\delta}(1) = f'$ since f' is the unique fibre product of hyperelliptic covers with branch loci B'_i for $1 \leq i \leq n$. Since $(H_{g,n,\vec{g}})_{\mathbb{C}}$ is path connected, it is also connected. □

We will now concentrate on the irreducible components of $\mathcal{H}_{g,n}$ rather than $H_{g,n}$. Suppose $\xi \in \mathcal{H}_{g,n}$ and let X be the corresponding curve which admits an action of G. The fibre $\tau^{-1}(\xi) \in H_{g,n}$ corresponds to the set of G-Galois covers $f : X \to \mathbb{P}^1$. There is a natural action of $\mathrm{Aut}(G) = \mathrm{GL}_n(\mathbb{Z}/2)$ on this fibre. The action of $\mathrm{Aut}(G)$ permutes the components $H_{g,n,\vec{g}}$ of $H_{g,n}$ since it permutes the hyperelliptic quotients of X. On the other hand, a permutation of the ordered set of hyperelliptic quotients of X by an element of $\mathrm{GL}_n(\mathbb{Z}/2)$ yields a permutation of this fibre. The irreducible components of $\mathcal{H}_{g,n}$ are indexed by the orbits of the tuples $\vec{g}$ by the action of $\mathrm{Aut}(G)$. In particular, when $n = 2$, $\mathrm{Aut}(G) \simeq S_3$ and thus the components of $\mathcal{H}_{g,2}$ are indexed by unordered triples $\{g_1, g_2, g_3\}$ such that $g_i \in \mathbb{N}$ and $g_1 + g_2 + g_3 = g$.

3. The geometry of $\mathcal{H}_{g,2}$

In this section, we restrict to the case $n = 2$ and investigate the geometry of the locus $\mathcal{H}_{g,2}$. We note that $\mathcal{H}_{g,2}$ is the union of components $\mathcal{H}_{g,2,\mathfrak{p}} = \tau(H_{g,2,\vec{g}})$ where $\mathfrak{p}$ ranges over the partitions $\{g_1, g_2, g_3\}$ of g. We assume without loss of generality that $g_1 \leq g_2 \leq g_3$. A priori, these components of $\mathcal{H}_{g,2}$ can intersect since the automorphism group of a curve X may contain several copies of $(\mathbb{Z}/2)^2$. In Section 3.3, we show that this in fact rarely happens.

3.1. *Number of irreducible components*

Lemma 3: *If g is even (resp. odd) then the number of non-empty irreducible components of $\mathcal{H}_{g,2}$ is $[(g+6)^2/48]$ (resp. $[(g+3)^2/48]+\lfloor (g+3)/4 \rfloor$).*

Here $[x]$ denotes the integer closest to x and one can check that the fractional parts of the expressions in Lemma 3 do not equal 1/2.

Proof: Let N_g denote the number of non-empty irreducible components $\mathcal{H}_{g,2,\mathfrak{p}}$ of $\mathcal{H}_{g,2}$. These components are indexed by partitions $\mathfrak{p} = \{g_1, g_2, g_3\}$ of g. In the next paragraph, we show that N_g equals the number of such partitions $\mathfrak{p}$ so that $g_3 \leq g_1 + g_2 + 1$.

Suppose $\mathfrak{p}$ is a partition so that $\mathcal{H}_{g,2,\mathfrak{p}}$ is non-empty and let $f : X \to \mathbb{P}^1$ denote a $(\mathbb{Z}/2)^2$-cover corresponding to a point of $\mathcal{H}_{g,2,\mathfrak{p}}$. By [3, Lemma 3.2], the branch loci of the three $\mathbb{Z}/2$-quotients of f satisfy $B_3 = (B_1 \cup B_2) \backslash (B_1 \cap B_2)$. It follows that $|B_3| \leq |B_1| + |B_2|$ and so $g_3 \leq g_1 + g_2 + 1$. Conversely, if $\mathfrak{p} = \{g_1, g_2, g_3\}$ is a partition of g so that $g_1 \leq g_2 \leq g_3$ and $g_3 \leq g_1 + g_2 + 1$, then $\mathcal{H}_{g,2,\mathfrak{p}}$ is non-empty. Namely, $\mathcal{H}_{g,2,\mathfrak{p}}$ contains a point corresponding to the fibre product of two hyperelliptic covers f_1 and f_2 whose branch loci are chosen with the following restrictions: $|B_1| = 2g_1 + 2$, $|B_2| = 2g_2 + 2$, and $|B_1 \cap B_2| = 2e$ where $e = g_1 + g_2 + 1 - g_3$.

If g is even, then $g_3 \leq g_1 + g_2 + 1$ implies that $g_3 \leq g_1 + g_2$. Thus N_g equals the number of (possibly degenerate) triangles with perimeter g and sides of integer length. By [1], $N_g = [(g+6)^2/48]$ (when g is even).

If g is odd, then $g_3 \neq g_1 + g_2$ and there are two cases to consider. If $g_3 \leq g_1 + g_2 - 1$ then $1 \leq g_1 \leq g_2 \leq g_3$. There is a bijective correspondence between these partitions of g and the partitions $\{a_1, a_2, a_3\}$ of $g - 3$ with $a_1 \leq a_2 \leq a_3$ and $a_3 \leq a_1 + a_2$ (by taking $a_i = g_i - 1$). It follows that these partitions correspond to the number of (possibly degenerate) triangles with sides of integer length and perimeter $g - 3$. By [1], there are $[(g+3)^2/48]$ such triples. In the other case, $g_3 = g_1 + g_2 + 1$. The number of these partitions equals the number of pairs $0 \leq g_1 \leq g_2$ so that $g = 2g_1 + 2g_2 + 1$. There are $\lfloor (g+3)/4 \rfloor$ such pairs. □

When $n > 2$, it is more difficult to count the number of nonempty strata, as the interactions between the numbers g_i pose additional restrictions on the partitions.

3.2. *Intersection of Components*

Suppose g is odd. Let $\mathfrak{p}_1 = \{1, (g-1)/2, (g-1)/2\}$ and $\mathfrak{p}'_1 = \{1, (g-3)/2, (g+1)/2\}$. These are the only two partitions $\{1, x, y\}$ of g satisfying the constraint $x, y \leq (g+1)/2$. They correspond to two irreducible components of $\mathcal{H}_{g,2}$. In this section we show that these two components intersect. In Section 3.3, we show that the irreducible components corresponding to $\mathfrak{p}_1$ and $\mathfrak{p}'_1$ are the only irreducible components of $\mathcal{H}_{g,2}$ which intersect.

Recall that the dihedral group D_{2c} has presentation $\langle s, r \mid s^2 = r^c = 1,\ srs = r^{-1}\rangle$. Suppose $G = D_{2c} \times \mathbb{Z}/2$ where r and s generate D_{2c} as above and τ is a generator of $\mathbb{Z}/2$.

Lemma 4: *Suppose $c, d \in \mathbb{N}^+$ with c even and d odd. Suppose $G = D_{2c} \times \mathbb{Z}/2$. There exists a $(d+1)$-dimensional family of curves X having an action by G so that $g(X) = 1 + cd$, $g(X/G) = g(X/\langle s, \tau\rangle) = g(X/\langle sr, \tau\rangle) = 0$, and $g(X/s) \neq g(X/sr)$.*

Proof: Let $B = \{0, \infty, 1, \lambda', \lambda_1, \ldots, \lambda_d\}$ be a set of $4 + d$ distinct points in $\mathbb{P}^1$. Let $\gamma_0 = \gamma_\infty = s$, $\gamma'' = sr$, $\gamma' = sr\tau$, and $\gamma_i = \tau$ for $1 \leq i \leq d$. We note that $\gamma_0\gamma_\infty\gamma''\gamma'\Pi_{i=1}^d\gamma_i = 1$ and that these elements generate G. By Riemann's Existence Theorem [10, Theorem 2.13], there exists a $D_{2c} \times \mathbb{Z}/2$ cover $f : X \to \mathbb{P}^1_k$ branched at B so that γ_i is the canonical generator of inertia at a point in the fibre above λ_i (resp. γ_0 above 0, γ_∞ above ∞, γ'' above 1, and γ' above λ'). Above 0 and ∞, the fibre of f consists of $2c$ points; for $1 \leq j \leq c/2$, there are four points in this fibre with inertia group $\langle sr^{2j}\rangle$. Above 1 (resp. λ'), the fibre of f consists of $2c$ points; for $1 \leq j \leq c/2$, there are four points in this fibre with inertia group $\langle sr^{2j-1}\rangle$ (resp. $\langle sr^{2j-1}\tau\rangle$). For $1 \leq i \leq d$, the fibre of f consists of $2c$ points each of which has inertia $\langle\tau\rangle$. By the Riemann-Hurwitz formula [4, IV.2.4], $g(X) = 1 + cd$.

Suppose σ has order 2 in G. Let $F_\sigma = \{P \in X | \sigma(P) = P\}$. By the Riemann-Hurwitz formula, $g(X/\sigma) = (g(X)+1)/2 - |F_\sigma|/4$. We calculate:

σ	s	$s\tau$	sr	$sr\tau$	τ
$\lvert F_\sigma\rvert$	8	0	4	4	$2cd$
$g(X/\sigma)$	$(g(X)-3)/2$	$(g(X)+1)/2$	$(g(X)-1)/2$	$(g(X)-1)/2$	1

It follows that $g(X/s) \neq g(X/sr)$. The fact that $g(X) = g(X/s) + g(X/\tau) + g(X/s\tau)$ implies that $g(X/\langle s, \tau\rangle) = 0$. One shows that $g(X/\langle sr, \tau\rangle) = 0$ similarly. □

We note that the group D_{2c} acts on the quotient X/τ which has genus 1. The element r acts on X/τ via translation by a point of order c while s is the hyperelliptic involution.

Proposition 5: *Let g be odd. The irreducible components $\mathcal{H}_{g,2,\mathfrak{p}_1}$ and $\mathcal{H}_{g,2,\mathfrak{p}'_1}$ intersect.*

Proof: Let $g - 1 = cd$ where d is an odd number and c is a power of 2. Applying Lemma 4, there exists a family of dimension $d + 1$ of curves X of genus g whose automorphism group contains two copies of the Klein four group (namely $\langle s, \tau \rangle$ and $\langle sr, \tau \rangle$) so that the two sets of quotients yield the two partitions $\mathfrak{p}_1$ and $\mathfrak{p}'_1$. □

We note that if $g \equiv 3 \bmod 4$, then $\mathcal{H}_{g,2,\mathfrak{p}_1} \cap \mathcal{H}_{g,2,\mathfrak{p}'_1}$ contains a family of dimension $(g + 1)/2$ whose generic point has automorphism group $(\mathbb{Z}/2)^3$.

3.3. *Nonintersection of Components*

In this section, we prove the converse of Proposition 5: the irreducible components corresponding to $\mathfrak{p}_1$ and $\mathfrak{p}'_1$ are the only irreducible components of $\mathcal{H}_{g,2}$ which intersect. We use this to count the number of connected components of $\mathcal{H}_{g,2}$.

To begin, we study the intersection of $\mathcal{H}_{g,2}$ with $\mathcal{H}_g$ in order to show that $\mathcal{H}_{g,2}$ is not connected for $g > 3$. Let $\mathfrak{p}_0 = \{0, g/2, g/2\}$ if g is even and $\mathfrak{p}_0 = \{0, (g - 1)/2, (g + 1)/2\}$ if g is odd. Then $\mathfrak{p}_0$ is the unique partition of g of the form $\{0, x, y\}$ where $y \leq x + 1$. As such, it corresponds to a non-empty irreducible component $\mathcal{H}_{g,2,\mathfrak{p}_0}$ of $\mathcal{H}_{g,2}$.

Proposition 6: *The component $\mathcal{H}_{g,2,\mathfrak{p}_0}$ equals $\mathcal{H}_{g,2} \cap \mathcal{H}_g$. It is a connected component of $\mathcal{H}_{g,2}$ for $g > 3$. The sublocus $\mathcal{H}_{g,2}$ of $\mathcal{M}_g$ is connected if and only if $g \leq 3$.*

Proof: By definition, $\mathcal{H}_{g,2,\mathfrak{p}_0} \subset \mathcal{H}_{g,2} \cap \mathcal{H}_g$. Conversely, suppose X corresponds to a point of $\mathcal{H}_{g,2} \cap \mathcal{H}_g$. Then $\mathrm{Aut}(X)$ contains a hyperelliptic involution σ as well as another involution τ. Also $\langle \sigma, \tau \rangle \simeq (\mathbb{Z}/2)^2$ since σ is in the center of $\mathrm{Aut}(X)$. It follows that $X \to X/\langle \sigma, \tau \rangle$ is a Klein four cover of the projective line with a $\mathbb{Z}/2$-quotient of genus 0. Thus X corresponds to a point of $\mathcal{H}_{g,2,\mathfrak{p}_0}$ and $\mathcal{H}_{g,2,\mathfrak{p}_0} = \mathcal{H}_{g,2} \cap \mathcal{H}_g$.

Next we show that $\mathcal{H}_{g,2,\mathfrak{p}_0}$ is a connected component of $\mathcal{H}_{g,2}$ for $g > 3$ since it does not intersect any other component of $\mathcal{H}_{g,2}$. Suppose X is a curve corresponding to a point of intersection of $\mathcal{H}_{g,2,\mathfrak{p}_0}$ and $\mathcal{H}_{g,2,\mathfrak{p}}$ for some $\mathfrak{p} \neq \mathfrak{p}_0$. We write $\mathfrak{p} = \{a, b, c\}$ with $0 < a \leq g/3$. Let σ (resp. τ) be the involution of X such that $X/\langle\sigma\rangle$ has genus 0 (resp. such that $X/\langle\tau\rangle$ has genus a). As above, $X \to X/\langle\sigma, \tau\rangle$ is a $(\mathbb{Z}/2)^2$-cover of the projective line. This implies that X also corresponds to a point of $\mathcal{H}_{g,2,\mathfrak{p}'}$ with partition $\mathfrak{p}' = \{0, a, g-a\}$. Then $\mathfrak{p}' = \mathfrak{p}_0$ since $\mathfrak{p}_0$ is the unique partition for $\mathcal{H}_{g,2}$ which contains 0. The conditions $a \in \{(g-1)/2, g/2, (g+1)/2\}$ and $a \leq g/3$ imply that such a curve X can exist only if $g \leq 3$. It follows that the intersection $\mathcal{H}_{g,2,\mathfrak{p}_0} \cap \mathcal{H}_{g,2,\mathfrak{p}}$ is empty for $g \geq 4$ and $\mathfrak{p} \neq \mathfrak{p}_0$.

If $g \leq 2$, then $\mathcal{H}_{g,2,\mathfrak{p}_0}$ is the unique component of $\mathcal{H}_{g,2}$. If $g = 3$, then the two components of $\mathcal{H}_{g,2}$ intersect by Lemma 4. For $g > 3$, $\mathcal{H}_{g,2}$ has at least one pair of non-intersecting components. It follows that $\mathcal{H}_{g,2}$ is connected if and only if $g \leq 3$. □

We now consider the intersection of other components of $\mathcal{H}_{g,2}$.

Hypotheses: Suppose $\mathfrak{p} \neq \tilde{\mathfrak{p}}$ are two partitions of g so that $\mathcal{H}_{g,2,\mathfrak{p}} \cap \mathcal{H}_{g,2,\tilde{\mathfrak{p}}} \neq \emptyset$. Let X be a curve corresponding to a point of the intersection. Let $H_1 \subset \mathrm{Aut}(X)$ and $H_2 \subset \mathrm{Aut}(X)$ be the two Klein four groups which lead to the partitions $\mathfrak{p}$ and $\tilde{\mathfrak{p}}$. In particular, $H_1 \neq H_2$ and $g(X/H_1) = g(X/H_2) = 0$. We write $H_1 = \langle\alpha_1, \alpha_2\rangle$ and $H_2 = \langle\sigma_1, \sigma_2\rangle$ where $g(X/\alpha_1) \leq g(X/\alpha_2) \leq g(X/\alpha_1\alpha_2)$ and $g(X/\sigma_1) \leq g(X/\sigma_2) \leq g(X/\sigma_1\sigma_2)$.

Lemma 7: *Under the hypotheses above, either* $\alpha_1 = \sigma_1$ *or* $\alpha_1\sigma_1 = \sigma_1\alpha_1$.

Proof: By hypothesis, $g(X/\alpha_1) \leq g(X)/3$ and $g(X/\sigma_1) \leq g(X)/3$. In fact, if $g \equiv 0 \bmod 3$ (resp. $g \equiv 1 \bmod 3$), then $g(X/\alpha_1)$ and $g(X/\sigma_1)$ cannot both equal $g/3$ (resp. $(g-1)/3$). The reason is that, if $g \equiv 0 \bmod 3$ (resp. $g \equiv 1 \bmod 3$), there is only one partition whose smallest entry is $g/3$ (resp. $(g-1)/3$) and the partitions $\mathfrak{p}$ and $\tilde{\mathfrak{p}}$ are different by hypothesis. It follows that $g(X/\alpha_1) + g(X/\sigma_1) \leq 2g(X)/3 - 1$. Let m be the order of $\alpha_1\sigma_1$. If $Y = X/\langle\alpha_1, \sigma_1\rangle$, then $g(X)+mg(Y) = g(X/\alpha_1)+g(X/\sigma_1)+g(X/\alpha_1\sigma_1)$, [5]. Thus, $g(X/\alpha_1\sigma_1) \geq g(X) + mg(Y) - (2g(X)/3 - 1) \geq g(X)/3 + 1$. By the Riemann-Hurwitz formula, $g(X/\alpha_1\sigma_1) \leq (g(X) + m - 1)/m$. Combining the two inequalities, along with the fact that m must be an integer, yields that $m \leq 2$ which implies that $\alpha_1 = \sigma_1$ or $\alpha_1\sigma_1 = \sigma_1\alpha_1$. □

Let X be a curve with an action by G. If γ_1 and γ_2 are conjugate in G, then the function fields of X/γ_1 and X/γ_2 are conjugate and so $g(X/\gamma_1) = g(X/\gamma_2)$. Recall that if c is odd, then all involutions in D_{2c} are conjugate. If c is even, then sr^i and sr^j are conjugate in D_{2c} if and only if i and j have the same parity.

Lemma 8: *Let X be a curve with an action by $G = D_{2c} \times \mathbb{Z}/2$ where $c \geq 2$ is even. Suppose that $g(X/G) = 0$. Then $g(X/\tau) = g(X/\langle\tau, r\rangle) + g(X/\langle\tau, s\rangle) + g(X/\langle\tau, sr\rangle)$.*

Proof: Consider the action of D_{2c} on X/τ. Any element of D_{2c} is in one of the subgroups $\langle r\rangle$ or $\langle sr^i\rangle$ for some $0 \leq i < c$. The intersection of any two of these subgroups has order 1. Applying [5] to the curve X/τ and using the fact that $g((X/\tau)/D_{2c}) = 0$, we see that $cg(X/\tau) = cg(X/\langle r, \tau\rangle) + \sum_{i=0}^{c-1} 2g(X/\langle\tau, sr^i\rangle)$. This simplifies to $g(X/\tau) = g(X/\langle r, \tau\rangle) + g(X/\langle\tau, s\rangle) + g(X/\langle\tau, sr\rangle)$. □

Theorem 9: The intersection of $\mathcal{H}_{g,2,\mathfrak{p}}$ and $\mathcal{H}_{g,2,\tilde{\mathfrak{p}}}$ is nonempty for $\mathfrak{p} \neq \tilde{\mathfrak{p}}$ if and only if $1 \in \mathfrak{p} \cap \tilde{\mathfrak{p}}$.

Proof: The reverse implication is true by Lemma 4. For the forward implication, suppose that $\mathcal{H}_{g,2,\mathfrak{p}} \cap \mathcal{H}_{g,2,\tilde{\mathfrak{p}}}$ is nonempty for some $\mathfrak{p} \neq \tilde{\mathfrak{p}}$ and let X be a curve corresponding to a point of the intersection. This situation satisfies the hypotheses above. By Lemma 7, either $\alpha_1 = \sigma_1$ or $\alpha_1\sigma_1 = \sigma_1\alpha_1$.

Suppose $\alpha_1 = \sigma_1$. Then $G = H_1H_2$ is isomorphic to $D_{2c} \times \mathbb{Z}/2$ for some c, where $\sigma_2 = s$ and $\alpha_2 = sr$. If c is odd, then $g(X/\alpha_2) = g(X/\sigma_2)$ which contradicts the hypothesis that $\mathfrak{p} \neq \tilde{\mathfrak{p}}$; so c is even. Note that by hypothesis $g(X/H_1) = g(X/H_2) = 0$ and so $g(X/G) = 0$. It follows from Lemma 8 that $g(X/\alpha_1) = g(X/\langle\alpha_1, \alpha_2\sigma_2\rangle)$. This implies that $g(X/\alpha_1) \leq 1$ by the Riemann-Hurwitz formula. Thus $1 \in \mathfrak{p} \cap \tilde{\mathfrak{p}}$.

Now suppose $\alpha_1\sigma_1 = \sigma_1\alpha_1$. Consider the group $G_1 = \langle\sigma_1, \alpha_2, \alpha_1\rangle$ which is isomorphic to $D_{2c_1} \times \mathbb{Z}/2$ and the group $G_2 = \langle\alpha_1, \sigma_2, \sigma_1\rangle$ which is isomorphic to $D_{2c_2} \times \mathbb{Z}/2$. Here c_1 (resp. c_2) is the order of $\sigma_1\alpha_2$ (resp. $\sigma_2\alpha_1$). If c_1 is odd, then $g(X/\sigma_1) = g(X/\alpha_2)$. If c_2 is odd, then $g(X/\sigma_2) = g(X/\alpha_1)$. Since $\mathfrak{p} \neq \tilde{\mathfrak{p}}$, it follows that c_1 and c_2 are not both odd.

Without loss of generality, we suppose that $g(X/\sigma_1) \leq g(X/\alpha_1)$. If c_1 is odd, then $g(X/\sigma_1) = g(X/\alpha_2)$. By hypothesis, $g(X/\alpha_2) \geq g(X/\alpha_1)$ which implies $g(\alpha_1) = g(\sigma_1)$ in this case. The implication of this is that (after possibly switching the roles of H_1 and H_2) we can simultaneously suppose that c_1 is even and $g(X/\sigma_1) \leq g(X/\alpha_1)$. Then $g(\sigma_1) \leq (g(X) - 2)/3$ by the hypothesis that $\mathfrak{p} \neq \tilde{\mathfrak{p}}$. Let $m = c_1/2$.

Consider the automorphism $r = \alpha_2\sigma_1$. Let H be the subgroup of $\mathrm{Aut}(X)$ generated by involutions α_1, α_2, and r^m. Then $H \simeq (\mathbb{Z}/2)^3$ and $g(X/H) =$

0. It follows from [3, Corollary 3.4] that $g(X)$ is odd. By hypothesis, $g(X) = g(\alpha_1)+g(\alpha_2)+g(\alpha_1\alpha_2)$. The Kani-Rosen theorem [5] applied to the action of H on X implies that $3g(X) = \sum_{\sigma\in H^*} g(X/\sigma)$. After simplifying, we see that

$$2g(X) = g(X/r^m) + g(X/r^m\alpha_1) + g(X/r^m\alpha_2) + g(X/r^m\alpha_1\alpha_2).$$

By the Riemann-Hurwitz formula, each of the terms on the righthand side of this equation is at most $(g(X)+1)/2$. It follows that $g(X/r^m\alpha_2) + g(X/r^m\alpha_1\alpha_2) \geq g(X) - 1$ and that $g(\alpha_2 r^m) \geq (g(X)-3)/2$.

Suppose m is even. In this case, $r^m\alpha_2$ and α_2 are conjugate in D_{2c_1}. Then $g(X) = g(\alpha_1) + g(X/r^m\alpha_2) + g(X/r^m\alpha_1\alpha_2)$. It follows that $g(\sigma_1) \leq g(\alpha_1) \leq 1$. By Proposition 6, $1 \in \mathfrak{p}\cap\tilde{\mathfrak{p}}$

Suppose m is odd. In this case, σ_1 and $r^m\alpha_2$ are conjugate in D_{2c_1}. Thus $(g(X)-2)/3 \geq g(\sigma_1) \geq (g(X)-3)/2$. This is only possible if $g(X) \leq 5$. The statement of the theorem is vacuous if $g = 1$ or $g = 3$. If $g = 5$, the only partitions are $\mathfrak{p}_0$, $\mathfrak{p}_1$ and $\mathfrak{p}_1'$ and so $1 \in \mathfrak{p}\cap\tilde{\mathfrak{p}}$ by Proposition 6. □

The following small generalization of Theorem 9 shows that any curve satisfying the hypotheses above is part of the family constructed in Lemma 4.

Corollary 10: *Suppose $X \in \mathcal{H}_{g,2,\mathfrak{p}} \cap \mathcal{H}_{g,2,\tilde{\mathfrak{p}}}$ for some $\mathfrak{p} \neq \tilde{\mathfrak{p}}$ and let H_1 and H_2 be the two corresponding Klein four sugroups of* Aut(X). *Then $H_1 \cap H_2$ contains an involution σ_1, H_1H_2 is of the form $D_{2c} \times \mathbb{Z}/2$ for some even c and $g(X/\sigma_1) = 1$.*

Proof: By earlier work in this section, the result will follow if $\sigma_1 = \alpha_1$. By Theorem 9, $g(X/\alpha_1) = g(X/\sigma_1) = 1$. Strengthening the equations in the proof of Lemma 7, we see that $(g(X)+m-1)/m \geq g(X/\alpha_1\sigma_1) \geq g(X)-2$ where m is the order of $\alpha_1\sigma_1$. If $g(X) > 5$, then this implies $m = 1$ and so $\sigma_1 = \alpha_1$. If $g(X) \leq 5$, the result follows from the approach of [6]. When $g = 3$, $H_1H_2 = (\mathbb{Z}/2)^3$ and when $g = 5$, $H_1H_2 = D_8 \times \mathbb{Z}/2$. □

Corollary 11: *The number of non-empty connected components of $\mathcal{H}_{g,2}$ equals $[(g+6)^2/48]$ if g is even and equals $[(g+3)^2/48] + \lfloor (g-1)/4 \rfloor)$ if g is odd.*

Proof: Theorem 9 implies that if g is even then no pair of irreducible components of $\mathcal{H}_{g,2}$ intersect and if g is odd then there is exactly one pair of irreducible components which intersect. The proof is then immediate from Lemma 3. □

4. The geometry of $\mathcal{H}_{g,3}$

In this section, we prove that $\mathcal{H}_{g,3}$ is not connected for g odd and $g \geq 5$. Recall that $\mathcal{H}_{g,3}$ is empty if g is even, [3, Corollary 3.4]. Again the method is to show that $\mathcal{H}_{g,3} \cap \mathcal{H}_g$ is a connected component of $\mathcal{H}_{g,3}$. We note that this approach will not work for $n \geq 4$ due to the following lemma.

Lemma 12: *The intersection $\mathcal{H}_{g,n} \cap \mathcal{H}_g$ is empty for $n \geq 4$.*

Proof: Suppose X is a curve corresponding to a point of intersection of $\mathcal{H}_{g,n} \cap \mathcal{H}_g$. Let σ be a hyperelliptic involution of X. Then $g(X/\sigma) = 0$. Since σ is in the center of $\mathrm{Aut}(X)$, it follows that there exists $H \simeq (\mathbb{Z}/2\mathbb{Z})^n \subset \mathrm{Aut}(X)$ so that $\sigma \in H$. Thus, $X/\sigma \in \mathcal{H}_{g,n-1}$ and $g(X/\sigma) \equiv 1 \bmod 2^{n-3}$ by [3, Corollary 3.4]. This is a contradiction unless $n \leq 3$. □

Let $\mathfrak{p}'_0 = \{0,0,0,j,j,j,j+1\}$ if $g = 4j+1$ and $\mathfrak{p}'_0 = \{0,0,0,j,j+1,j+1,j+1\}$ if $g = 4j+3$.

Lemma 13: *The component $\mathcal{H}_{g,3,\mathfrak{p}'_0}$ of $\mathcal{H}_{g,3}$ is non-empty and equals $\mathcal{H}_{g,3} \cap \mathcal{H}_g$.*

Proof: First, note that $\mathcal{H}_{g,3} \cap \mathcal{H}_g$ is non-empty. In particular, consider three hyperelliptic covers with branch loci $B_1 = \{0, \infty\}$, $B_2 = \{0,1\}$, and B_3 so that $|\cup_{i=1}^{3} B_i| = (g+7)/2$. The normalized fibre product of these covers yields a curve in $\mathcal{H}_{g,3} \cap \mathcal{H}_g$. It remains to show that the only partition $\mathfrak{p}$ of g for which $\mathcal{H}_{g,3,\mathfrak{p}}$ contains hyperelliptic curves is $\mathfrak{p}'_0$.

Suppose that X corresponds to a point of $\mathcal{H}_{g,3} \cap \mathcal{H}_g$ and let $\mathfrak{p}$ be its partition. Let σ_1 be such that $g(X/\sigma_1)$ has genus 0. There exists a subgroup $H \simeq (\mathbb{Z}/2)^3 \subset \mathrm{Aut}(X)$ containing σ_1. There are three hyperelliptic quotients of $X \to X/H$ dominated by X/σ_1. Each of these must have genus 0. Thus $\mathfrak{p} = \{0,0,0,g_4,g_5,g_6,g_7\}$ and $\sum_{i=4}^{7} g_i = g$. Note that $g_i \leq (g+3)/4$ by the Riemann-Hurwitz formula.

Let $Y \to X/H$ be a Klein four quotient of $X \to X/H$. Since $Y \to X/H$ is not disjoint from $X \to X/H$, it follows that Y corresponds to a point of $\mathcal{H}_{g(Y),2} \cap \mathcal{H}_{g(Y)}$. By Proposition 6, the three hyperelliptic quotients of $Y \to X/H$ have genera $\{0,x,y\}$ where $y \leq x+1$. The numerical constraints then imply that $\mathfrak{p} = \mathfrak{p}'_0$. □

Proposition 14: *The locus $\mathcal{H}_{g,3,\mathfrak{p}'_0}$ is a connected component of $\mathcal{H}_{g,3}$. The sublocus $\mathcal{H}_{g,3}$ of $\mathcal{M}_g$ is connected if and only if $g \leq 5$.*

Proof: Suppose X corresponds to a point of $\mathcal{H}_{g,3,\mathfrak{p}'_0}$ and let σ_1 be such that $g(X/\sigma_1) = 0$. Recall that σ is in the center of $\mathrm{Aut}(X)$. By Lemma 12, $H_{g,4}$ is empty. Thus there does not exist a subgroup $H \simeq (\mathbb{Z}/2)^3 \subset \mathrm{Aut}(X)$ which is disjoint from σ_1. It follows that X does not correspond to a point of $\mathcal{H}_{g,3,\mathfrak{p}}$ for any $\mathfrak{p} \neq \mathfrak{p}'_0$.

For $g = 1$ and $g = 3$, $\mathfrak{p}'_0$ is the only partition for which $\mathcal{H}_{g,3,\mathfrak{p}}$ is non-empty. It follows that $\mathcal{H}_{1,3}$ and $\mathcal{H}_{3,3}$ are connected.

It now suffices to show for $g \geq 5$ that there exists a partition $\mathfrak{p}$ other than $\mathfrak{p}'_0$ for which $\mathcal{H}_{g,3,\mathfrak{p}}$ is non-empty. Let X' be the normalized fibre product of the three hyperelliptic curves branched at B_1, B_2, and B_3 as follows. If $g = 4j + 1$, let $B_1 = \{\lambda_1, \ldots, \lambda_{2j}, 0, \infty\}$, $B_2 = \{0, 1\}$, and $B_3 = \{\infty, \mu\}$. Then X' corresponds to a point of $\mathcal{H}_{g,3,\{j,j,j,j,1,0,0\}}$. If $g = 4j + 3$, let $B_1 = \{\lambda_1, \ldots, \lambda_{2j}, 0, \infty\}$, $B_2 = \{\lambda_1, \ldots, \lambda_{2j}, 0, 1\}$, and $B_3 = \{\lambda_1, \ldots, \lambda_{2j}, \mu_1, \mu_2\}$. Then X' corresponds to a point of $\mathcal{H}_{g,3,\{j+1,j,j,j,1,1,0\}}$. □

References

1. Andrews, G., A note on partitions and triangles with integer sides, Amer. Math. Monthly, The American Mathematical Monthly, 86, 1979, 6, 477–478,
2. Cornalba, M., On the locus of curves with automorphisms, Ann. Mat. Pura Appl. (4), Annali di Matematica Pura ed Applicata. Serie Quarta, 149, 1987, 135–151,
3. Glass, D. and Pries, R., Hyperelliptic Curves with Prescribed p-Torsion, accepted by Manuscripta
4. Hartshorne, R., Algebraic geometry, Graduate Texts in Mathematics, No. 52, Springer-Verlag, New York, 1977, xvi+496,
5. Kani, E. and Rosen, M., Idempotent relations and factors of Jacobians, Math. Ann., Mathematische Annalen, 284, 1989, 2, 307–327
6. Magaard, K. and Shaska, T. and Shpectorov, S. and Völklein, H., The locus of curves with prescribed automorphism group, Communications in arithmetic fundamental groups (Kyoto, 1999/2001), Sūrikaisekikenkyūsho Kōkyūroku, Sūrikaisekikenkyūsho Kōkyūroku, 1267, 2002, 112–141,
7. Magaard, Kay and Shpectorov, Sergey and Völklein, Helmut, A GAP package for braid orbit computation and applications, Experiment. Math., Experimental Mathematics, 12, 2003, 4, 385–393
8. Popp, H., The singularities of the moduli schemes of curves, J. Number Theory, 1, 1969, 90–107
9. Serre, Jean-Pierre, Géométrie algébrique et géométrie analytique, Ann. Inst. Fourier, Grenoble, 6, 1955–1956, 1–42
10. Völklein, H., Groups as Galois groups, Cambridge Studies in Advanced Mathematics, 53, Cambridge University Press, Cambridge, 1996, xviii+248
11. Wewers, S., Construction of Hurwitz spaces, Thesis

Field of moduli and field of definition for curves of genus 2

Gabriel Cardona & Jordi Quer*

Dept. Ciències Matemàtiques i Informàtica,
Universitat de les Illes Balears,
Campus UIB, Ed. Anselm Turmeda, Desp. 241.
Carretera Valldemossa, km. 7.5,
07122 – Palma de Mallorca, Spain
gabriel.cardona@uib.es

Dept. Matemàtica Aplicada II,
Universitat Politècnica de Catalunya,
Campus Nord, Ed. Omega, Desp. 438.
Jordi Girona, 1–3,
08034 – Barcelona, Spain
jordi.quer@upc.edu

Let $\mathcal{M}_2$ be the moduli space that classifies genus 2 curves. If a curve C is defined over a field k, the corresponding moduli point $P = [C] \in \mathcal{M}_2$ is defined over k. In [7], Mestre solves the converse problem for curves with $\mathrm{Aut}(C) \simeq C_2$. Given a moduli point defined over k, Mestre finds an obstruction to the existence of a corresponding curve defined over k, that is an element in $\mathrm{Br}_2(k)$ not always trivial. In this paper we prove that for all the other possibilities of $\mathrm{Aut}(C)$, every moduli point defined over k is represented by a curve defined over k. We also give an explicit construction of such a curve in terms of the coordinates of the moduli point.

1. Preliminaries on genus 2 curves

Curves of genus 2, sextic forms and invariants. Let us fix k a perfect field of characteristic different from 2, $\overline{k}$ an algebraic closure of k, and

*Supported by grants BFM-2003-06768-C02-01 and 2002SGR 00148

$G_k = \mathrm{Gal}(\overline{k}/k)$ the absolute Galois group of k. If C/k is a genus 2 curve, then it always admits an affine model

$$y^2 = f(x),$$

where $f(x) \in k[x]$ is a polynomial of degree 5 or 6 without multiple roots. We will call this model a hyperelliptic model for C. Isomorphisms between curves of genus 2, given by hyperelliptic models, can be written as

$$(x,y) \mapsto \left(\frac{mx+n}{px+q}, \frac{y(mq-np)}{(px+q)^3}\right),$$

associated to a uniquely determined $M = \left(\begin{smallmatrix} m & n \\ p & q \end{smallmatrix}\right) \in \mathrm{GL}_2(\overline{k})$. In particular, automorphisms of curves of genus 2 can be represented by matrices, and the mapping $\mathrm{Aut}(C) \to \mathrm{GL}_2(\overline{k})$ is a morphism of G_k-groups. The canonical hyperelliptic involution $\imath$, given by $(x,y) \mapsto (x,-y)$ in terms of hyperelliptic models, is represented by the matrix $\left(\begin{smallmatrix} -1 & 0 \\ 0 & -1 \end{smallmatrix}\right)$.

The classification of genus 2 curves up to isomorphism was performed by Clebsch and Bolza (cf. [4] and [1]) using the classification of binary sextic forms up to linear equivalence. Given a genus 2 curve with hyperelliptic model $y^2 = f(x)$, the associated sextic form is given by

$$F(x_1, x_2) = x_2^6 f(x_1/x_2).$$

An isomorphism associated to $M = \left(\begin{smallmatrix} m & n \\ p & q \end{smallmatrix}\right) \in \mathrm{GL}_2(\overline{k})$ traslates into the language of binary sextic forms as the linear transformation

$$(x_1; x_2) \mapsto (mx_1 + nx_2, px_1 + qx_2).$$

The classification of binary sextic forms uses Hilbert's theory of algebraic invariants (cf. [5]). An invariant is a polynomial expression $I \in k[a_0, \ldots, a_6]$ in the coefficients of a generic sextic form

$$F(x_1, x_2) = a_0 x_1^6 + a_1 x_1^5 x_2 + \cdots + a_5 x_1 x_2^5 + a_6 x_2^6,$$

that under a change of variables associated to any $M \in \mathrm{GL}_2(\overline{k})$ changes by a fixed power of $\det M$. Covariants are polynomial expressions in $k[a_0, \ldots, a_6, x_1, x_2]$ with the same property. The homogeneous degree in x_i is called the order of the covariant, while the degree in a_i is called the degree of the covariant; in particular, invariants are just covariants of order zero. Given an invariant I and a genus 2 curve C or a sextic form F, we will denote by $I(C)$ or $I(F)$ the value that takes the invariant when evaluated at this curve or form.

In terms of invariants, the classification goes as follows. Two sextic forms F and F' are linearly equivalent if, and only if, there exists $r \in \overline{k}^*$ such that for every invariant I, one has $I(F) = r^d I(F')$, where d is the degree of I. As a consequence, the fact that an invariant annihilates when evaluated to a certain curve depends only on the isomorphism class of the curve.

The algebra generated by the invariants of sextic forms can be generated by five of them, with degrees 2, 4, 6, 10 and 15; this implies that the condition of linear equivalence of forms needs only to be tested for this generating set. Among these generators, there is, up to equivalence, a single algebraic relation which involves terms of degree 30; namely, if R is a generator of degree 15, then this relation states that R^2 is a polynomial expression in the four generators of even degree. As a consequence, the condition of equivalence needs not to be tested for the invariant R. Indeed, since the even degree invariants determine R up to sign, after replacing r with $-r$ if necessary, the condition holds for R if it holds for the even degree generators.

Therefore, one can reduce the former condition on linear equivalence to test it only for the four generator invariants of even degree. In the literature appear several such sets of generators: Clebsch invariants, that we will note as c_2, c_4, c_6, c_{10}, and are noted A, B, C, D in [4] and [7], Igusa invariants, that we will note as I_2, I_4, I_6, I_{10}, and are noted A, B, C, D in [6], and the last set we will mention is that of Igusa arithmetic invariants J_2, J_4, J_6, J_{10} introduced by Igusa in [6]. All these generators have rational coefficients when expressed as polynomials in the coefficients of a generic form. However, while the invariants J_i reduce well in any characteristic, the I_i fail to work in characteristics 2 and 3, and the c_i in characteristics 2, 3 and 5.

The invariant $I_{10} = 2^{12} J_{10}$ is just the discriminant of the form F. Therefore, forms F corresponding to genus 2 curves are those with $J_{10}(F) \neq 0$.

Absolute invariants and the moduli space. Absolute invariants are quotients of invariants of equal degree. The classification of genus 2 curves up to isomorphism can be stated in terms of absolute invariants: Two curves C and C' are $\overline{k}$-isomorphic if, and only if, for every absolute invariant $\jmath$, one has $\jmath(C) = \jmath(C')$. From this fact, Igusa constructed the variety $\mathcal{M}_2$ that classifies genus 2 curves up to isomorphism as a 3-dimensional affine variety over $\mathbb{Z}$ (cf. [6]).

In this moduli space, the classes of curves for which all invariants of degree 2 don't annihilate, which is equivalent to have $J_2(C) \neq 0$, form an open set and we can take as coordinates for the moduli point the absolute invariants

$$j'_1 = \frac{J_2^5}{J_{10}}, \qquad j'_2 = \frac{J_2^3 J_4}{J_{10}}, \qquad j'_3 = \frac{J_2^2 J_6}{J_{10}}.$$

Similarly, for the set of curves for which all invariants of degree 2 annihilate, but there exists a non-vanishing invariant of degree 4, that is $J_2 = 0$ and $J_4 \neq 0$, one can take

$$j''_1 = 0, \qquad j''_2 = \frac{J_4^5}{J_{10}^2}, \qquad j''_3 = \frac{J_4 J_6}{J_{10}}.$$

Finally, for the remaining points, one can take

$$j'''_1 = 0, \qquad j'''_2 = 0, \qquad j'''_3 = \frac{J_6^5}{J_{10}^3}.$$

In must be noted that a point $P = [C] \in \mathcal{M}_2$ is defined over a field k if, and only if, its coordinates are in k, which amounts to say that all absolute invariants, when particularized to C, are in k. This way, the set of points on $\mathcal{M}_2(k)$ defined over k is in bijection with the set of 3-tuples (j_1, j_2, j_3), $j_i \in k$, where

$$j_i = \begin{cases} j'_i, & \text{if } j_1 \neq 0, \\ j''_i, & \text{if } j_1 = 0,\ j_2 \neq 0, \\ j'''_i, & \text{if } j_1 = j_2 = 0. \end{cases} \qquad i = 1, 2, 3.$$

At first sight, it could seem that to have absolute invariants defined over k does not guarantee that one can find a set of corresponding generator invariants $J_2, J_4, J_6, J_{10}, R \in k$ giving rise to them. The following lemma proves the existence of such invariants.

Lemma 1: *For every point $P \in \mathcal{M}_2(k)$, there exist invariants defined over k that give the absolute invariants which are coordinates for this point.*

Proof: One can take

$$(J_2, J_4, J_6, J_{10}) = \begin{cases} (j_1, j_1 j_2, j_1^2 j_3, j_1^4), & \text{if } j_1 \neq 0, \\ (0, j_2, j_2 j_3, j_2^2), & \text{if } j_1 = 0,\ j_2 \neq 0, \\ (0, 0, j_3^2, j_3^3), & \text{if } j_1 = j_2 = 0, \end{cases}$$

and verify that with this choices, we get the values for the absolute invariants as defined above.

We can also make a choice in such a way that R is defined over k. Since R^2 can be written as a polynomial expression in the invariants J_i, we have that R is of the form $R = \sqrt{d}R_0$ with $R_0 \in k$. We can then take $r = \sqrt{d}$ and construct another set of invariats

$$\begin{aligned}(J_2', J_4', J_6', J_{10}', R') &= (r^2 J_2, r^4 J_4, r^6 J_6, r^{10} J_{10}, r^{15} R)\\ &= (dJ_2, d^2 J_4, d^3 J_6, d^5 J_{10}, d^8 R_0),\end{aligned}$$

equivalent to the former and all of them defined over k. □

Group of automorphisms. In [1], Bolza gives the different possibilities for the reduced group of automorphisms $\mathrm{Aut}'(C) = \mathrm{Aut}(C)/\langle \imath \rangle$ of a genus 2 curve defined over a field of characteristic zero. In [2], 8 the corresponding structures for the full group $\mathrm{Aut}(C)$ are given.

If char k is different from 2, 3 and 5, the possible automorphism groups of genus 2 curves are

$$C_2, \quad V_4, \quad D_8, \quad D_{12}, \quad 2D_{12}, \quad \tilde{S}_4, \quad C_{10}.$$

The generic case corresponds to $\mathrm{Aut}(C) \simeq C_2$. The curves having an involution different from the hyperelliptic involution form a surface in $\mathcal{M}_2$; the curves with group of automorphisms isomorphic to D_8 or D_{12} describe two curves in this surface, and they intersect in the single point corresponding to curves with $\mathrm{Aut}(C) \simeq 2D_{12}$; the curve corresponding to D_8 has another distinguished point corresponding to the case $\tilde{S}_4$. Finally, the curves with group of automorphisms cyclic of order 10 correspond to an isolated point outside the surface of curves with non-hyperelliptic involutions.

As for the remaining cases, in characteristic 3, there is no curve with group of automorphisms isomorphic to $2D_{12}$. In characteristic 5, the three isolated points reduce to a single one, for which the reduced group of automorphisms is isomorphic to $\tilde{S}_5$. We refer the reader to [3] for the case of characteristic 2.

The subvariety of curves admitting a non-hyperelliptic involution is characterized by the condition $R = 0$, while the single moduli point of curves with group of automorphisms isomorphic to C_{10} is characterized by the condition $J_2 = J_4 = J_6 = 0$, that is, $(\jmath_1, \jmath_2, \jmath_3) = (0, 0, 0)$.

Field of moduli and field of definition. Let C be any curve defined over $\overline{k}$. We say that k is a field of moduli for C if $C \simeq {}^{\sigma}C$ for every $\sigma \in G_k$. In the case of genus 2 curves, this condition is equivalent to have the absolute invariants defined over k and, therefore, to say that the point of moduli $P = [C] \in \mathcal{M}_2$ is defined over k.

If C is defined over k, then $P = [C]$ is defined over k. The problem we want to solve is the converse: Given a point $P \in \mathcal{M}_2(k)$, decide whether there is a curve representing P defined over k. In an equivalent formulation, we are interested in the following Galois descent problem: Given a genus 2 curve defined over $\overline{k}$ having k as a field of moduli, decide if there is an isomorphic curve defined over k.

The answer to this question is summarized in the following theorem, the first part of which is proved by Mestre in [7],

Theorem 2: Let $P = [C] \in \mathcal{M}_2(k)$.

(1) (Mestre) If $\mathrm{Aut}(C) \simeq C_2$, there exists an obstruction to the existence of a curve C' defined over k and isomorphic to C, which is an element in $\mathrm{Br}_2(k)$ not necessarily trivial.
(2) If $\mathrm{Aut}(C) \not\simeq C_2$, there always exists C'/k with $C \simeq C'$.

The remaining sections of the paper contain the proof of the second part of this theorem, split into several cases corresponding to different structures for $\mathrm{Aut}(C)$. We remark that all proofs are constructive in the sense that they give an algorithm for constructing a curve defined over k starting from the coordinates of a moduli point $P \in \mathcal{M}_2(k)$ corresponding to curves with non-trivial reduced group of automorphisms.

2. The case $\mathrm{Aut}(C) \simeq V_4$

If one wants to study genus 2 curves with $\mathrm{Aut}(C) \simeq V_4$ over an algebraically closed field, one can use the following result, that goes back to Bolza (cf. [1]).

Lemma 3: *Let C be a genus 2 curve defined over k with $\mathrm{Aut}(C) \simeq V_4$. Then, C is $\overline{k}$-isomorphic to a curve defined over $\overline{k}$ of the form*

$$y^2 = x^6 + \alpha_1 x^4 + \alpha_2 x^2 + 1, \tag{1}$$

for some $\alpha_1, \alpha_2 \in \overline{k}$.

It has to be noted that not all equations of the form (1) correspond to curves of genus 2 with group of automorphisms isomorphic to V_4. The following lemma, that also appears in [8], gives conditions for this.

Lemma 4: *An equation of the form*

$$y^2 = x^6 + \alpha_1 x^4 + \alpha_2 x^2 + 1$$

corresponds to a curve C of genus 2 with $\mathrm{Aut}(C) \simeq V_4$ if, and only if,

$$(u^2 - 4v + 18u - 27)(v^2 - 4u^3)(4v - u^2 + 110u - 1125) \neq 0,$$

where $u = \alpha_1\alpha_2$ and $v = \alpha_1^3 + \alpha_2^3$.

Proof: Such an equation corresponds to a genus 2 curve if, and only if, the discriminant of the polynomial $x^6 + \alpha_1 x^4 + \alpha_2 x^2 + 1$ is non-zero, which is equivalent to

$$u^2 - 4v + 18u - 27 \neq 0.$$

If the previous condition holds, then (1) defines a genus 2 curve, and the group of automorphisms of this curve contains V_4. As we will prove in the next section, those curves having $V_4 \subsetneq \mathrm{Aut}(C)$ are those isomorphic to either

$$y^2 = x^5 + x^3 + tx$$

or

$$y^2 = x^6 + x^3 + t$$

for some $t \in k \setminus \{0, 1/4\}$. Computing the corresponding parameters u and v one finds that they satisfy the relations

$$v^2 - 4u^3 = 0$$

or

$$4v - u^2 + 110u - 1125 = 0,$$

respectively. Note also that if the characteristic of k is 3, then one needs to replace the equation $y^2 = x^6 + x^3 + t$ with

$$y^2 = \frac{1}{t}x^6 + x^4 + x^2 + 1,$$

but also in this case holds that $4v - u^2 + 110u - 1125 = 0$. □

The following proposition shows that, given C with $\mathrm{Aut}(C) \simeq V_4$, the corresponding parameters introduced in the previous lemma:

$$u = \alpha_1\alpha_2, \qquad v = \alpha_1^3 + \alpha_2^3,$$

can be obtained as absolute invariants of C.

Proposition 5: *Let C be a genus 2 curve with* $\mathrm{Aut}(C) \simeq V_4$*. Then the parameters u, v can be obtained as the values that take at C the absolute invariants*

$$j_u = \frac{A_u}{B_u}, \qquad j_v = \frac{A_v}{B_v},$$

where

$$\begin{aligned}
A_u &= J_2^5 J_4^2 - 64 J_2^3 J_4^3 + 1024 J_2 J_4^4 + 3 J_2^6 J_6 - 202 J_2^4 J_4 J_6 + 4014 J_2^2 J_4^2 J_6 - \\
&\quad - 20160 J_4^3 J_6 + 666 J_2^3 J_6^2 - 20520 J_2 J_4 J_6^2 + 48600 J_6^3 - 30 J_2^4 J_{10} + \\
&\quad + 2800 J_2^2 J_4 J_{10} - 192000 J_4^2 J_{10} - 360000 J_2 J_6 J_{10} \\
B_u &= 2 J_2^4 J_4 J_6 - 114 J_2^2 J_4^2 J_6 + 1344 J_4^3 J_6 + 6 J_2^3 J_6^2 + 216 J_2 J_4 J_6^2 - 3240 J_6^3 + \\
&\quad + 18 J_2^4 J_{10} - 1040 J_2^2 J_4 J_{10} + 12800 J_4^2 J_{10} + 4800 J_2 J_6 J_{10} \\
A_v &= J_2^6 J_4^2 - 96 J_2^4 J_4^3 + 3072 J_2^2 J_4^4 - 32768 J_4^5 + 3 J_2^7 J_6 - 164 J_2^5 J_4 J_6 + \\
&\quad + 1250 J_2^3 J_4^2 J_6 + 29760 J_2 J_4^3 J_6 + 858 J_2^4 J_6^2 - 22680 J_2^2 J_4 J_6^2 - \\
&\quad - 172800 J_4^2 J_6^2 + 81000 J_2 J_6^3 + 1176 J_2^5 J_{10} - 79600 J_2^3 J_4 J_{10} + \\
&\quad + 1344000 J_2 J_4^2 J_{10} - 72000 J_2^2 J_6 J_{10} - 12960000 J_4 J_6 J_{10} - 134400000 J_{10}^2 \\
B_v &= 3 J_2^3 J_4^2 J_6 - 160 J_2 J_4^3 J_6 + J_2^4 J_6^2 - 36 J_2^2 J_4 J_6^2 + 3456 J_4^2 J_6^2 - 1188 J_2 J_6^3 + \\
&\quad + 24 J_2^3 J_4 J_{10} - 1280 J_2 J_4^2 J_{10} + 160 J_2^2 J_6 J_{10} + 105600 J_4 J_6 J_{10} + \\
&\quad + 640000 J_{10}^2,
\end{aligned}$$

and J_i are Igusa's arithmetic invariants.

Proof: From Lemma 3, and since j_u and j_v are absolute invariants, the equalities $j_u(C) = u = \alpha_1\alpha_2$ and $j_v(C) = v = \alpha_1^3 + \alpha_2^3$ need only to be tested for equations of the form (1). It follows from a simple computation that these equalities really hold. Note also that the denominators B_u and B_v evaluated at a curve of the form (1) get the value

$$\begin{aligned}
B_u &= -8(v^2 - 4u^3)(4v - u^2 + 110u - 1125) \\
B_v &= 64(v^2 - 4u^3)(4v - u^2 + 110u - 1125)
\end{aligned}$$

and, from Lemma 4, they don't vanish. □

As a direct consequence of the previous lemma, we have that if C is defined over a field k, then the corresponding parameters u, v are elements in k. Conversely, the following result provides a model defined over k for every possible choice of $(u, v) \in k^2$.

Proposition 6: *Let* $(u, v) \in k^2$ *such that*

$$(u^2 - 4v + 18u - 27)(v^2 - 4u^3)(4v - u^2 + 110u - 1125) \neq 0.$$

Then, the curve of genus 2 *defined over* k *given by*

$$y^2 = a_0x^6 + a_1x^5 + a_2x^4 + a_3x^3 + ta_2x^2 + t^2a_1x + t^3a_0,$$

where, if $u \neq 0$,

$$\begin{aligned} t &= v^2 - 4u^3, \\ a_0 &= v^2 + u^2v - 2u^3, \\ a_1 &= 2(u^2 + 3v)(v^2 - 4u^3), \\ a_2 &= (15v^2 - u^2v - 30u^3)(v^2 - 4u^3), \\ a_3 &= 4(5v - u^2)(v^2 - 4u^3)^2, \end{aligned}$$

or, if $u = 0$,

$$\begin{aligned} t &= 1, \\ a_0 &= 1 + 2v, \\ a_1 &= 2(3 - 4v), \\ a_2 &= 15 + 14v, \\ a_3 &= 4(5 - 4v), \end{aligned}$$

has u, v *as absolute invariants* j_u, j_v.

Proof: The result follows easily evaluating the absolute invariants j_u, j_v at the given curves. Note also that the invariants B_u, B_v appearing in the denominator, evaluated at these curves get the values

$$\begin{aligned} B_u &= -2^{57}u^{54}(v^2 - 4u^3)^{28}(4v - u^2 + 110u - 1125)^2 \\ B_v &= 2^{66}u^{60}(v^2 - 4u^3)^{31}(4v - u^2 + 110u - 1125)^2 \end{aligned}$$

in the case $u \neq 0$ and

$$\begin{aligned} B_u &= -2^{57}v^{11}(4v - 1125)^2 \\ B_v &= 2^{66}v^{12}(4v - 1125)^2 \end{aligned}$$

in the case $u = 0$. Since we assume

$$(u^2 - 4v + 18u - 27)(v^2 - 4u^3)(4v - u^2 + 110u - 1125) \neq 0,$$

they never vanish. This same remark holds for the discriminant, which is equal to

$$-2^{36}u^{30}(v^2 - 4u^3)^{15}(u^2 - 4v + 18u - 27)^2$$

in the case $u \neq 0$ and to

$$-2^{36}v^5(-4v-27)^2$$

in the case $u = 0$. □

Remark 7: The group of automorphisms of the curves introduced in the previous lemma is generated by

$$\imath = \begin{pmatrix} -1 & 0 \\ 0 & -1 \end{pmatrix}, \qquad U = \begin{pmatrix} 0 & \sqrt{t} \\ 1/\sqrt{t} & 0 \end{pmatrix}.$$

In particular, all automorphisms are defined over $k(\sqrt{t})$.

These previous lemmas allow us to give an explicit parametrization of the moduli points in $\mathcal{M}_2(k)$ corresponding to curves with group of automorphisms isomorphic to V_4.

Proposition 8: *Let k be a field of characteristic different from 2. The set of points in $\mathcal{M}_2(k)$ corresponding to curves with group of automorphisms isomorphic to V_4 is in bijection with*

$$\{(u,v) \in k^2 \mid (u^2-4v+18u-27)(v^2-4u^3)(4v-u^2+110u-1125) \neq 0\}.$$

An explicit bijection is obtained by evaluating the absolute invariants j_u, j_v given in Proposition 5 at a given moduli point and, conversely, associating to a pair (u,v) the curve given by Proposition 6.

3. The case $\mathrm{Aut}(C) \simeq D_8, D_{12}$

In this section we give a parametrization of curves of genus 2 with dihedral group of automorphisms, that is, isomorphic to the dihedral group of 8 or 12 elements. The following two propositions provide such parametrizations, which are completely constructive, giving representatives defined over the field k for any suitable parameter in k.

Proposition 9: *Let k be a field of characteristic different from 2. The $\overline{k}$-isomorphism classes of curves C/k of genus 2 with $\mathrm{Aut}(C) \simeq D_8$ are classified by the open subset of the affine line*

$$k \setminus \{0, 1/4, 9/100\}.$$

An explicit bijection is obtained by evaluating at a given curve C the absolute invariant given by

$$t = \begin{cases} 1 + \dfrac{J_4}{J_2^2}, & \text{if } \operatorname{char} k = 5, \\ \dfrac{-J_2^2}{J_4}, & \text{if } \operatorname{char} k = 3, \\ \dfrac{8c_6(6c_4 - c_2^2) + 9c_{10}}{900c_{10}}, & \text{otherwise.} \end{cases}$$

and associating to an element $t \in k \smallsetminus \{0, 1/4, 9/100\}$ the curve with equation

$$y^2 = x^5 + x^3 + t\,x.$$

Proof: The polynomial $x^5 + x^3 + t\,x$ has multiple roots for exactly the two values $t = 0, 1/4$, hence the equation $y^2 = x^5 + x^3 + t\,x$ defines an hyperelliptic curve C_t of genus 2 for every $t \neq 0, 1/4$. One checks that the group of automorphisms of this curve contains the group isomorphic to D_8 generated by the two matrices

$$U = \begin{pmatrix} -\sqrt{-1} & 0 \\ 0 & \sqrt{-1} \end{pmatrix}, \qquad V = \begin{pmatrix} 0 & t^{1/4} \\ t^{-1/4} & 0 \end{pmatrix},$$

and the equality $\operatorname{Aut}(C) = \langle U, V \rangle$ holds if $t \neq 9/100$. For this value of t, the group of automorphisms of the curve is isomorphic to $2D_{12}$. Conversely, it is easily seen that every curve $C/\overline{k}$ with $\operatorname{Aut}(C) \simeq D_8$ is $\overline{k}$-isomorphic to a curve C_t for some $t \in \overline{k}$.

Using the formulas expressing the Clebsch and Igusa arithmetic invariants of a curve as polynomials on the coefficients of a polynomial of degree 5 or 6 in an hyperelliptic equation defining the curve, one checks that $t \in \overline{k}$ is the absolute invariant of the curve C_t given in the statement of the proposition.

Since t is an absolute invariant of the curve C_t, it follows that C_t and $C_{t'}$ are isomorphic if, and only if, $t = t'$.

As for the field of definition, if $t \in k$ the curve C_t is defined over k. Conversely, if a curve C is defined over k and C_t is a curve isomorphic to it, then $t \in k$, since t is an absolute invariant of C. □

Proposition 10: *Let k be a perfect field of characteristic different from 2. The $\overline{k}$-isomorphism classes of curves C/k of genus 2 with $\operatorname{Aut}(C) \simeq D_{12}$ are classified by the open subset of the affine line*

$$k \smallsetminus \{0, 1/4, -1/50\}.$$

An explicit bijection is obtained by evaluating at a given curve C the absolute invariant given by

$$t = \begin{cases} -1 - \dfrac{J_4}{J_2^2}, & \text{if} \quad \operatorname{char} k = 5, \\ \dfrac{-J_2^3}{J_6}, & \text{if} \quad \operatorname{char} k = 3, \\ \dfrac{3c_4c_6 - c_{10}}{50c_{10}}, & \text{otherwise.} \end{cases}$$

and associating to an element $t \in k \setminus \{0, 1/4, -1/50\}$ *the curve with equation*

$$y^2 = x^6 + x^3 + t,$$

if $\operatorname{char} k \neq 3$, *or*

$$y^2 = t_*^{-1}x^6 + x^4 + x^2 + 1,$$

if $\operatorname{char} k = 3$, *where* $t_* \in k$ *is such that* $t_*^3 = t$.

Proof: If $\operatorname{char} k \neq 3$, the proof goes the same way as the previous one taking as generators for the group of automorphisms

$$U = \begin{pmatrix} -\zeta_6 & 0 \\ 0 & \zeta_6^2 \end{pmatrix}, \quad V = \begin{pmatrix} 0 & t^{1/6} \\ t^{-1/6} & 0 \end{pmatrix},$$

where ζ_6 is a primitive 6-th root of unity. Notice that for $t = -1/50$, the group of automorphisms of the curve is isomorphic to $\tilde{S}_4$.

For the case $\operatorname{char} k = 3$, one takes

$$U = \begin{pmatrix} 1 & 0 \\ -1 & 1 \end{pmatrix}, \quad V = \begin{pmatrix} 1 & 0 \\ 2 & 1 \end{pmatrix},$$

and checks that they generate the group of automorphisms of the curve $y^2 = t_*^{-1}x^6 + x^4 + x^2 + 1$, whose invariant t, given in the statement of the proposition, is t_*^3. □

4. The isolated cases

The remaining cases correspond to three single points in $\mathcal{M}_2$. For the sake of completeness, we give below an explicit equation defined over k for each of these moduli points.

Proposition 11: *Let C be a genus 2 curve with group of automorphisms isomorphic to* $2D_{12}$, $\tilde{S}_4$, $\tilde{S}_5$ *or* C_{10}. *Then, C is* $\overline{k}$*-isomorphic to the curve* *given by*

$$
\begin{aligned}
y^2 &= x^6 - 1, && \textit{if } \mathrm{Aut}(C) \simeq 2D_{12} \ (\mathrm{char}\, k \neq 3, 5), \\
y^2 &= x^5 - x, && \textit{if } \mathrm{Aut}(C) \simeq \tilde{S}_4 \ (\mathrm{char}\, k \neq 5) \textit{ or } \tilde{S}_5 \ (\mathrm{char}\, k = 5), \\
y^2 &= x^5 - 1, && \textit{if } \mathrm{Aut}(C) \simeq C_2 \times C_5 \ (\mathrm{char}\, k \neq 5).
\end{aligned}
$$

References

1. O. Bolza. On binary sextics with linear transformations into themselves. *Amer. J. Math*, 10:47–70, 1888.
2. G. Cardona, J. González, J. C. Lario, and A. Rio. On curves of genus 2 with Jacobian of GL_2-type. *Manuscripta Math.*, 98(1):37–54, 1999.
3. G. Cardona, E. Nart, and J. Pujolás. Curves of genus two over fields of even characteristic. *Math. Zeitschrift*, to appear, DOI: 10.1007/s00209-004-0750-0, 2005.
4. A. Clebsch. *Theorie der Binärien Algebraischen Formen*. Verlag von B. G. Teubner, Leipzig, 1872.
5. D. Hilbert. *Theory of Algebraic Invariants*. Cambridge University Press, 1993.
6. J. I. Igusa. Arithmetic variety of moduli for genus two. *Annals of Mathematics*, 72(3):612–649, 1960.
7. J. F. Mestre. Construction de courbes de genre 2 à partir de leurs modules. In *Effective methods in algebraic geometry (Castiglioncello, 1990)*, pages 313–334. Birkhäuser Boston, Boston, MA, 1991.
8. T. Shaska and H. Völklein. Elliptic subfields and automorphisms of genus 2 function fields. *Algebra, Arithmetic and Geometry with Applications. Papers from Shreeram S. Abhyankar's 70th Birthday Conference*, pg. 687 - 707, Springer (2004).

Explicit computation of Hurwitz spectra

R. Vogeler

Department of Mathematics,
Ohio State University
231 W 18th Ave,
Columbus, Ohio, 43210, USA
vogeler@math.ohio-state.edu

We describe an algorithmic method of computing the length spectrum of the $(2,3,7)$ triangle group G in exact arithmetic form. From G descend all the Hurwitz surfaces, and their spectra can in turn be computed.

1. Introduction

Much of the geometry [1] of a Riemann surface S of genus $g \geq 2$ can be approached by way of the closed geodesics [7] on S, where S is endowed with its conformally unique metric of constant curvature -1. The sequence of lengths of these curves is called the *length spectrum* of S. It is natural to think of the length spectrum as also belonging to a uniformizing Fuchsian group associated to S.

Frequently it is convenient to work with the trace of a curve, rather than its length. More precisely, this means the trace of a matrix $M \in SL_2(\mathbb{R})$ representing the covering transformation obtained by lifting the closed curve, with M acting on the upper half-plane $\mathbb{H}$ as a linear fractional transformation. The relation between the length and the trace is given by

$$\text{trace} = \pm 2 \cosh\left(\frac{\text{length}}{2}\right).$$

The ambiguity in the sign of the trace is unavoidable, in general, and is a consequence of the lifting from $\text{Isom}^+(\mathbb{H}) \cong PSL_2(\mathbb{R})$ to $SL_2(\mathbb{R})$.

In this report we describe an explicit computation of the trace sequence for a large initial portion of the length spectrum of G, the $(2,3,7)$ triangle group. This begins with a combinatorial technique, developed in previous work [10] which exploits the high degree of regularity present in triangle groups. Section 2 summarizes this part of the computation. In Section 3 we explain the computation of the traces in exact, algebraic form, which is possible because G is an arithmetic group. Finally, in Section 4 we describe in more detail the structure of the trace sequence, which is closely connected to the fact that G is derived from a quaternion algebra.

Our interest in G began with an investigation of closed curves on Hurwitz surfaces [2], [4], the very special Riemann surfaces for which the automorphism group attains the maximal size of $84(g-1)$; see [3]. Each of these surfaces is uniformized by a normal, finite-index subgroup of G. Our original computations of curve lengths on these surfaces used floating-point approximations. As more and more curves of apparently equal length arose, it began to seem important to determine how much of the apparent multiplicity was genuine. This, of course, is resolved by computing the exact trace values.

2. The combinatorial technique

G is the group of orientation-preserving symmetries of the tiling of $\mathbb{H}$ by regular 7-gons with angles of $\frac{2\pi}{3}$; this tiling is denoted by the Schlafli symbol $\{7,3\}$, and is depicted (using the disk model) in Figure 1. The torsion elements of G are elliptic transformations, or rotations, having orders 2, 3 and 7, and falling into just a few conjugacy classes. The remaining elements are hyperbolic transformations, or translations, each having a characteristic translation distance d along a unique oriented geodesic axis. Such an axis can be closely approximated by a path restricted to the 1-skeleton of the tiling and formed by simply stringing together successive edges. Since the 1-skeleton is trivalent, this edge-path can be encoded by a doubly-infinite sequence in the letters R and L, each letter indicating a right or left turn, respectively, at the junction of successive edges in the path. Due to the periodic nature of the action of a translation on the tiling, the approximating edge-path can be chosen to have a periodic structure, too. As a result, the doubly-infinite R-L sequence can be reduced to a finite sequence corresponding to a single period and having length roughly proportional to d. We call this finite sequence a *label* for the underlying translation. It is worth noting that a given label applies not just to a single translation g, but in fact to the entire conjugacy class of g in G.

There are many different labels corresponding to any given translation,

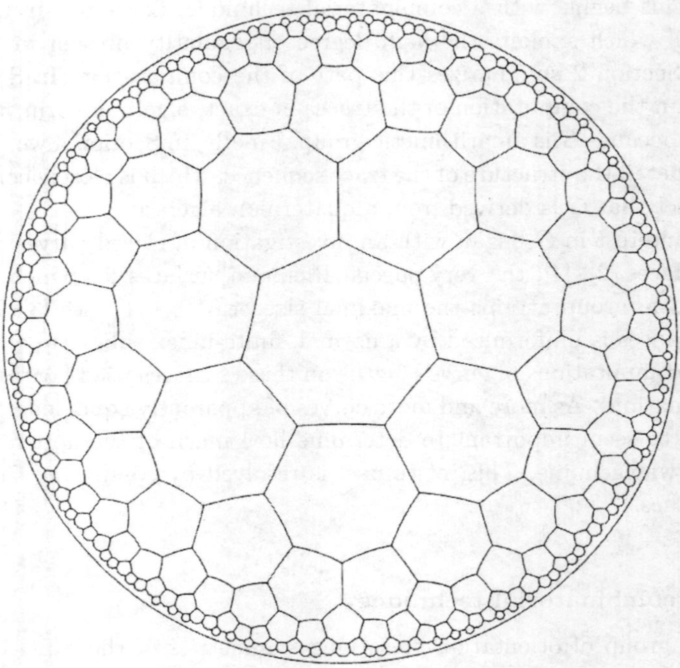

Fig. 1. The $\{7,3\}$ tiling of $\mathbb{H}$

though it would be desirable to have just one. It is possible to force this uniqueness by restricting to a special subset of labels called *reduced* labels. There is then a bijective correspondence between the set of reduced labels and the conjugacy classes of translations in G.

At this point one can begin writing down a list of reduced labels representing the length spectrum of G. This procedure makes use of the fact that any label can be checked algorithmically to determine whether it is reduced or not. Thus for each positive integer n one simply generates all possible labels of length n, checks each one, and keeps those that are reduced. The resulting list (as far as it goes) contains the conjugacy classes of translations in G, without repetition, and ordered roughly (though not precisely) according to the translation distance.

For example, the shortest translation in G has an axis that can be approximated by an edge-path encoded as ...LRLRLR... , which leads to the single-period label RL. This label is reduced, and thus appears as the first entry on the list of labels representing the length spectrum.

3. Computing the traces

In this section we explain how the traces can be computed in a very direct way from reduced labels. The first step is to develop matrices $R, L \in SL_2(\mathbb{R})$ which correspond to the R's and L's in the labels.

We arrange the $\{7,3\}$ tiling of $\mathbb{H}$ so that one 7-gon has a vertex v at i and its center c at bi, where $b = e^h$ and h is the hyperbolic distance between v and c. Basic trigonometric formulas give

$$\cosh h = (\cot \frac{\pi}{3})(\cot \frac{\pi}{7}),$$

and we note for future use the formula

$$b + \frac{1}{b} = e^h + e^{-h} = 2\cosh h = \frac{2}{\sqrt{3}} \cot \frac{\pi}{7}.$$

The group G of orientation-preserving tiling symmetries is now given by the presentation

$$G = \langle A, B \mid A^3, B^7, (AB)^2 \rangle,$$

where the generators are

$$A = \begin{pmatrix} \cos\frac{\pi}{3} & \sin\frac{\pi}{3} \\ -\sin\frac{\pi}{3} & \cos\frac{\pi}{3} \end{pmatrix} = \begin{pmatrix} \frac{1}{2} & \frac{\sqrt{3}}{2} \\ -\frac{\sqrt{3}}{2} & \frac{1}{2} \end{pmatrix},$$

which is a $2\pi/3$ counter-clockwise rotation about i, and

$$B = \begin{pmatrix} \cos\frac{\pi}{7} & b\sin\frac{\pi}{7} \\ -\frac{1}{b}\sin\frac{\pi}{7} & \cos\frac{\pi}{7} \end{pmatrix},$$

which is a $2\pi/7$ counter-clockwise rotation about bi.

Next, we define matrices $R = A^{-1}B$ and $L = B$. Multiplication gives

$$R = \begin{pmatrix} \frac{1}{2}\cos\frac{\pi}{7} + \frac{\sqrt{3}}{2b}\sin\frac{\pi}{7} & \frac{b}{2}\sin\frac{\pi}{7} - \frac{\sqrt{3}}{2}\cos\frac{\pi}{7} \\ \frac{\sqrt{3}}{2}\cos\frac{\pi}{7} - \frac{1}{2b}\sin\frac{\pi}{7} & \frac{1}{2}\cos\frac{\pi}{7} + \frac{\sqrt{3}}{2}b\sin\frac{\pi}{7} \end{pmatrix}.$$

Taking traces, we find

$$Tr(L) = 2\cos\frac{\pi}{7}$$

and also

$$Tr(R) = \cos\frac{\pi}{7} + \frac{\sqrt{3}}{2}(b + \frac{1}{b})\sin\frac{\pi}{7} = 2\cos\frac{\pi}{7}.$$

Matrices R and L generate right and left turns at vertex i in an edge-path entering i from the left. It follows that any given label corresponds to the translation obtained by reinterpreting the label as a product of matrices. Hence the trace could, at least in principle, be computed by first performing the full matrix multiplication and then taking the trace. However, this computation is extremely costly as it leads to huge, complicated expressions. A more efficient method uses the following identity, which is easily verified for all matrices $u, v \in SL_2(\mathbb{R})$:

$$Tr(uv) = Tr(u)Tr(v) - Tr(u^{-1}v).$$

Using this, along with the fact that trace is invariant under conjugation, we have

$$Tr(RL) = Tr(R)Tr(L) - Tr(R^{-1}L) = 4\cos^2\frac{\pi}{7} - Tr(B^{-1}AB) = 1 + 2\cos\frac{2\pi}{7}.$$

Before computing more traces, we now change the numerical representation from trigonometric to algebraic form. Letting Θ be the seventh root of unity given by

$$\Theta = \cos\frac{2\pi}{7} + i\sin\frac{2\pi}{7},$$

we set

$$X = \Theta + \Theta^6 \approx 1.25, Y = \Theta^2 + \Theta^5 \approx -0.45, Z = \Theta^3 + \Theta^4 \approx -1.80;$$

these are the roots of $x^3 + x^2 - 2x - 1$, and are algebraic integers in the number field $\mathbb{Q}[X]$. Multiplicative relations such as $XY = X + Z$ and $X^2 = Y + 2 = -2X - Y - 2Z$ are easily worked out, and show that all elements of the ring $\mathbb{Z}[X, Y, Z]$ can be expressed in the form $aX + bY + cZ$ with $a, b, c \in \mathbb{Z}$. Numbers in this form can be denoted by the abbreviated expression $[a, b, c]$, which we adopt from now on as the standard form for all traces. Casting the traces computed above into this new form gives

$$Tr(R) = Tr(L) = [0, 0, -1] \quad \text{and} \quad Tr(RL) = [0, -1, -1].$$

In general we need to compute the trace of a product of matrices R and L as represented by an RL label. To do so, we recursively apply the trace identity in order to express the desired trace in terms of traces of "shorter" products.

There are potentially many different ways that such a reduction algorithm could proceed. A technique that we have used is to first remove all occurrences of double letters. For instance, the label RLRRL contains a double R. We move RR to the beginning by cyclic permutation, set $u = R$ and $v = RLRL$, and then reduce:

$$Tr(RRLRL) = Tr(R)Tr(RLRL) - Tr(LRL).$$

A product containing no double letters (even under cyclic permutation) must be of the form $(RL)^n$. In this case, assuming $n \geq 2$, we set $u = RL$ and $v = (RL)^{n-1}$ to obtain

$$Tr((RL)^n) = Tr(RL)Tr((RL)^{n-1}) - Tr((RL)^{n-2}).$$

Since products of this form occur repeatedly, it is convenient to compute the trace of $(RL)^n$ just once (for many small values of n) and store the results in a lookup table.

It is now evident from the reduction algorithm and from the few traces already computed that all the traces of G must be algebraic integers of the form specified above. Some results are shown in the table, which gives the first twenty terms of the length spectrum of G.

To find the length spectra of the Hurwitz surfaces, the closed curves are first determined by doing group calculations involving the list of reduced labels for G; see [10]. This gives a new list of reduced labels for the surface, and the traces are then computed just as for G.

4. Structure of the trace sequence

From the number theoretic viewpoint, G is considered an arithmetic group derived from a quaternion algebra; see [5]. Takeuchi [8] has given a useful characterization of such groups, which has strong implications for the nature of the values that can occur as traces; one aspect of this is stated below as a theorem. (See also Schmutz [6] for more on arithmetic length spectra, and Takeuchi [9] for the determination of all arithmetic triangle groups.)

There are three embeddings of the number field $\mathbb{Q}[X]$ into $\mathbb{R}$. One of them, ϕ_1, is obtained by extending the permutation $X \to Y \to Z \to X$. Another, ϕ_2, is the second power of ϕ_1, and lastly ϕ_3 is the identity embedding. There is a restriction on the traces of elements of G in terms of ϕ_1 and ϕ_2.

Theorem: (Takeuchi) Suppose an arithmetic Fuchsian group Γ, with trace set $Tr(\Gamma)$, is derived from a quaternion algebra. Then $\phi(\mathbb{Q}[Tr(\Gamma)] \subset [-2, 2]$ for every non-identity embedding $\phi : \mathbb{Q}[Tr(\Gamma)] \to \mathbb{C}$.

In order to easily visualize the situation as it applies to G, we may consider traces having the form $[a, b, c]$ as elements of the integer lattice

Length	Label	Trace
0.98398656	RL	[0, -1, -1]
1.73600575	RRLL	[-1, -1, -2]
1.96797312	RLRL	[1, 0, -1]
2.13110501	RLRLL	[-1, -2, -2]
2.66193075	RLRRLL	[0, -1, -2]
2.89814945	RLLRRLL	[0, -2, -2]
2.95195969	RLRLRL	[-1, -1, -3]
3.15482446	RLRLRLL	[-1, -2, -3]
3.47201150	RRLLRRLL	[0, -1, -3]
3.54271043	RLRRLRLL	[-2, -3, -4]
3.62731599	RLRLRRLL	[0, -2, -3]
3.80470389	RLRRLRRLL	[-1, -2, -4]
3.93594625	RLRLRLRL	[-1, -3, -4]
3.93594625	RLRLLRRLL	[-1, -3, -4]
4.15197165	RLRLRLRLL	[0, -2, -4]
4.20180693	RLLRRLRRLL	[-2, -4, -5]
4.26221003	RLRLLRLRLL	[0, -3, -4]
4.39145967	RLRRLLRRLL	[-1, -3, -5]
4.48925712	RLRLRRLRLL	[-1, -4, -5]
4.60473270	RLRLRLRRLL	[-2, -4, -6]

$\mathbb{Z}^3$ in $\mathbb{R}^3$. Then the embeddings ϕ_i $(i = 1, 2, 3)$ extend naturally to linear functionals $\hat{\phi}_i : \mathbb{R}^3 \to \mathbb{R}$. Let ℓ be the line of intersection of the zero sets of $\hat{\phi}_1$ and $\hat{\phi}_2$; it turns out that ℓ is spanned by the vector $[X-2, Y-2, Z-2]$. Then Takeuchi's characterization says that all the traces of G are clustered around ℓ, lying specifically within the region bounded by the planes $\hat{\phi}_1, \hat{\phi}_2 = \pm 2$.

It appears that even more may be true. Let T be the tube-like region bounded by the planes mentioned above, and limited by the additional constraint $\hat{\phi}_3 > 2$. Let S be the set of integral lattice points contained in T, ordered by the value of $\hat{\phi}_3$ (which is nearly the same as ordering by distance from the origin). Then the first thousand points of S correspond precisely to the first thousand terms of the length spectrum of G. We are not certain whether *every* point of S matches a trace of G, but it seems reasonable to formally state our suspicion.

Conjecture: $\hat{\phi}_3(S) = Tr(G)$.

We now describe the structure of S in more detail. This is really just an exercise in linear algebra, but the results are rather pleasing. We use the terms *jump sequence* for the sequence of successive vector differences in S, and *gap sequence* for the sequence obtained from the jump sequence by applying $\hat{\phi}_3$, i.e., the sequence of gaps in the conjectured trace sequence of G.

By inspecting the values of $\hat{\phi}_i$ over a small portion of the integer lattice, one easily finds that there are just nine vectors that can occur as elements of the jump sequence, hence there are also just nine values that can occur in the gap sequence. These data are shown in Figure 2, arranged in order of increasing gap size and labeled A through I.

The square diagram in Figure 2 contains nine rectangles also labeled A through I. Each rectangle has one vertex at a corner of the square, and a label that is located just inside the rectangle at the opposite vertex. The square is the target space $[-2,2]\times[-2,2]$ of the mapping $x \mapsto (\hat{\phi}_1(x), \hat{\phi}_2(x))$ for points $x = [a,b,c] \in T$. Each rectangle is the image of the points of T whose sum with the associated vector also lies in T. We need not go beyond rectangle I because the square is completely covered by these first nine rectangles.

For example, suppose lattice point $[a,b,c] \in S$ maps into rectangle A. Then $[a,b,c]+[-1,1,-1]$ also lies in S, and the associated trace values differ by gap size A, or about 0.1099. Clearly $[a,b,c]$ also maps into rectangle E (since it encloses rectangle A), so $[a,b,c] + [-1,0,-1]$ lies in S and the associated traces differ by gap size E. However, these lattice points are not consecutive in S so there is no corresponding term in the gap sequence.

The jumping from one point of S to another by adding the appropriate vector can be completely carried over to the square diagram. The required add-on vector is now simply the arrow extending from the labeled vertex of the rectangle to the opposite corner of the square, or in other words, the translation that carries the rectangle across diagonally to the opposite corner of the square. The iteration of jumps is thus easily visualized in the square. This makes it easy to view the iteration of jumps as a type of dynamical system taking place in the square.

In Figure 3, the square diagram has been modified by subtracting from each rectangle all the rectangles that precede it in the labeling order. Rectangle H thereby vanishes, so there are just eight different values that actually occur in the gap sequence. Each point of S now maps to just one of the eight regions; by adding the corresponding vector, the next point of S is obtained. By starting with lattice point $[-2,-2,-2]$ (or, equivalently,

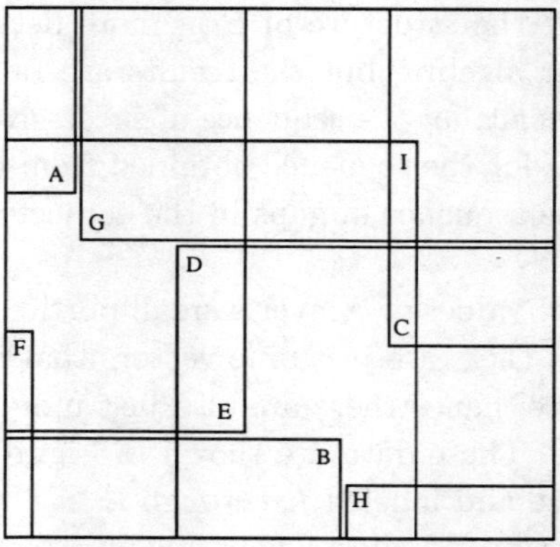

	[a, b, c]	gap size
A	-1, 1, -1	0.1099
B	-2, -2, -1	0.1981
C	2, 1, 1	0.2470
D	0, -1, 0	0.4450 = B+C
E	-1, 0, -1	0.5550 = A+B+C
F	-3, -2, -2	0.7530 = A+2B+C
G	1, 1, 0	0.8019 = A+B+2C
H	0, -2, 0	0.8901 = 2B+2C
I	-1, -1, -1	1.0000 = A+2B+2C

Fig. 2. Gap sequence data

the upper right corner of the square diagram), the entire gap sequence can be generated by this iteration scheme.

The areas of the modified regions correspond to the frequency with which the corresponding gap sizes occur in the gap sequence. These data appear in the lower part of Figure 3. For example, region A fills about 4.4% of the total area of the square, hence gap size A ought to occur in 4.4% of the terms of the gap sequence. This amounts to 51.3 expected occurrences among the first 1161 terms, while the actual number of occurrences is 51. (Why 1161? It takes that many terms to reach the first occurrence of gap size F.)

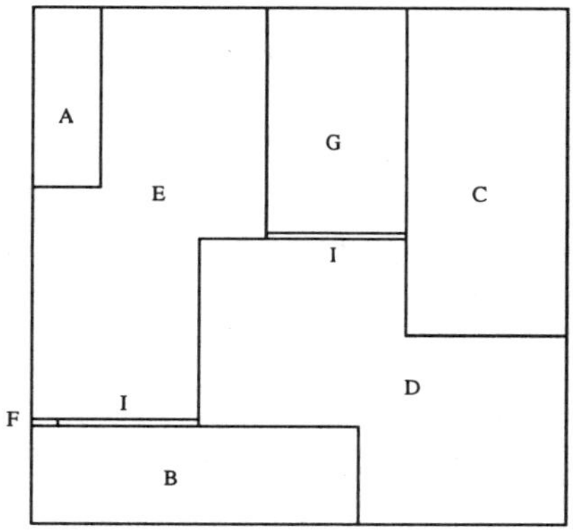

	area %	First 1161 gaps: expected	actual
A	4.4	51.3	51
B	11.5	133.6	133
C	19.1	222.1	220
D	26.5	308.2	310
E	26.2	304.3	304
F	0.06	0.7	1
G	11.5	133.4	135
H	0	0	0
I	0.64	7.4	7

Fig. 3. Gap sequence statistics

Other structural facts can easily be found by examining the diagram. For example, every occurrence of gap size A must be followed immediately by gap size D; likewise, F is always followed by C. Also, gap size D is the only value that can occur at two consecutive terms of the gap sequence.

References

1. P. Buser, Geometry and Spectra of Compact Riemann Surfaces, Progress in Mathematics, Vol. 106, Birkhäuser, 1992.
2. M. Conder, Hurwitz groups: a brief survey, Bull. Amer. Math. Soc., Vol.

23, No. 2, 1990.
3. A. HURWITZ, Über algebraische Gebilde mit eindeutigen Transformationen in sich, Math. Ann., Vol. 41, 1893.
4. A.M. MACBEATH, Hurwitz groups and surfaces, The Eightfold Way, MSRI Publications, Vol. 35, 1998.
5. C. MACLACHLAN, Introduction to arithmetic Fuchsian groups, London Math. Soc. Lecture Note Series 287, 2001.
6. P. SCHMUTZ, Arithmetic groups and the length spectrum of Riemann surfaces, Duke Math. J., Vol. 84, No. 1, 1996.
7. P. SCHMUTZ SCHALLER, Geometry of Riemann surfaces based on closed geodesics, Bull. Amer. Math. Soc., Vol. 35, No. 3, 1998.
8. K. TAKEUCHI, A characterization of arithmetic Fuchsian groups, J. Math. Soc. Japan, Vol. 27, No. 4, 1975.
9. K. TAKEUCHI, Arithmetic triangle groups, J. Math. Soc. Japan, Vol. 29, No. 1, 1977.
10. R. VOGELER, Combinatorics of curves on Hurwitz surfaces, Annales Academiæ Scientiarum Fennicæ Mathematica Dissertationes, No. 137, 2004.

Non-normal Belyi p-gonal surfaces

Aaron Wootton*

Department of Mathematics,
University of Arizona,
Tucson, AZ 85721, USA
awootton@math.arizona.edu

We show that there are only a small number of non-normal Belyi p-gonal surfaces, most of which are well known. For each such surface, we find the full automorphism group and a defining equation. A consequence is the discovery of an explicit family of surfaces which show that Accola's bound is sharp for p-gonal surfaces for all primes $p \geq 5$.

1. Introduction

A Belyi surface is a compact Riemann surface that is a ramified covering of the sphere with at most three critical values. One of the most important properties of Belyi surfaces is that any such surface will a possess a defining equation over a finite extension of $\mathbb{Q}$. For this reason, a study of Belyi surfaces can provide valuable insight into the inverse Galois problem or aid in a search for non-trivial elements in the absolute Galois group $Gal(\bar{\mathbb{Q}}/\mathbb{Q})$. This suggests that a central problem in the theory of Belyi surfaces is to find unique defining equations for each such surface. Such a program presently seems unreasonable as even for low genus, knowledge regarding Belyi surfaces is scarce. Therefore, the results of this paper focus on families of Belyi surfaces with additional properties.

A family of Belyi surfaces for which such a program has been completed are normal p-gonal Belyi surfaces, see Wootton [10], defined as:

Definition 1: A normal Belyi p-gonal surface X is a compact Riemann

*The author would like to express his thanks to the editor for his support

surface of genus $g \geq 2$ such that the full group of automorphisms G of X has the following properties:

(1) X/G has genus 0.
(2) The map $\pi\colon X \to X/G$ is branched over three points.
(3) There exists a proper normal cyclic prime p-subgroup of G, C_p such that X/C_p has genus 0.

These results were developed through the use of Fuchsian groups, hyperbolic geometry and Galois theory. One of the problems encountered was that there exist surfaces which share many properties with normal Belyi p-gonal surfaces, but are not themselves normal Belyi p-gonal. They are defined as follows:

Definition 2: A non-normal Belyi p-gonal surface X is a compact Riemann surface of genus $g \geq 2$ which admits a group of automorphisms L with the following properties:

(1) X/L has genus 0.
(2) The map $\pi_L\colon X \to X/L$ is branched over three points.
(3) L has a proper normal cyclic prime subgroup C_p of order p such that X/C_p has genus 0.
(4) C_p is not normal in G, the full automorphism group of X.

Therefore, before the problem of finding defining equations for normal Belyi p-gonal surfaces can be solved, one must first find all non-normal Belyi p-gonal surfaces.

In this paper we shall classify all non-normal Belyi p-gonal surfaces of genus $g \geq 2$. For each such surface we will find its full automorphism group and a defining equation using the results developed in Wootton [10].

2. Fuchsian Groups and Uniformization

To classify non-normal Belyi p-gonal surfaces, we shall start by developing the necessary theory regarding uniformization and Fuchsian groups - discrete subgroups of $PSL(2,\mathbb{R})$. The primary reason for this is that Belyi p-gonal surfaces have a very useful description when regarded as quotients of the upper half plane. Let X denote a compact Riemann surface of genus $g \geq 2$ and G a group of automorphisms of X.

Uniformization implies that X is conformally equivalent to a quotient of the upper half plane $\mathbb{H}$ by a torsion free Fuchsian group Λ called a surface group or a surface kernel for X. Under such a realization, a group

G is a group of automorphisms of X if and only if $G = \Gamma/\Lambda$ for some Fuchsian group Γ containing Λ as a normal subgroup, see Macbeath [5]. It follows that the full automorphism group Aut (X) of X is the quotient group $N(\Lambda)/\Lambda$. Uniformization also provides a way of determining conformal equivalence of two surfaces. Specifically, two surfaces X_1 and X_2 are conformally equivalent if and only if any surface group for X_1 is conjugate in $PSL(2, \mathbb{R})$ to any surface group for X_2.

Since G is a group of automorphisms acting on X, we can form the quotient space X/G which can be imposed with a unique structure making the map $\pi_G \colon X \to X/G$ a holomorphic map between compact Riemann surfaces. The group Γ is also a group of biholomorphic maps acting on $\mathbb{H}$, so we can form the quotient space $\mathbb{H}/\Gamma$ with a complex structure so the map $\pi_\Gamma \colon \mathbb{H} \to \mathbb{H}/\Gamma$ is holomorphic. Let $\pi_\Lambda \colon \mathbb{H} \to \mathbb{H}/\Lambda = X$ denote smooth cover of X by $\mathbb{H}$. Then X/G can be identified with $\mathbb{H}/\Gamma$ and after identification, we have $\pi_\Gamma = \pi_G \circ \pi_\Lambda$:

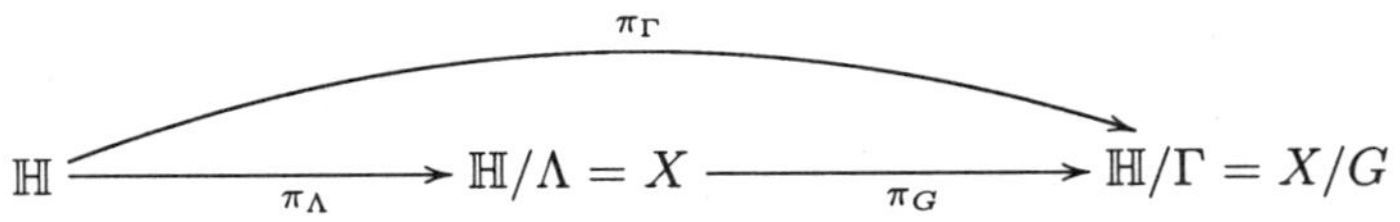

Fig. 1. Holomorphic Quotient Maps and Surface Identifications

For a Fuchsian group Γ with compact orbit space $\mathbb{H}/\Gamma$ of genus g, a presentation for Γ is:

$$\Gamma = \langle a_1, b_1, \ldots, a_g, b_g, c_1, \ldots, c_r | c_1^{m_1}, \ldots, c_r^{m_r}, \prod_{i=1}^{r} c_i \prod_{j=1}^{g} [a_j, b_j] \rangle$$

where the quotient map π_Γ branches over r points with ramification indices m_i for $1 \leq i \leq r$. The presentation of such a group is described by the tuple $(g; m_1, \ldots, m_r)$ called the signature of Γ, g the orbit genus of Γ and $m_1, \ldots, m_r$ the periods of Γ. In the special case $g = 0$ and $r = 3$, we call Γ a triangle group and we shall sometimes denote such a group by Δ. It is a fact that triangle groups are unique up to conjugation in $PSL(2, \mathbb{R})$, see Singerman [9]. Notice that if Γ is a surface group for a surface of genus g, since it is torsion free, it must have signature $(g; -)$.

We now interpret this information into the situation we are considering. If X is a non-normal Belyi p-gonal surface, let L and C_p denote the groups of automorphisms as specified in the definition and if Λ is a surface group for

X, let Γ and Γ_p denote the Fuchsian groups with $\Gamma/\Lambda = L$ and $\Gamma_p/\Lambda = C_p$. Since C_p is normal in L, it follows that the group $K = L/C_p$ acts by automorphism on the quotient space X/C_p. Let π_K denote the quotient map of the space X/C_p by K. After appropriate identifications, we get the following tower of Galois covers.

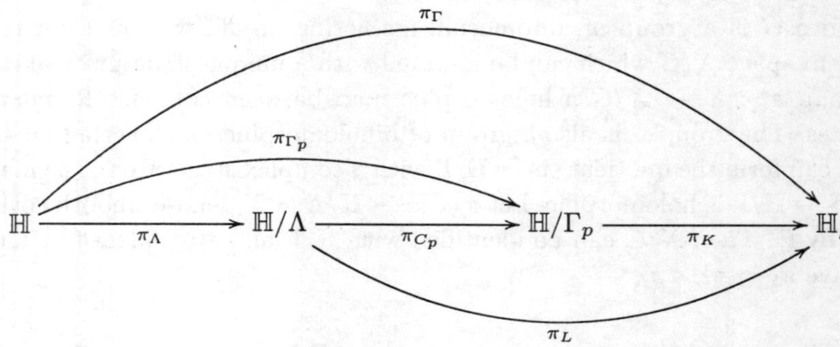

Fig. 2. Holomorphic Quotient Maps and Surface Identifications

Since π is unramified, our analysis implies that Γ must be a Fuchsian triangle group and that Γ_p has signature $(0; \underbrace{p, \ldots, p}_{r \text{ times}})$ where r is the number of branch points of the map π_{C_p}. Our next task is determine specifically the signatures for Γ and Γ_p.

3. Finding Signatures

The map π_K is a Galois map from the Riemann sphere to itself. The properties we require of such maps are common knowledge, so we summarize them in Table 1. The branching data is a vector whose length is the number of branch points of the quotient map π_K and whose entries are the ramification indices of ramification points above these branch points.
To find the possible signatures for Γ, we shall use that fact that we know complete branching data of the map π_K and partial branching data for the map π_{C_p}.

Proposition 3: *Let K be the group Γ/Γ_p.*

(1) If $K \neq C_n$, the signature of Γ is $(0; m_1\tilde{p_1}, m_2\tilde{p_2}, m_3\tilde{p_3})$ where $\tilde{p_i} = 1$ or p depending upon whether the branch points of π_{C_p} coincide with ram-

Table 1. Groups of Automorphisms of the Riemann Sphere and Branching Data.

Group	Branching Data
C_n	(n, n)
D_n	$(2, 2, n)$
A_4	$(2, 3, 3)$
S_4	$(2, 3, 4)$
A_5	$(2, 3, 5)$

ification points of π_K. *For such a* Γ, *the signature of* Γ_p *is* $(0; \underbrace{p \ldots, p}_{r\ times})$ *where*

$$r = \Sigma_{i=1}^{3} |\tilde{K}|/m_i$$

and $|\tilde{K}| = 0$ *if* $\tilde{p_i} = 1$ *or* $|\tilde{K}| = |K|$ *if* $\tilde{p_i} = p$.

(2) If $K = C_n$, *the signature of* Γ *is* $(0; n\tilde{p_1}, n\tilde{p_2}, p)$ *where* $\tilde{p_i} = 1$ *or* p *depending upon whether the branch points of* π_{C_p} *coincide with ramification points of* π_K . *For such a* Γ, *the signature of* Γ_p *is* $(0; \underbrace{p, \ldots, p}_{r\ times})$ *where*

$$r = n + \Sigma_{i=1}^{2} |\tilde{K}|/m_i$$

with $|\tilde{K}|/m_i$ *as defined in* (i).

Proof: Since $\mathbb{H}/\Gamma$ has genus 0 and the map π_L is branched over three points, we know that Γ must be a triangle group with signature $(0; m_1, m_2, m_3)$ where the m_i's are the ramification indices of the map π_L. Therefore, to find the signature of Γ, we need to find the values of the m_i.

Assume $K \neq C_n$ (the proof is almost identical when $K = C_n$ so we omit it). By assumption, the map $\pi_L \colon X \to \mathbb{H}/\Gamma$ will be branched over three points. Since $\pi_K \colon \mathbb{H}/\Gamma_p \to \mathbb{H}/\Gamma$ is branched over only three points and $\pi_L = \pi_K \circ \pi_{C_p}$, it follows that π_L must be branched over the same points.

Let α be a ramification point of the map π_L. Our remarks above imply that the ramification index of π_K at $\pi_{C_p}(\alpha)$ is m_i for $i = 1$, 2 or 3. Without loss of generality, assume it is m_1. To calculate th ramification index of π_L at α, we just need to determine if α is a ramification point of π_{C_p}. If it is, since π_{C_p} is a degree p Galois cover, it follows that the ramification index of π_L at α is $m_1 p$. Else, the ramification index of π_L at α is m_1.

To find the signature of Γ_p, we examine the orbits of branch points of π_{C_p} under the action of K. The length of an orbit of of a branch point x of π_{C_p} will be the equal to $|K|/|\text{Stab}_K(x)|$. It is easy to show that $|\text{Stab}_K(x)| = m_i$ where m_i is the ramification index of π_K at x, so the result follows. □

There are a large number of signatures which satisfying Proposition 3. We shall not write down all possibilities as we are only concerned with those which could contain a normal surface subgroup whose quotient $\mathbb{H}/\Lambda$ is a non-normal Belyi p-gonal surface X. We examine the additional conditions imposed by this assumption.

If Λ is a surface kernel for X, then Λ will be normal in a Fuchsian group Γ with signature matching one derived using Proposition 3. In addition, there will be some overgroup Δ of Γ in which Λ is normal, but Γ_p is not. Therefore, to progress, we need to find all signatures for overgroups of Fuchsian groups with signature as given in Proposition 3. For general Fuchsian groups, such a task is not possible without additional information. However, for triangle groups, the possible overgroups are completely signature dependent. We summarize in Table 2, for details see Singerman [9].

To proceed, we examine each possible signature which arises using Proposition 3 and compare them to those in Singerman's list, Table 2 to see which ones have overgroups. For each such case, we then check to see if the necessary conditions are satisfied. Our results make essential use of the following result.

Proposition 4: *(Sah [7]) Let Γ be a Fuchsian group with signature $(0; m_1, \ldots, m_r)$ and $\Gamma_1 \leq \Gamma$ a normal subgroup of finite index N such that $c_i\Gamma_1$ has order t_i in the quotient group Γ/Γ_1 and the quotient space $\mathbb{H}/\Gamma_1$ has genus 0. Then,*

$$-1 = -N + \tfrac{N}{2} \sum_{i=1}^{r} (1 - \tfrac{1}{t_i}),$$

and the entries in the signature of Γ_1 are $f_{i,j} = \frac{m_i}{t_i}$, $1 \leq j \leq \frac{N}{t_i}$, $1 \leq i \leq r$, where $f_{i,j} = 1$ are deleted.

We illustrate by examining the case when Γ has signature $(0; n, n, p)$ in detail.

Table 2. Singerman's List for Triangle Groups.

Case	Signature Γ	Signature Δ	[Δ : Γ]
N6	(0;t,t,t),$t \geq 4$	(0;3,3,t)	3
N7	(0;t,t,t),$t \geq 4$	(0;2,3,2t)	6
N8	(0;t,t,u),$t \geq 3$,$t+u \geq 7$	(0;2,t,2u)	2
T1	(0;7,7,7)	(0;2,3,7)	24
T2	(0;2,7,7)	(0;2,3,7)	9
T3	(0;3,3,7)	(0;2,3,7)	8
T4	(0;4,8,8)	(0;2,3,8)	12
T5	(0;3,8,8)	(0;2,3,8)	10
T6	(0;9,9,9)	(0;2,3,9)	12
T7	(0;4,4,5)	(0;2,4,5)	6
T8	(0;n,4n,4n),$n \geq 2$	(0;2,3,4n)	6
T9	(0;n,2n,2n),$n \geq 3$	(0;2,4,2n)	4
T10	(0;3,n,3n),$n \geq 3$	(0;2,3,3n)	4
T11	(0;2,n,2n),$n \geq 4$	(0;2,3,2n)	3

Example 5: Suppose that Γ has signature $(0; n, n, p)$. Proposition 3 implies that Γ_p will have signature $(0; \underbrace{p. \ldots, p}_{n \text{ times}})$. Checking through Singerman's list, there are a number of choices for values of n and p which would be the signature of a Fuchsian group Γ necessarily contained in a Fuchsian overgroup Δ. We tabulate them in Table 3.

For the groups with signatures $(0; 4p, 4p, p)$, $(0; 2p, 2p, p)$ and $(0; n, n, p)$ with $n \neq p$, the group Γ_p is the unique subgroup of Γ with its defining properties. It is easy to show for each such case there also exists a unique normal subgroup of Δ also contained in Γ with these properties, so Γ_p must be normal in the overgroup Δ. This means that in each of these cases, Δ is not a possible overgroup signature for a surface group of a non-normal Belyi p-gonal surface.

When Γ has signature $(0; p, p, p)$, there are three subgroups with the defining properties of Γ_p. For $\Delta \geq \Gamma$ with signature $(0; 2, p, 2p)$, one of these groups will be simultaneously normal in Δ. Therefore, Δ is not a possible overgroup signature for a surface group of a non-normal Belyi p-gonal surface.

For the remaining cases, to show Γ_p is not normal in Δ, we simply apply Proposition 4 to show that no normal subgroup exists with the same

Table 3. Signatures for the Case $(0; n, n, p)$.

n	p	Γ	Δ	$\Gamma_p \lhd \Delta$?
n	p	(0;n,n,p)	(0;2,n,2p)	Yes
p	p	(0;p,p,p)	(0;3,3,p)	No
p	p	(0;p,p,p)	(0;2,3,2p)	No
7	7	(0;7,7,7)	(0;2,3,7)	No
7	2	(0;7,7,2)	(0;2,3,7)	No
3	7	(0;3,3,7)	(0;2,3,7)	No
8	3	(0;3,8,8)	(0;2,3,8)	No
4p	p	(0;4p,4p,p)	(0;2,3,4p)	Yes
2p	p	(0;2p,2p,p)	(0;2,4,2p)	Yes

signature as Γ_p.

For all other signatures, the process we follow is the same. We start with a signature and examine all the possible choices of primes p (and choices for n when appropriate), to find which signatures are signatures for Fuchsian groups with non-trivial overgroups. Following this, we filter the list to contain only those in which Γ_p is not normal. To avoid tedious and repetitious calculations, we summarize the results in Table 3.

Table 4 supplies us with a lot of information. The different cases $1-4$ are for the different possible signatures from Proposition 3, the subcases being the different choices for n and p. Specifically, Case 1 is for Γ with signature $(0; n, n, p)$, Case 2 for Γ with signature $(0; n, np, p)$, Case 3 for Γ with signature $(0; 2, 2p, n)$ and Case 4 for Γ with signature $(0; 2p, 2p, n)$. The columns for Δ and Γ are the signatures for Δ and Γ respectively, and the column for Γ_p is the number of periods of Γ_p (all of order p). The groups L were found using the results of Wootton [10]. When L is a semi-direct product, the column Ker $(\mathcal{A})$ denotes the kernel of the action. Finally, the group K is the quotient group Γ/Γ_p and the last column is the choice for p.

4. Finding Non-Normal Belyi Surfaces

There are twenty two different signature types we need to consider which may be overgroups for the surface kernel of a non-normal Belyi p-gonal surface. We examine a small number of them on a a case by case basis finding the full automorphism groups. The arguments are similar for all other cases, so we omit the details and summarize at the end of the section.

Table 4. Pairs for Non-Normal Belyi cyclic p-gonal Surfaces.

Case	Δ	Γ	Γ_p	L	Ker ($\mathcal{A}$)	K	$[\Delta:\Gamma]$	p
1A	(0,3,3,p)	(0;p,p,p)	p	$C_p \times C_p$		C_p	3	p
1B	(0,2,3,2p)	(0;p,p,p)	p	$C_p \times C_p$		C_p	6	p
1C	(0,2,3,7)	(0;7,7,7)	7	$C_7 \times C_7$		C_7	24	7
1D	(0,2,3,7)	(0;3,3,7)	3	$C_7 \rtimes C_3$		C_3	8	7
1E	(0,2,3,8)	(0;3,8,8)	8	$C_3 \rtimes C_8$		C_8	10	3
1F	(0;2,4,5)	(0;4,4,5)	4	$C_5 \rtimes C_4$	C_2	C_4	6	5
1G	(0;2,4,5)	(0;4,4,5)	4	$C_5 \rtimes C_4$	1	C_4	6	5
1H	(0,2,3,4p)	(0;p,4p,4p)	4p	$C_p \times C_{4p}$		C_{4p}	6	p
1I	(0,2,3,4p)	(0;p,4p,4p)	4p	$C_p \rtimes C_{4p}$	C_2	C_{4p}	6	p
1J	(0,2,3,4p)	(0;p,4p,4p)	4p	$C_p \rtimes C_{4p}$	C_4	C_{4p}	6	p
1K	(0,2,4,2p)	(0;p,2p,2p)	2p	$C_p \rtimes C_{2p}$		C_{2p}	4	p
2A	(0,2,3,3n)	(0;3,n,3n)	n+1	C_{3n}		C_3	4	3
2B	(0,2,3,2n)	(0;2,n,2n)	n+1	C_{2n}		C_2	3	2
3A	(0;2,4,2p)	(0;2,2p,2p)	2p	$C_p \times D_{2p}$		D_{2p}	2	p
3B	(0;2,4,2p)	(0;2,2p,2p)	2p	$C_p \rtimes D_{2p}$		D_{2p}	2	p
3C	(0;2,3,2p)	(0;2,p,2p)	p	$C_p \times D_p$		D_p	3	p
3D	(0;2,3,4p)	(0;2,2p,4p)	4p	$C_p \times D_{4p}$		D_{4p}	3	p
3E	(0;2,3,4p)	(0;2,2p,4p)	4p	$C_p \rtimes D_{4p}$		D_{4p}	3	p
4A	(0;2,3,4p)	(0;2p,2p,2p)	4p	$C_p \times D_{2p}$		D_{2p}	6	p
4B	(0;3,3,2p)	(0;2p,2p,2p)	4p	$C_p \times D_{2p}$		D_{2p}	3	p
4C	(0;2,4,5)	(0;4,4,5)	10	$C_2 \times D_5$		D_5	6	2
4D	(0;2,4,2p)	(0;2p,2p,p)	2p	$C_p \times D_p$		D_p	4	p

For those which are normal p-gonal, we eliminate them from the list. For those which are not normal p-gonal, we shall find the full automorphism group of each and defining equations. As we shall see, many of these cases are examples of well known surfaces. For some cases when the genus is small enough, we refer to Breuer's list developed in Breuer [2] and stored in GAP [4]. For the possible automorphism groups and methods to find the number of surface kernels, we refer to Wootton [10], Chapter 5 and Chapter 6. We shall also use Wootton [10], Chapter 5 to find defining equations for each of these surfaces. All other cases follow similar reasoning to the ones outlined below. We present complete results following these examples. The following two results will be very useful in our calculations.

Theorem 6: (Accola [1]) Suppose X is a compact Riemann surface of genus g which admits a cyclic p cover of the Riemann sphere for p a prime. If $g > (p-1)^2$, then the corresponding cyclic prime subgroup is normal in the full automorphism group of X.

Theorem 7: (Conder et al. [3]) Let G be a finite group which is the image of a Fuchsian group Γ with signature as specified below under a surface kernel epimorphism $\rho\colon \Gamma \to G$ acting on the surface $\mathbb{H}/\mathrm{Ker}\ (\rho)$ with genus g. Let a, b and c denote the images in G of the elliptic generators c_1, c_2 and c_3 of Γ respectively under ρ.

(1) If Γ has signature $(0; t, t, t)$ and the assignment $a \mapsto b$, $b \mapsto c$ and $c \mapsto a$ is an automorphism of G, then $Ker(\rho)$ is normal in Δ with signature $(0; 3, 3, t)$.
(2) If Γ has signature $(0; t, t, u)$ and the assignment $a \mapsto b$, $b \mapsto a$ and $c \mapsto bcb^{-1}$ is an automorphism of G, then $Ker(\rho)$ is normal in Δ with signature $(0; 2, t, 2u)$.

Example 8: The group Γ with signature $(0; p, p, p)$ contains one surface kernel Λ with quotient group $C_p \times C_p$ (we assume that $p \neq 7$ and examine this following the more general case). This implies Λ will be normal in any group Δ of which Γ is a normal subgroup. There are three overgroup signatures for Γ, and these are $(0; 2, 3, 2p)$, $(0; 3, 3, p)$ and $(0; 2, p, 2p)$. As we noted above, Λ will be normal in each of these groups, so we only need to consider the maximal overgroup of Γ, the Fuchsian group Δ with signature $(0; 2, 3, 2p)$.

Let $X = \mathbb{H}/\Lambda$. Our reasoning implies that the full automorphism group of X is the group $G = \Delta/\Lambda$ for Δ with signature $(0; 2, 3, 2p)$. It is not hard to show that $G = (C_p \times C_p) \rtimes S_3$. In particular, X will not be normal p-gonal for any choice of p. Before we move on to the next case, using the results of Wootton [10], a defining equation for X will be:

$$y^p = \Pi_{i=1}^{p}(x - \zeta^i)$$

where ζ is a primitive p-th root of unity. Letting $y' = -y$, and simplifying, we obtain the following equation also as a model for X:

$$(y')^p + x^p = 1$$

This is a model for the Fermat curve of genus $(p-1)(p-2)/2$.

When $p = 7$, there is are additional overgroups. In particular, there will be an overgroup Δ_2 with signature $(0; 2, 3, 7)$. The reasoning above implies that the surface kernel Λ of Γ with signature $(0; 7, 7, 7)$ must be normal in Δ with signature $(0; 2, 3, 14)$. However, Δ is finitely maximal, so Λ cannot be simultaneously normal in Δ_2, so our previous reasoning also holds when $p = 7$.

Example 9: Let Γ be a Fuchsian group with signature $(0; 3, 3, 7)$ which has overgroups Δ_1 with signature $(0; 2, 3, 14)$ and Δ_2 with signature $(0; 2, 3, 7)$. The only possible quotient group of Γ which has order 21 is the semi-direct product $C_7 \rtimes C_3$. There will be two surface kernels Λ_1 and Λ_2 with this quotient group. We need to determine if either of these groups are normal in Δ_1 or Δ_2.

Using Theorem 7, it is easy to show that neither of the surface kernels will be normal in Δ_1. Since Γ is normal in Δ_1, this implies that Λ_1 and Λ_2 are conjugate by an element in $\Delta_1 \backslash \Gamma$. In particular, the corresponding surfaces will be conformally equivalent. Therefore, we just consider one of the surface kernels. Denote this surface kernel by Λ.

Applying the Riemann-Hurwitz formula, the surface X has genus 3. Therefore, if Λ were normal in Δ_2, there would exist a genus 3 curve with an automorphism group of order 168 and conversely. Klein's genus 3 surface is a curve with such properties so must be conformally equivalent to $\mathbb{H}/\Lambda$. This implies that X is not normal 7-gonal.

Using the results of Wootton [10], Klein's curve has the following affine singular models:

$$y^7 = (x-1)(x-\zeta)^2(x-\zeta^2)^4$$

$$y^7 = x(x-1)^2$$

where ζ is a primitive third root of unity.

Example 10: Suppose that X is a trigonal curve uniformized by a surface kernel Λ normal in a Fuchsian group Γ with signature $(0; 3, 8, 8)$. Using the Riemann-Hurwitz formula, the genus of X is 6. Since $6 > (3-1)^2 = 4$, Theorem 6 implies that X must be normal trigonal.

Example 11: Suppose X is a surface uniformized by a surface subgroup of a group Γ with signature $(0; 4, 4, 5)$. There are two possible quotient groups, both non-trivial semi-direct products, Cases $1F$ and $1G$ in Table 4. Γ has index 6 in a group Δ_1 with signature $(0;2,4,5)$. We need to determine if any of the corresponding surface kernels are normal in Γ. We examine these two different possibilities separately.

For Case $1F$, the action of C_4 on the elements of C_5 is inversion. This means that the surface kernel is unique. If Λ denotes this surface kernel, similar to previous examples, uniqueness implies that Λ will be normal in Δ_2 with signature $(0;2,4,10)$. Checking Singerman's list, there are no overgroups of the Fuchsian group Δ_2 with signature $(0;2,4,10)$. Therefore, the full automorphism group of the surface $X = \mathbb{H}/\Lambda$ will be Δ_2/Λ. In particular, since Γ_5 is normal in Δ_2, this surface will be normal 5-gonal.

For Case $1G$, there are 2 different surface kernels. Theorem 7 implies that neither of these surface kernels are normal in Δ_2 with signature $(0;2,4,10)$, so they must be conjugate by an element in $\Delta_2\backslash\Gamma$. Therefore, we are just considering one surface X, up to conformal equivalence, with these properties. Let X denote this surface and Λ a surface group for X.

The genus of X is 4, so we examine Breuer's list of groups and signatures. For genus 4, the only group which is a quotient of a Fuchsian group with signature $(0;2,4,5)$ is the Symmetric group S_5. The group S_5 has subgroups of order 20 isomorphic to the group L. There is only one signature for a Fuchsian group in the genus 4 list with quotient group L, and that is the signature for Γ. Therefore, X must have full automorphism group S_5 and so it will not be normal 5-gonal.

The curve of genus 4 with automorphism group isomorphic to S_5 is called Bring's curve, see Edge [4]. This curve admits a cyclic 5 cover of the Riemann sphere but is not normal 5-gonal. Using the results developed in Wootton [10], an equation for Bring's curve is:

$$y^5 = (x^2-1)(x^2+1)^4.$$

Example 12: The next set of signatures we consider are for Γ with signature $(0;3,n,3n)$ as a subgroup of a Fuchsian group Δ with signature $(0;2,3,3n)$ for different choices of n. Using the Riemann-Hurwitz formula, the genus of such a surface, if it exists, is $g = n-1$. Since X is trigonal, Theorem 6 implies that we only need consider surfaces of genus $g \leq 4$. For this, we study Breuer's lists.

For $n = 3$ and $g = 2$, there exists no such curve so there is nothing to show. For $n = 4$ and $g = 3$, Γ has signature $(0;3,4,12)$ and Δ has signature $(0;2,3,12)$. There exists a unique normal surface kernel in Γ of genus 3 with quotient group C_{12}. Let Λ be this surface kernel and X the surface $\mathbb{H}/\Lambda$.

In Breuer's list for genus 3, there is only one group G which is a quotient of Δ with signature $(0;2,3,12)$. This group has group identity $[48,33]$ in the library of small groups from GAP [4]. The group G does contains cyclic subgroups of order 12, so we need to check which group signatures for genus 3 are for Fuchsian groups with quotient group C_{12}. Checking Breuer's list, there are two signatures of Fuchsian groups for genus 3 which have quotient group C_{12}, the other being $(0;2,12,12)$. Using Theorem 1 of Singerman [8], only Γ is a possible subgroup of Δ. Therefore the surface X is not trigonal and has full automorphism group G with group identification $[48,33]$. A defining equation for X is

$$y^3 = x^2(x^2-1)(x^2+1)$$

For $n=5$ and $g=4$, Γ has signature $(0;3,5,15)$ and Δ has signature $(0;2,3,15)$. In Breuer's list for genus 4, there are no groups G which are a quotient of Δ_1 with signature $(0;2,3,15)$ so we have nothing to show.

Example 13: For the signatures $(0;2,n,2n)$ and $(0;4,4,5)$ with $p=2$ and $n=5$, Theorem 6 implies they are normal 2-gonal, so there is nothing to show.

We summarize our results.

Theorem 14: There are two infinite families of non-normal Belyi p-gonal surfaces and three other such surfaces. Table 5 *shows the genus, signatures for Δ and Γ, defining equations and the full automorphism group of each of these surfaces.*

Table 5. Defining Equations for non-normal Belyi p-gonal Surfaces.

Genus	Γ	Γ_1	Equation	Aut (X)
$\frac{(p-1)(p-2)}{2}$	(0;p,p,p)	(0;2,3,2p)	$x^p + y^p = 1$	$(C_p \times C_p) \rtimes S_3$
$(p-1)^2$	(0;p,2p,2p)	(0;2,4,2p)	$y^p = (x^p-1)(x^p+1)^{p-1}$	$(C_p \times C_p) \rtimes D_4$
3	(0;3,3,7)	(0;2,3,7)	$y^7 = x(x-1)^2$	$PSL(2,7)$
4	(0;4,4,5)	(0;2,4,5)	$y^5 = (x^2-1)(x^2+1)^4$	S_5
3	(0;2,4,12)	(0;2,3,12)	$y^3 = x^2(x^2-1)(x^2+1)$	$[48,33]$

Our results imply the following interesting Corollary.

Corollary 15: *For every odd prime, there exists a non-normal Belyi p-gonal surface of genus $(p-1)^2$. In particular, Accola's bound given in Theorem 6 is sharp.*

References

1. R. D. Accola. *Strongly Branched Covers of Closed Riemann Surfaces.* Proc. of the AMS, Vol. 26, Issue 2, 315-322.
2. T. Breuer. *Characters and Automorphism Groups of Compact Riemann Surfaces.* Cambridge University Press (2001).
3. E. Bujalance, F.J. Cirre, M.D.E. Conder. *On Extendability of Group Actions on Compact Riemann Surfaces.* Trans. Amer. Math. Soc. 355 (2003), 1537-1557.
4. Edge, W. L. *Bring's curve.* J. London Math. Soc. (2) 18 (1978), no. 3, 539–545.
5. A. M. Macbeath. *On a Theorem of Hurwitz.* Proceedings of the Glasgow Mathematical Association, Vol. 5 (1961) pp. 90-96.
6. M. Schönert et al. GAP, Groups, Algorithms and Programming, Lehrstuhl D. fur Mathematik, RWTH, Aachen, 4.0 ed 2003
7. C. H. Sah. *Groups Related to Compact Riemann Surfaces.* Acta Math. 123 (1969) 13-42.
8. David Singerman. *Subgroups of Fuchsian Groups and Permutation Groups.* Bull. London Math. Soc., 2(1970), 319-323.
9. David Singerman. *Finitely Maximal Fuchsian Groups.* J. London Math. Soc., (2), 6(1972),29-38.
10. A. Wootton. *Defining Algebraic Polynomials for Cyclic Prime Covers of the Riemann Sphere.* Dissertation.

Hyperelliptic curves of genus 3 with prescribed automorphism group

Jaime Gutierrez*
Dpt. of Math., Stat. and Comp. Univ. of Cantabria, 39071 Santander, Spain
jaime.gutierrez@unican.es

D. Sevilla†
Dpt. of Comp. Sci. and Software Eng., Concordia University
1455 de Maisonneuve W., Montreal QC, H3G 1M8 Canada
sevillad@gmail.com

T. Shaska‡
Dep. of Mathematics, Oakland University, Rochester, MI, 48309-4485.
shaska@oakland.edu

We study genus 3 hyperelliptic curves which have an extra involution. The locus $\mathcal{L}_3$ of these curves is a 3-dimensional subvariety in the genus 3 hyperelliptic moduli $\mathcal{H}_3$. We find a birational parametrization of this locus by affine 3-space. For every moduli point $\mathfrak{p} \in \mathcal{H}_3$ such that $|\text{Aut}(\mathfrak{p})| > 2$, the field of moduli is a field of definition. We provide a rational model of the curve over its field of moduli for all moduli points $\mathfrak{p} \in \mathcal{H}_3$ such that $|\text{Aut}(\mathfrak{p})| > 4$. This is the first time that such a rational model of these curves appears in the literature.

1. Introduction

Let $\mathcal{X}_g$ be an irreducible, smooth, projective curve of genus $g \geq 3$, defined over the complex field $\mathbb{C}$. We denote by $\mathcal{M}_g$ the coarse moduli space of

*Partially supported by project MTM2004-07086 of the Spanish Ministry of Science and Technology
†Partially supported by Spanish Ministry of Science Project MTM2004-07086.
‡Supported by the NSA grant R1-05-0129

smooth curves of genus g and by $\mathcal{H}_g$ the hyperelliptic locus in $\mathcal{M}_g$. It is well known that dim $\mathcal{M}_g = 3g - 3$ and $\mathcal{H}_g$ is a $(2g-1)$ – dimensional subvariety of $\mathcal{M}_g$. A curve $\mathcal{X}_g$ is called *bielliptic* if it admits a degree 2 morphism $\pi : \mathcal{X}_g \to E$ onto an elliptic curve. This morphism is called a *bielliptic structure* in $\mathcal{X}_g$; see [3]. The locus $\mathcal{M}_g^b$ of bielliptic curves is a $(2g-2)$ – dimensional subvariety of $\mathcal{M}_g$.

From the Castelnouvo-Severi inequality it follows that $\mathcal{X}_g$ admits precisely one bielliptic structure for $g \geq 6$, but if $g \leq 5$ then there are curves which admit more than one bielliptic structure. Since every bielliptic structure corresponds to an involution in the automorphism group Aut$(\mathcal{X}_g)$ of the curve, then these results can be obtained easily if the list of groups that occur as automorphism groups is known for a given genus g. Lately, algorithms have been developed to determine such lists of groups for reasonably small g; see [23].

A bielliptic curve of genus $g \geq 4$ can not be hyperelliptic (Castelnouvo-Severi inequality), but for $g = 3$ this can be the case. The bielliptic (non-hyperelliptic) curves of genus 3 were studied in [3]; see also [31]. In this paper we will focus in the hyperelliptic case. Such curves are known in the literature also as hyperelliptic curves with *extra involutions*. This extends our previous work in [32], [10], [25].

In the second section we give a brief description of the dihedral invariants and how they are used to describe the loci of curves with fixed automorphism groups. Such invariants can be helpful in determining the automorphism group of the curve and determining its field of moduli. For further details on dihedral invariants we refer to [10]. The reader can also check [32] for the case $g = 2$. Further in this section we briefly define the classical invariants of binary forms.

In section three, we focus on genus 3 hyperelliptic curves. We define the invariants of binary octavics and compute them explicitly in terms of the coefficients of the curve for curves with extra involutions. The locus $\mathcal{L}_3$ of genus 3 hyperelliptic curves with extra involutions is a 3-dimensional subvariety of $\mathcal{H}_3$. Using such explicit expressions for the invariants of the binary forms we find a birational parametrization of the locus $\mathcal{L}_3$ via dihedral invariants. Further, we make use of such invariants to study certain subvarieties of the moduli space of hyperelliptic curves of genus 3. The list of groups that occur as automorphism groups of genus 3 hyperelliptic curves is described. Then, for each group in the list we describe algebraic relations that define the corresponding locus.

If $\mathcal{X} \in \mathcal{L}_3$ then $V_4 \hookrightarrow$ Aut$(\mathcal{X})$. Let G be a group which occurs as an automorphism group of hyperelliptic curves of genus 3 and such that $V_4 \hookrightarrow G$. We describe each locus of curves with automorphism group G in terms of the dihedral invariants and prove that the field of moduli of such curves is a field of definition. If $|G| > 4$ then a rational model of the curve is provided over its field of moduli. As far as we are aware, this is the first time that such rational models are known for genus 3 curves.

Notation: We denote a hyperelliptic curve of genus 3 by $\mathcal{X}_3$. D_n denotes the dihedral group of order $2n$ and V_4 the Klein 4-group. Further, $\mathbb{Z}_n$ denotes the cyclic group of order n.

2. Dihedral invariants of hyperelliptic curves

In this section we give a brief review of some of the basic results on hyperelliptic curves with extra involutions and their dihedral invariants. For details see [10].

Let $\mathcal{X}_g$ be a genus g hyperelliptic curve defined over $\mathbb{C}$ and $\text{Aut}(\mathcal{X}_g)$ its automorphism group. We say that $\mathcal{X}_g$ has an *extra involution* when there is a non-hyperelliptic involution in $\text{Aut}(\mathcal{X}_g)$. If the fixed field of such an extra involution is an elliptic field then sometimes these are called *elliptic involutions.* The hyperelliptic involution $\alpha_0 \in \text{Aut}(\mathcal{X}_g)$ is in the center of $\text{Aut}(\mathcal{X}_g)$. We denote $\overline{\text{Aut}}(\mathcal{X}_g) := \text{Aut}(\mathcal{X}_g)/\langle \alpha_0 \rangle$ and call it the *reduced automorphism group* of $\mathcal{X}_g$.

Let $\mathcal{X}_g$ be a genus g hyperelliptic curve with an extra involution $\alpha_1 \in \text{Aut}(\mathcal{X}_g)$. Then, $\mathcal{X}_g$ is isomorphic to a curve given by an equation

$$Y^2 = X^{2g+2} + a_g X^{2g} + \cdots + a_1 X^2 + 1.$$

Such equation is called the *normal equation* of the curve $\mathcal{X}_g$. There is a degree 2 map

$$\phi_1 : \mathcal{X}_g \to C_1$$

where C_1 is the hyperelliptic curve with equation

$$Y^2 = X^{g+1} + a_g X^g + \cdots + a_1 X + 1$$

and genus $g_1 = \left[\frac{g}{2}\right]$. Since every hyperelliptic curve $\mathcal{X}_g$ has the hyperelliptic involution $\alpha_0 \in \text{Aut}(\mathcal{X}_g)$, then the extra involutions come in pairs $(\alpha_1, \alpha_2 = \alpha_0\alpha_1)$. The extra involution α_2 determines another degree 2 covering

$$\phi_2 : \mathcal{X}_g \to C_2$$

where C_2 is the hyperelliptic curve with equation

$$Y^2 = X(X^{g+1} + a_g X^g + \cdots + a_1 X + 1)$$

and genus $g_2 = \left[\frac{g+1}{2}\right]$. The curve $\mathcal{X}_g$ is called *bielliptic* if C_1 is an elliptic curve. This always happens if $g = 2$ or 3.

The Jacobian $J_{\mathcal{X}_g}$ of $\mathcal{X}_g$ is isogenous to $J_{C_1} \times J_{C_2}$. We say that $J_{\mathcal{X}_g}$ splits. Our goal is to determine the locus of such curves $\mathcal{X}_g$ in the variety of moduli. The locus of genus g hyperelliptic curves with an extra involution is an irreducible g-dimensional subvariety of $\mathcal{H}_g$ which we denote by $\mathcal{L}_g$. The following

$$u_i := a_1^{g-i+1}\, a_i + a_g^{g-i+1}\, a_{g-i+1}, \quad 1 \leq i \leq g$$

are called *dihedral invariants* of genus g. The next theorem shows that $\mathcal{L}_g$ is a rational variety; see [10].

Theorem 1: Let $g \geq 2$ and $(u_1, \ldots, u_g)$ be the g-tuple of dihedral invariants. Then, $k(\mathcal{L}_g) = k(u_1, \ldots, u_g)$.

A generic point $\mathfrak{p} \in \mathcal{L}_g$ has automorphism group $\mathrm{Aut}(\mathfrak{p}) \cong V_4$. Singular points of $\mathcal{L}_g$ have more than one tuple of dihedral invariants and therefore more than one conjugacy class of involutions in $\overline{\mathrm{Aut}}(\mathfrak{p})$. For curves with automorphism group isomorphic to V_4 we have the following:

Corollary 2: *Let $\mathcal{X}_g$ and $\mathcal{X}_g'$ be genus g hyperelliptic curves with automorphism groups isomorphic to V_4, and $(u_1, \ldots, u_g)$, $(u_1', \ldots, u_g')$ their respective dihedral invariants. Then,*

$$\mathcal{X}_g \cong \mathcal{X}_g' \iff (u_1, \ldots, u_g) = (u_1', \ldots, u_g').$$

Theorem 3: Let $\mathcal{X}_g$ be a genus g hyperelliptic curve with an extra involution and $(u_1, \ldots, u_g)$ its corresponding dihedral invariants.

i) If $V_4 \hookrightarrow \overline{\mathrm{Aut}}(\mathcal{X}_g)$ then $2^{g-1}\, u_1^2 = u_g^{g+1}$.

ii) Moreover, if g is odd then $V_4 \hookrightarrow \overline{\mathrm{Aut}}(\mathcal{X}_g)$ implies that

$$\left(2^r\, u_1 - u_g^{r+1}\right)\, \left(2^r\, u_1 + u_g^{r+1}\right) = 0$$

where $r = \frac{g-1}{2}$. The first factor corresponds to the case when involutions of $V_4 \hookrightarrow \overline{\mathrm{Aut}}(\mathcal{X}_g)$ lift to involutions in $\mathrm{Aut}(\mathcal{X}_g)$. The second factor corresponds to the case when one of the two involutions lifts to an element of order 4 in $\mathrm{Aut}(\mathcal{X}_g)$.

For the proofs of these statements see [10].

From a computational point of view, determining the normal equation of a given curve with an extra automorphism can be done simply by solving a system of equations; this is quite efficient both theoretically and in practice. If an extra involution $\alpha = X \to \frac{aX+b}{cX+d}$ is known explicitly, one can easily find σ such that $\alpha^\sigma = X \to -X$, then the equation after applying σ^{-1} has the form $Y^2 = F(X^2)$. One more substitution $X \to \lambda X$ for certain λ will provide the normal equation.

2.1. *Genus 2 case*

The case $g = 2$ has been studied in [32]. Every point in $\mathcal{M}_2$ is a triple (i_1, i_2, i_3) of absolute invariants, see [32] for details. The curve of genus 2 with extra involutions has equation

$$Y^2 = X^6 + a_2 X^4 + a_1 X^2 + 1.$$

We denote its dihedral invariants by

$$\mathfrak{u} := a_1^3 + a_2^3, \quad \textit{and} \quad \mathfrak{v} := 2a_1 a_2. \tag{1}$$

The following lemma is proved in [32 .

Lemma 4: *Let $\mathcal{X}_2$ be a genus 2 curve such that $G := \mathrm{Aut}(\mathcal{X}_2)$ has an extra involution and $(\mathfrak{u}, \mathfrak{v})$ its dihedral invariants. Then,*

a) $G \cong \mathbb{Z}_3 \rtimes D_8$ if and only if $(\mathfrak{u}, \mathfrak{v}) = (0, 0)$ or $(\mathfrak{u}, \mathfrak{v}) = (6750, 450)$.
b) $G \cong GL_2(3)$ if and only if $(\mathfrak{u}, \mathfrak{v}) = (-250, 50)$.
c) $G \cong D_{12}$ if and only if $\mathfrak{v}^2 - 220\mathfrak{v} - 16\mathfrak{u} + 4500 = 0$ for $\mathfrak{v} \neq 18,\ 140 + 60\sqrt{5},\ 50$.
d) $G \cong D_8$ if and only if $2\mathfrak{u}^2 - \mathfrak{v}^3 = 0$, for $\mathfrak{v} \neq 2, 18, 0, 50, 450$. Cases $\mathfrak{v} = 0,\ 450$ and $\mathfrak{u} = 50$ are reduced to cases a) and b) respectively.

Notice that the parameters $u = \dfrac{\mathfrak{v}}{2}$ and $v = \mathfrak{u}$ instead of $\mathfrak{u}, \mathfrak{v}$ are used in [32]. The mapping

$$\Phi : (\mathfrak{u}, \mathfrak{v}) \to (i_1, i_2, i_3)$$

gives a birational parametrization of $\mathcal{L}_2$. The dihedral invariants $\mathfrak{u}, \mathfrak{v}$ are given explicitly as rational functions of i_1, i_2, i_3, see [32]. For $g = 2$, the curve $Y^2 = X^6 - X$ is the only genus 2 curve (up to isomorphism) which has extra automorphisms and is not in $\mathcal{L}_2$. The automorphism group in this case is $\mathbb{Z}_{10}$. Relations between elliptic subcovers of such $\mathcal{X}_2$ were studied in detail in [32].

2.2. *Invariants of binary forms*

In this section we define the action of $GL_2(k)$ on binary forms and discuss the basic notions of their invariants. Let $k[X, Z]$ be the polynomial ring in two variables and let V_d denote the $(d+1)$-dimensional subspace of $k[X, Z]$ consisting of homogeneous polynomials

$$f(X, Z) = a_0X^d + a_1X^{d-1}Z + ... + a_dZ^d$$

of degree d. Elements in V_d are called *binary forms* of degree d. We let $GL_2(k)$ act as a group of automorphisms on $k[X, Z]$ as follows:

$$\text{for} \quad \begin{pmatrix} a\ b \\ c\ d \end{pmatrix} \in GL_2(k)\ , \quad M \begin{pmatrix} X \\ Z \end{pmatrix} = \begin{pmatrix} aX + bZ \\ cX + dZ \end{pmatrix}. \tag{2}$$

This action of $GL_2(k)$ leaves V_d invariant and acts irreducibly on V_d.

Let $A_0,\ A_1, \dots, A_d$ be coordinate functions on V_d. Then the coordinate ring of V_d can be identified with $k[A_0, \dots, A_d]$. For any $I \in k[A_0, \dots, A_d]$ and $M \in GL_2(k)$, we define $I^M \in k[A_0, \dots, A_d]$ as follows:

$$I^M(f) := I(M(f))$$

for all $f \in V_d$. Then this equation and $I^{MN} = (I^M)^N$ define an action of $GL_2(k)$ on $k[A_0, \dots, A_d]$. A homogeneous polynomial $I \in k[A_0, \dots, A_d, X, Z]$ is called a *covariant* of index s if

$$I^M(f) = \delta^s I(f)$$

where $\delta = \det(M)$. The homogeneous degree in $a_1, \dots, a_n$ is called the *degree* of I, and the homogeneous degree in X, Z is called the *order* of I. A covariant of order zero is called *invariant*. An invariant is a $SL_2(k)$-invariant on V_d.

We will use the symbolic method of classical theory to construct covariants of binary forms. Let

$$f(X, Z) := \sum_{i=0}^{n} \binom{n}{i} a_i X^{n-i}\, Z^i, \qquad g(X, Z) := \sum_{i=0}^{m} \binom{m}{i} b_i X^{n-i}\, Z^i$$

be binary forms of degree n and m respectively with coefficients in k. We define the *r-transvection*

$$(f, g)^r := d \sum_{k=0}^{r} (-1)^k \binom{r}{k} \cdot \frac{\partial^r f}{\partial X^{r-k}\, \partial Z^k} \cdot \frac{\partial^r g}{\partial X^k\, \partial Z^{r-k}}$$

where $d = \frac{(m-r)!\,(n-r)!}{n!\,m!}$. Then $(f, g)^r$ is a homogeneous polynomial in $k[X, Z]$ and therefore a covariant of order $m + n - 2r$ and degree 2. In general, the r-transvection of two covariants of order m, n (resp. degree p, q) is a covariant of order $m + n - 2r$ (resp. degree $p + q$). See [2], [6], [9], [8] for details.

3. Hyperelliptic curves of genus three

In this section we study hyperelliptic curves of genus 3 with extra involutions. Let $\mathcal{X}_3$ be such a curve. Then, $\mathcal{X}_3$ has normal equation

$$Y^2 = X^8 + a_3\,X^6 + a_2\,X^4 + a_1\,X^2 + 1,$$

see [10]. The dihedral invariants of $\mathcal{X}_3$ are

$$u_1 = a_1^4 + a_3^4 \quad u_2 = (a_1^2 + a_3^2)\,a_2 \quad u_3 = 2\,a_1\,a_3.$$

If $a_1 = a_3 = 0$, then $u_1 = u_2 = u_3 = 0$. In this case $w := a_2^2$ is invariant. Thus, we define

$$\mathfrak{u}(\mathcal{X}_3) = \begin{cases} w & \text{if} \quad a_1 = a_3 = 0, \\ (u_1, w, u_3) & \text{if} \quad a_1^2 + a_3^2 = 0 \text{ and } a_2 \neq 0, \\ (u_1, u_2, u_3) & \text{otherwise.} \end{cases} \tag{3}$$

To have an explicit way of describing a point in the moduli space of hyperelliptic curves of genus 3 we need the generators of the field of invariants of binary octavics. These invariants are described in terms of covariants of binary octavics. Such covariants were first constructed by van Gall who showed that the graded ring of covariants is generated by 70 covariants and explicitly constructed them, see [14].

Let $f(X,Y)$ be the binary octavic

$$f(X,Y) = \sum_{i=0}^{8} a_i X^i Y^{8-i}.$$

We define the following covariants:

$$\begin{aligned} &g = (f,f)^4, \quad k = (f,f)^6, \quad h = (k,k)^2, \quad m = (f,k)^4, \\ &n = (f,h)^4, \quad p = (g,k)^4, \quad q = (g,h)^4. \end{aligned} \tag{4}$$

Then the following

$$\begin{aligned} &J_2 = (f,f)^8, \quad J_3 = (f,g)^8, \quad J_4 = (k,k)^4, \\ &J_5 = (m,k)^4, \quad J_6 = (k,h)^4, \quad J_7 = (m,h)^4 \end{aligned} \tag{5}$$

are $SL_2(k)$-invariants. Shioda has shown that the ring of invariants is a finitely generated module of $k[J_2, \dots, J_7]$, see [14]. The expressions of these covariants are very large in terms of the coefficients of the curve and difficult to compute. However, in terms of the dihedral invariants u_1, u_2, u_3 these expressions are smaller. Analogously, J_{14} is the discriminant of the octavic. We define $M := 2u_1 + u_3^2$ and assume $M \neq 0$.

$$
\begin{aligned}
J_2 &= \frac{1}{M}\,(560u_1 + 280u_3^2 + 10u_3u_1 + 5u_3^3 + 2u_2^2) \\
J_3 &= \frac{u_2}{a_2M^2}\,(12u_2^3 + 4200u_1^2 + 4200u_1u_3^2 + 1050u_3^4 - 110u_3u_2u_1 - 55u_3^3u_2 \\
&\quad + 7840u_2u_1 + 3920u_2u_3^2) \\
J_4 &= \frac{32}{M^2}\,(2u_2^4 - 1568u_2^2u_1 - 784u_2^2u_3^2 + 1008u_2u_1^2 + 1008u_2u_1u_3^2 + 252u_2u_3^4 \\
&\quad + 8u_1^2u_3^2 + 307328u_1^2 + 307328u_1u_3^2 + 76832u_3^4 + 62u_3u_2^2u_1 + 31u_3^3u_2^2 \\
&\quad - 784u_3u_1^2 + 8u_1u_3^4 + 2u_3^6 - 784u_3^3u_1 - 196u_3^5) \\
J_5 &= -\frac{16u_2}{a_2M^3}\,(104u_2u_1^2u_3^2 - 614656u_2u_1u_3^2 - 41160u_3^6 - 614656u_2u_1^2 - 153664u_2u_3^4 \\
&\quad - 246960u_1u_3^4 - 2296u_2^2u_1u_3^2 + 104u_2u_1u_3^4 - 41552u_3u_2u_1^2 - 41552u_3^3u_2u_1 \\
&\quad + 26u_2^3u_3u_1 + 1568u_2^3u_3^2 + 13u_2^3u_3^3 + 26u_2u_3^6 - 2296u_2^2u_1^2 - 574u_2^2u_3^4 - 10388u_3^5u_2 \\
&\quad + 1120u_3u_1^3 + 840u_3^5u_1 - 4u_2^5 - 329280u_1^3 + 140u_3^7 - 493920u_1^2u_3^2 + 3136u_2^3u_1 \\
&\quad + 1680u_3^3u_1^2) \\
J_6 &= -\frac{256}{M^3}\,(2u_3u_1 + u_3^3 - 392u_1 - 196u_3^2 + u_2^2)(-2u_2^4 - 8u_1^2u_3^2 - 8u_1u_3^4 - 2u_3^6 \\
&\quad + 154u_3u_2^2u_1 + 77u_3^3u_2^2 + 1568u_2^2u_1 + 784u_2^2u_3^2 + 3024u_2u_1^2 + 3024u_2u_1u_3^2 \\
&\quad + 756u_2u_3^4 + 10192u_3u_1^2 + 2548u_3^5 - 307328u_1^2 - 307328u_1u_3^2 \\
&\quad - 76832u_3^4 + 10192u_3^3u_1) \qquad (6) \\
J_7 &= \frac{64u_2}{a_2M^4}\,(129077760u_3^6u_1 - 481890304u_2u_1^3 + 516311040u_1^3u_3^2 \\
&\quad + 387233280u_1^2u_3^4 + 921984u_2^3u_3^4 + 7299040u_3^7u_2 - 90u_2^5u_3^3 - 14896u_2^4u_1^2 \\
&\quad - 3724u_2^4u_3^4 + 3360u_3^6u_1^2 + 1120u_3^8u_1 + 141120u_2u_1^4 + 4480u_1^3u_3^4 + 16134720u_3^8 \\
&\quad + 2086u_3^7u_2^2 + 345u_2^3u_3^6 + 38u_3^9u_2 + 3687936u_2^3u_1^2 - 68600u_3^9 - 25480u_2u_3^8 \\
&\quad + 5180672u_2^2u_1^3 + 647584u_2^2u_3^6 - 1097600u_3u_1^4 + 8u_2^7 + 140u_3^{10} + 258155520u_1^4 \\
&\quad - 1646400u_3^5u_1^2 - 548800u_3^7u_1 - 9408u_2^5u_1 - 4704u_2^5u_3^2 - 722835456u_2u_1^2u_3^2 \\
&\quad + 16688u_3u_2^2u_1^3 + 304u_3^3u_2u_1^3 + 456u_3^5u_2u_1^2 - 199920u_2u_1^2u_3^4 + 7840u_2u_1^3u_3^2 \\
&\quad - 14896u_2^4u_1u_3^2 + 25032u_3^3u_2^2u_1^2 + 43794240u_3^5u_2u_1 + 87588480u_3^3u_2u_1^2 \\
&\quad + 228u_3^7u_2u_1 + 1380u_2^3u_3^4u_1 - 78400u_2^3u_3u_1^2 + 58392320u_3u_2u_1^3 + 1380u_2^3u_3^2u_1^2 \\
&\quad - 135240u_2u_1u_3^6 + 3885504u_2^2u_1u_3^4 + 3687936u_2^3u_3^2u_1 - 60236288u_2u_3^6 \\
&\quad + 12516u_3^5u_2^2u_1 - 78400u_2^3u_3^3u_1 - 361417728u_2u_1u_3^4 + 2240u_1^4u_3^2 \\
&\quad + 7771008u_2^2u_1^2u_3^2 - 19600u_2^3u_3^5 - 180u_2^5u_3u_1 - 2195200u_3^3u_1^3)
\end{aligned}
$$

$$
\begin{aligned}
J_{14} = \frac{16}{M^4} & (-1024u_3^4 - 64u_2^4 - 4096u_1^2 - 4096u_1u_3^2 - 2304u_2u_1u_3^2 + 6u_3^6 + 384u_3^5 \\
& + 1024u_2^2u_1 + 512u_2^2u_3^2 - 2304u_2u_1^2 - 576u_2u_3^4 + 456u_1^2u_3^2 + 132u_1u_3^4 \\
& + 160u_3^3u_2^2 + 1536u_3u_1^2 + 1536u_3^3u_1 - 2u_2^2u_1u_3^2 + 320u_3u_2^2u_1 - 144u_3u_2u_1^2 \\
& - 144u_3^3u_2u_1 + 32u_2^3u_1 + 16u_2^3u_3^2 - u_2^2u_3^4 - 36u_3^5u_2 + 8u_3^3u_1^2 + 8u_3^5u_1 \\
& + 432u_1^3 + 2u_3^7)^2
\end{aligned} \tag{7}
$$

The next theorem is a direct corollary of Theorem 4. However, we provide a computational proof.

Theorem 5: $k(\mathcal{L}_3) = k(u_1, u_2, u_3)$.

Proof: Notice that J_3, J_5, J_7 have a_2 as a factor. However, this does not contradict Theorem 4. The function field $k(\mathcal{L}_3)$ is generated by absolute invariants. To define such absolute invariants one must raise J_3, J_5, J_7 to some power and therefore absolute invariants will have only $a_2^2 = \frac{2u_2^2}{2u_1+u_3^2}$ as a factor. Hence, computationally we have shown that $i_1, \dots, i_5 \in k(u_1, u_2, u_3)$, as expected. Let

$$
i_j = \frac{p_j(u_1, u_2, u_3)}{q_j(u_1, u_2, u_3)}
$$

for $j = 1, \dots, 5$ and certain polynomials p_i, q_i. Then, we have the system of equations

$$
i_j \cdot q_j(u_1, u_2, u_3) - p_j(u_1, u_2, u_3) = 0, \quad j = 1, \dots, 5.
$$

We can solve for u_1, u_2, u_3 and express them as rational functions in i_1, i_2, i_3. Therefore, $k(u_1, u_2, u_3) = k(\mathcal{L}_3)$. □

Remark 6: The expressions of u_1, u_2, u_3 as rational functions in i_1, i_2, i_3 are rather large and we don't display them. Using the above equations, one can find explicit equations of the locus $\mathcal{L}_3$ in terms of $i_1, \dots, i_5$. However, such equations are very large and not practical to use. Instead, using the dihedral invariants u_1, u_2, u_3 is much more convenient.

3.1. *The locus of genus 3 hyperelliptic curve with prescribed automorphism group*

In this section we describe the locus of genus 3 hyperelliptic curves in terms of dihedral invariants or classical invariants. First we briefly describe the list of groups that occur as automorphism groups of genus 3 hyperelliptic curves. This list has been computed by many authors; we refer to [23] for the correct result and a complete list of references.

We denote by U_6, V_8 the following groups:

$$U_6 := \langle x, y \mid x^2, y^6, x\,y\,x\,y^4\rangle, \qquad V_8 := \langle x, y \mid x^4, y^4, (x\,y)^2, (x^{-1}y)^2\rangle.$$

In Table 1 we list the automorphism groups of genus 3 hyperelliptic curves. The first column is the case number, in the second column the groups which occur as full automorphism groups are given, and the third column indicates the reduced automorphism group for each case. The dimension δ of the locus and the equation of the curve are also given in the next two columns. The last column is the GAP identity of each group in the library of small groups in GAP.

	$\mathrm{Aut}(\mathcal{X}_g)$	$\overline{G}$	δ	equation $y^2 = f(x)$	Id.
1	$\mathbb{Z}_2$	$\{1\}$	5	$x(x-1)(x^5+ax^4+bx^3+cx^2+dx+e)$	$(2,1)$
2	$\mathbb{Z}_2\times\mathbb{Z}_2$	$\mathbb{Z}_2$	3	$x^8+a_3x^6+a_2x^4+a_1x^2+1$	$(4,2)$
3	$\mathbb{Z}_4$	$\mathbb{Z}_2$	2	$x(x^2-1)(x^4+ax^2+b)$	$(4,1)$
4	$\mathbb{Z}_{14}$	$\mathbb{Z}_7$	0	x^7-1	$(14,2)$
5	$\mathbb{Z}_2^3$	D_4	2	$(x^4+ax^2+1)(x^4+bx^2+1)$	$(8,5)$
6	$\mathbb{Z}_2\times D_8$	D_8	1	x^8+ax^4+1	$(16,11)$
7	$\mathbb{Z}_2\times\mathbb{Z}_4$	D_4	1	$(x^4-1)(x^4+ax^2+1)$	$(8,2)$
8	D_{12}	D_6	1	$x(x^6+ax^3+1)$	$(12,4)$
9	U_6	D_{12}	0	$x(x^6-1)$	$(24,5)$
10	V_8	D_{16}	0	x^8-1	$(32,9)$
11	$\mathbb{Z}_2\times S_4$	S_4	0	x^8+14x^2+1	$(48,48)$

Remark 7: Note that $\mathbb{Z}_2$, $\mathbb{Z}_4$ and $\mathbb{Z}_{14}$ are the only groups which don't have extra involutions. Thus, curves with automorphism group $\mathbb{Z}_2$, $\mathbb{Z}_4$ or $\mathbb{Z}_{14}$ do not belong to the locus $\mathcal{L}_3$.

We want to describe each of the above subloci and study inclusions among them. In order to study such inclusions the lattice of the list of groups needs to be determined. The lattice of the groups for genus 3 is given in Fig. 1. Each group is presented by its GAP identity. Each level contains cases with the same dimension (i.e., the bottom level correspond to the 0-dimensional families). The boxed entries correspond to groups with extra involutions.

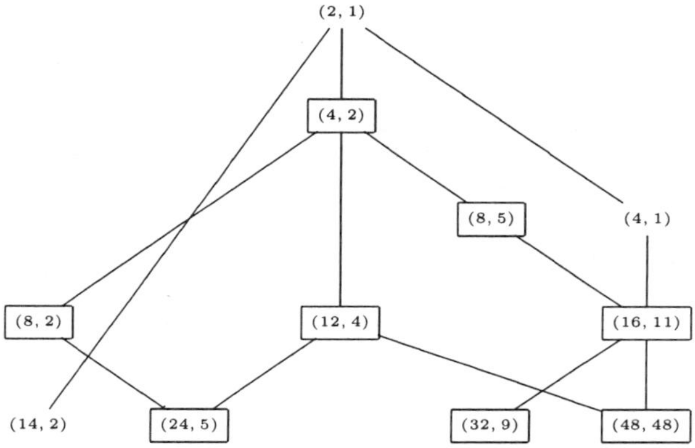

Fig. 1. The lattice of automorphism groups

Next we determine the algebraic relations among dihedral invariants for each case in the diagram.

Theorem 8: The algebraic relations of dihedral invariants for each case of Figure 1 are presented in Figure 2.

Proof: If $\overline{\mathrm{Aut}}(\mathcal{X}_3) \cong D_4$, then from Theorem 3 we have $4\,u_1^2 = u_3^4$. This can be checked easily by computing the dihedral invariants. If $2u_1^2 = u_3^2$ then $\mathrm{Aut}(\mathcal{X}_3) \cong \mathbb{Z}_2 \times \mathbb{Z}_2 \times \mathbb{Z}_2$. If $2u_1^2 = -u_3^2$ then $\mathrm{Aut}(\mathcal{X}_3) \cong \mathbb{Z}_2 \times \mathbb{Z}_4$, see the remark after Theorem 3.

Let $\mathcal{X}_3$ be a curve with equation $Y^2 = X\,(X^6 + aX^3 + 1)$, where $a \neq 0, \pm 2$. By a transformation $X \to \frac{X+1}{X-1}$, $\mathcal{X}_3$ has equation

$$Y^2 = X^8 + (9\lambda - 7)X^6 + 15(\lambda - 1)X^4 + (7\lambda - 9)X^2 - \lambda$$

where $\lambda = \frac{a-2}{a+2}$, $\lambda \neq 0, \pm 1$. Then, by another transformation $X \to \sqrt[8]{-\lambda}\,X$,

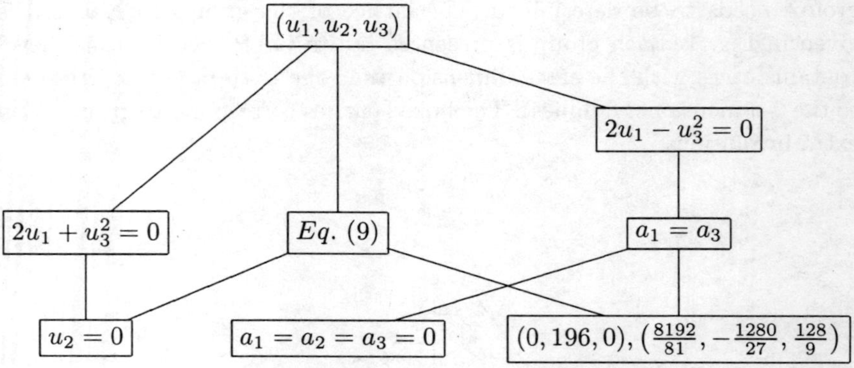

Fig. 2. Corresponding relations of dihedral invariants

we compute the dihedral invariants:

$$\begin{aligned} u_1 &= -\frac{1}{\lambda^3}\,(6561\lambda^6 - 20412\lambda^5 + 26215\lambda^4 - 24694\lambda^3 \\ &\quad + 26215\lambda^2 - 20412\lambda + 6561), \\ u_2 &= 15\frac{(\lambda-1)^2(81\lambda^2 - 94\lambda + 81)}{\lambda^2}, \\ u_3 &= -2\frac{(7\lambda - 9)(9\lambda - 7)}{\lambda}. \end{aligned} \tag{8}$$

Eliminating λ, we get $\lambda = -126\frac{1}{u_3 - 260}$ and

$$\begin{aligned} u_2 &= \frac{5}{588}\,(u_3 - 8)(9u_3 - 1024), \\ u_1 &= \frac{9}{2744}u_3^3 - \frac{873}{686}u_3^2 + \frac{149504}{3087}u_3 - \frac{1048576}{3087}. \end{aligned} \tag{9}$$

Notice that $u_3 \neq 260$, otherwise $a = 2$.

If $u_2 = 0$ then we have $81\lambda^2 - 94\lambda + 81 = 0$ and

$$(u_1, u_2, u_3) = (8, 0, -32) \text{ or } \left(-\frac{524288}{81}, 0, \frac{1024}{9}\right).$$

Both of these triples satisfy the relation $2u_1 = -u_3^2$ so both D_{12} and $\mathbb{Z}_2 \times \mathbb{Z}_4$ are embedded in $\mathrm{Aut}(\mathcal{X}_3)$. Hence, $\mathrm{Aut}(\mathcal{X}_3) \cong U_6$.

If $2\,u_1^2 = u_3^2$ then $81\,\lambda^2 - 1568\lambda + 81 = 0$. Then,

$$(u_1, u_2, u_3) = (0, 196, 0) \text{ or } \left(\frac{8192}{81}, -\frac{1280}{27}, \frac{128}{9}\right).$$

Both triples determine the same isomorphism class of curves and correspond to the case $\mathrm{Aut}(\mathcal{X}_3) \cong \mathbb{Z}_2 \times S_4$. The proof is complete. □

The above loci can be easily determined in terms of $J_2, \dots, J_7$. This would be beneficial because we don't have to find the normal decomposition form of the curve in order to determine the automorphism group. However, we don't display such equations in terms of $J_2, \dots, J_7$. It is worth mentioning that, if $\mathrm{Aut}(\mathcal{X}_3) \cong \mathbb{Z}_2 \times \mathbb{Z}_4$, U_6 or V_8, then $J_3 = J_5 = J_7 = 0$.

4. Field of moduli of genus 3 hyperelliptic curves

In this section we study the field of moduli of genus 3 hyperelliptic curves with extra automorphisms. Let $\mathcal{X}$ be a curve defined over $\mathbb{C}$. A field $F \subset \mathbb{C}$ is called a *field of definition* of $\mathcal{X}$ if there exists $\mathcal{X}'$ defined over F such that $\mathcal{X}'$ is isomorphic to $\mathcal{X}$ over $\mathbb{C}$.

Definition 9: The *field of moduli* of $\mathcal{X}$ is a subfield $F \subset \mathbb{C}$ such that for every automorphism $\sigma \in \mathrm{Aut}(\mathbb{C})$ the following holds: $\mathcal{X}$ is isomorphic to $\mathcal{X}^\sigma$ if and only if $\sigma_F = id$.

We will use $\mathfrak{p} = [\mathcal{X}] \in \mathcal{M}_g$ to denote the corresponding *moduli point* and $\mathcal{M}_g(\mathfrak{p})$ the residue field of $\mathfrak{p}$ in $\mathcal{M}_g$. The field of moduli of $\mathcal{X}$ coincides with the residue field $\mathcal{M}_g(\mathfrak{p})$ of the point $\mathfrak{p}$ in $\mathcal{M}_g$. The notation $\mathcal{M}_g(\mathfrak{p})$ (resp. $M(\mathcal{X})$) will be used to denote the field of moduli of $\mathfrak{p} \in \mathcal{M}_g$ (resp. $\mathcal{X}$). If there is a curve $\mathcal{X}'$ isomorphic to $\mathcal{X}$ and defined over $M(\mathcal{X})$, we say that $\mathcal{X}$ has a *rational model over its field of moduli.* As mentioned above, the field of moduli of curves is not necessarily a field of definition, see [14] for examples of such families of curves.

Lemma 10: *Let $\mathfrak{u}_0 \in \mathcal{L}_3(k)$ such that $|\mathrm{Aut}(\mathfrak{u}_0)| > 4$. Then, there exists a genus 3 hyperelliptic curve $\mathcal{X}_3$ defined over k such that $\mathfrak{u}(\mathcal{X}_3) = \mathfrak{u}_0$ as defined in Eq (3). Moreover, the equation of $\mathcal{X}_3$ over its field of moduli is given by:*

i) If $|\mathrm{Aut}(\mathcal{X}_3)| = 16$ then

$$Y^2 = wX^8 + wX^4 + 1.$$

ii) If $\mathrm{Aut}(\mathcal{X}_3) \cong D_{12}$ then

$$\begin{aligned} Y^2 = &(u_3 - 260)X^8 - 7(u_3 - 98)X^6 + 15(u_3 - 134)X^4 \\ &- 9(u_3 - 162)X^2 + 126. \end{aligned} \tag{10}$$

iii) If $\mathrm{Aut} \cong \mathbb{Z}_2 \times \mathbb{Z}_4$ then

$$Y^2 = u_3^4 X^8 + u_3^4 X^6 + 8u_3 X^2 - 16.$$

iv) If $\mathrm{Aut}(\mathfrak{u}) \cong \mathbb{Z}_2^3$ *then*

$$Y^2 = u_1 X^8 + u_1 X^6 + u_2 X^4 + u_3 X^2 + 2.$$

Proof: The proof in all cases consists of simply computing the dihedral invariants. It is easy to check that these dihedral invariants satisfy the corresponding relations for $\mathrm{Aut}(\mathcal{X}_3)$ given above. □

Corollary 11: *Let* $\mathfrak{p} \in \mathcal{H}_3$ *such that* $|\mathrm{Aut}(\mathfrak{p})| > 2$. *Then the field of moduli of* $\mathfrak{p}$ *is a field of definition.*

Proof: There is only one hyperelliptic curve of genus 3 which has no extra involutions and whose automorphism group has more than 4 elements, see [23]. This curve is $Y^2 = X^7 - 1$ and its field of moduli is $\mathbb{Q}$. The result follows from the above Lemma for all groups of order > 4. Let $\mathcal{X}_3$ be a curve such that its automorphism group $\mathrm{Aut}(\mathcal{X}_3)$ has order 4. Then $\mathrm{Aut}(\mathcal{X}_3)$ is isomorphic to $\mathbb{Z}_4$ or V_4. In both cases the quotient curve $\mathcal{X}_3$ $\mathrm{Aut}(\mathcal{X}_3)$ is a conic which contains a non-trivial rational point. The result follows from [5]. □

Remark 12: An interesting problem would be to find an algorithm which finds a rational model over the field of moduli for curves with automorphism group $\mathbb{Z}_2$. In the case of genus 2 this has been done by Mestre; see [10].

For all the computations in this paper we have used Maple; see [1]. The above results have been implemented in a Maple package which is available upon request. In this package we can compute the field of moduli of any genus 3 hyperelliptic curve and provide a rational model for curves which have more than 4 automorphisms. Further, the automorphism group can be computed and all invariants of binary octavics defined in Eq.(5).

References

1. Maple 9. *Maplesoft Inc.*, 2004.
2. R. Alagna. Le relazioni fra gl'invarianti d'una forma qualunque d'ottavo ordine. *Rendiconti del Circolo Matematico di Palermo*, 6:77–99, 1892.
3. F. Bardelli and A. D. Centina. Bielliptic curves of genus three: Canonical models and moduli space. *Indag. Mathem., N.S.*, 10(2):183–190.
4. A. Clebsch. *Theorie der Binären Algebraischen Formen.* Verlag von B.G. Teubner, Leipzig, 1872.
5. P. Debes and M. Emsalem. On the field of moduli of curves. *J. Algebra*, 211(1): 42–56, 1999.
6. J. Gutierrez and T. Shaska. Hyperelliptic curves with extra involutions. *LMS J. of Comput. Math.*, 8 (2005), 102-115.

7. D. Hilbert. *Theory of Algebraic Invariants.* Cambridge University Press, London, 1993.
8. V. Krishnamorthy, T. Shaska, and H. Völklein. Invariants of binary forms. *Developments in Mathematics*, 12: 101–122, 2004.
9. K. Magaard, T. Shaska, S. Shpectorov, and H. Völklein. The locus of curves with prescribed automorphism group. volume Communications in arithmetic fundamental groups, pages 112–141, 2002.
10. P. Mestre. Construction de courbes de genre 2 á partir de leurs modules. volume Effective methods in algebraic geometry, pages 313–334, April 1990.
11. D. Sevilla and T. Shaska. Hyperelliptic curves with reduced automorphism group A_5. *(preprint).*
12. T. Shaska and J. Thompson. On the generic curves of genus 3. Contemporary. Math., Vol. **369**, pg. 233-244, AMS, 2004.
13. T. Shaska and H. Völklein. Elliptic subfields and automorphisms of genus two fields. volume Algebra, Arithmetic and Geometry with Applications, pages 687 – 707, 2004.
14. T. Shioda. On the graded ring of invariants of binary octavics. *Amer. J. Math.*, 89:1022–1046, 1967.

Curves over finite fields with many points: an introduction

John Voight*

Department of Mathematics,
University of California,
Berkeley, California, 94720, USA
jvoight@math.berkeley.edu

The number of points on a curve defined over a finite field is bounded as a function of its genus. In this introductory article, we survey what is known about the maximum number of points on a curve of genus g defined over $\mathbb{F}_q$, including an exposition of upper bounds, lower bounds, known values of this maximum, and briefly indicate some methods of constructing curves with many points, providing many references to the literature.

1. Introduction

By the Hasse-Weil bound (also known as the Riemann hypothesis for curves over finite fields), the number of points on a smooth, geometrically integral projective curve X of genus $g = g_X$ over a finite field $\mathbb{F}_q$ satisfies

$$\#X(\mathbb{F}_q) \leq q + 1 + 2g\sqrt{q}$$

and is therefore bounded as a function of g and q. It is natural to inquire about the sharpness of this upper bound, and so we consider the quantity

$$N_q(g) = \max_{\substack{X/\mathbb{F}_q \\ g_X = g}} \#X(\mathbb{F}_q)$$

which is the maximum number of points on a curve of genus g defined over $\mathbb{F}_q$, as well as the asymptotic quantity

$$A(q) = \limsup_{g \to \infty} \frac{N_q(g)}{g}.$$

*The author wishes to thank Michael Zieve, Bjorn Poonen, Hendrik Lenstra, and an anonymous reader for their helpful comments.

Recently, these quantities have seen a great deal of study, motivated in part by applications in coding theory and cryptography and because it is an enticing problem. This article surveys the results in this area, including an exposition of upper bounds, lower bounds, known values of $N_q(g)$, and a brief indication of methods of constructing curves with many points.

2. Goppa codes

We begin with a brief motivation coming from coding theory. In 1981, Goppa introduced a way to associate an error-correcting code to a curve over a finite field [6, 7]—good codes require curves with a large number of points relative to their genus, as we will see. For more information, see [22, §§II.2, VII.4] and [16, §5.5].

Messages are transmitted using a finite *alphabet* which we will take to be the finite field $\mathbb{F}_q$; the entire message is split into *words* which are blocks of fixed size and *encoded*: that is, provided with redundant letters to identify errors that were introduced during transmission. Therefore encoding is an injective map $E : \mathbb{F}_q^k \to \mathbb{F}_q^n$. The *code* C is the image of E, the *length* of the code is n.

If no errors occur, then the receiver can recover the original k letters using knowledge of E. If a word $v \notin C$ occurs, an error has occurred and assuming that the error made was small, the decoder looks for a word $\widetilde{v}$ in C that is as close to v as possible, i.e. such that the distance $|\widetilde{v} - v|$ is minimized, where the metric $|w|$ is defined to be the number of nonzero coordinates of w, called the *Hamming weight.* The *minimal distance* d is the minimal Hamming distance between distinct words in the code; since every ball of radius $< d/2$ contains at most one code word, one can correct up to $\lfloor (d-1)/2 \rfloor$ errors in each word. Therefore if possible we would like to simultaneously maximize k/n (the *transmission rate*) and d/n for a good code, and so we identify codes by the array $[k, d, n]$.

Let X be a curve, i.e. a smooth, geometrically integral projective variety of dimension 1, of genus g, defined over the finite field $\mathbb{F}_q$. Let D be a divisor on X defined over $\mathbb{F}_q$ and let $P_1, \ldots, P_n$ be a collection of points in $X(\mathbb{F}_q)$. We assume that $P_1, \ldots, P_n$ do not occur in the support of D, but this assumption can be removed with a little extra effort. We define the map

$$\begin{aligned} \theta : L(D) &\to \mathbb{F}_q^n \\ f &\mapsto (f(P_1), \ldots, f(P_n)), \end{aligned}$$

and in this way obtain the code C as the image of θ. Note that the Goppa code defines a linear subspace of $\mathbb{F}_q^n$.

We can estimate the parameters of this code using standard facts about the geometry of curves (see e.g. [22]). The code has length n. If the image of $f \in L(D)$ has weight d, then f vanishes in $n-d$ points, so $\deg D-(n-d) \geq 0$ (a divisor has at least as many poles as zeros), i.e. $d \geq n - \deg D$, and hence we assume that $\deg D < n$ so that d is positive. With this assumption, we also have $\ker \theta = L(D - \sum_{i=1}^n P_i) = \{0\}$ since $\deg(D - \sum_{i=1}^n P_i) < 0$, and the dimension k of the code is by Riemann-Roch

$$k = \dim L(D) \geq \deg D + (1-g).$$

In particular, using the Singleton bound $k + d \leq n + 1$ (a code of dimension k has a codeword with at least $k-1$ coordinates equal to 0 and hence has weight at most $n - (k-1)$, so $d \leq n + 1 - k$), we find

$$1 + \frac{1-g}{n} \leq \frac{k}{n} + \frac{d}{n} \leq 1 + \frac{1}{n};$$

Therefore a good code (one with k/n and d/n large) is one which arises from a curve of small genus with a large number of points.

Example 1: As an example, we can consider the Klein quartic X defined by the equation

$$x^3y + y^3z + xz^3 = 0$$

in the projective plane over $\mathbb{F}_2$ (see [17, Example (4.1)]). Since X is smooth and is defined by an equation of degree 4, the curve X has genus $g = (4-1)(4-2)/2 = 3$. One has the three points

$$(0:0:1),\ (0:1:0),\ (1:0:0) \in X(\mathbb{F}_2),$$

and these are the only points $(x:y:z)$ on X over any field $\mathbb{F}_{2^k}$ $(k \geq 1)$ with $xyz = 0$. Let us now consider the curve over $\mathbb{F}_8$. We can represent $\mathbb{F}_8 = \mathbb{F}_2(\alpha)$ where $\alpha^3+\alpha+1 = 0$; then α generates the multiplicative group $\mathbb{F}_8^*$. If $(x:y:z) \in X(\mathbb{F}_8)$ with $xyz \neq 0$, then we may assume $z = 1$, $y = \alpha^i$, and $x = \alpha^{3i}\beta$ for some $\beta \in \mathbb{F}_8$, so that

$$(\alpha^{3i}\beta)^3 + \alpha^{3i} + \alpha^{3i}\beta = \alpha^{3i}(\beta^3 + \beta + 1) = 0$$

hence $\beta^3 + \beta + 1 = 0$, so $\beta \in \{\alpha, \alpha^2, \alpha^2 + \alpha\}$. It follows that $\#X(\mathbb{F}_8^*) = 21$ and so

$$\#X(\mathbb{F}_8) = 24.$$

Let P be a $\mathbb{F}_8$-rational point and let D be the divisor $D = 10P$. If we take the sum of the other 23 points on X as the rational points divisor, then this code has length 23, dimension $10 - g + 1 = 8$ and minimum distance $\geq 23 - \deg(D) = 13$. If one adds a parity-check row and views $\mathbb{F}_8$ as a dimension 3 vector space over $\mathbb{F}_2$, one obtains a $[92, 24, \geq 26]$-code which is very nearly an optimal code with these parameters.

Asymptotically, the quality of these codes came as quite a surprise, exceeding the so-called Varshamov-Gilbert bound for $q \geq 7^2$; for more on this, we refer the reader to [15].

3. Upper bounds

With this as our motivation, we begin by finding upper bounds on $N_q(g)$, the maximum number of points on a curve of genus g over $\mathbb{F}_q$.

3.1. *Hasse-Weil bound*

For a motivated introduction to the material in this section, see [8, Appendix C], and for an elementary proof of the Weil conjectures for curves, see [22, §V.1] or [16, Chapter 3].

To study the number of points $N_r = \#X(\mathbb{F}_{q^r})$ on X, we form the zeta function of X:

$$Z(T) = Z_X(T) = \exp\left(\sum_{r=1}^{\infty} \#X(\mathbb{F}_{q^r})\frac{T^r}{r}\right) \in \mathbb{Q}[[t]].$$

Let a_d be the number of closed points of degree d of X. Then equivalently, we claim that we may write $Z(T)$ as the product

$$Z(T) = \prod_{d=1}^{\infty} \frac{1}{(1 - T^d)^{a_d}}.$$

Since $N_r = \sum_{d|r} da_d$ (a closed point of degree d corresponds to a Galois-orbit of d points defined over $\mathbb{F}_{q^d}$ and hence $\mathbb{F}_{q^r}$ whenever $d \mid r$), taking the logarithmic derivative we find

$$\begin{aligned}\frac{d}{dT}\log\prod_{d=1}^{\infty}\frac{1}{(1-T^d)^{a_d}} &= \sum_{d=1}^{\infty}\frac{da_dT^{d-1}}{1-T^d} = \frac{1}{T}\sum_{d=1}^{\infty} da_d(T^d + T^{2d} + \dots) \\ &= \frac{1}{T}\sum_{r=1}^{\infty}\left(\sum_{d|r} da_d\right)T^r = \frac{1}{T}\sum_{r=1}^{\infty} N_rT^r = \frac{d}{dT}\log Z(T).\end{aligned}$$

The zeta function satisfies the following properties.

Theorem 2: (Weil conjectures for curves) We have

$$Z(T) = \frac{P(T)}{(1-T)(1-qT)}$$

where

$$P(T) = \prod_{i=1}^{2g} (1-\alpha_i T) \in \mathbb{Z}[T]$$

is a polynomial of degree $2g$ with reciprocal roots $\alpha_i \in \mathbb{C}$ that are algebraic integers satisfying $|\alpha_i| = \sqrt{q}$.

The zeta function $Z(T)$ satisfies the functional equation

$$Z(1/(qT)) = \frac{Z(T)}{(qT^2)^{g-1}}.$$

As a corollary, we have a bound on $\#X(\mathbb{F}_{q^r})$.

Corollary 3: (Hasse-Weil bound) *We have*

$$N_r = \#X(\mathbb{F}_{q^r}) = q^r + 1 - \sum_{i=1}^{2g} \alpha_i^r \leq q^r + 1 + \lfloor 2gq^{r/2} \rfloor.$$

Proof: Taking the logarithmic derivative and multiplying by T, we have

$$\begin{aligned} \sum_{r=1}^{\infty} N_r T^{r-1} &= \frac{1}{1-T} + \frac{q}{1-qT} - \sum_{i=1}^{2g} \frac{\alpha_i}{1-\alpha_i T} \\ &= \sum_{r=0}^{\infty} (1+q^{r+1})T^r - \sum_{r=1}^{\infty} \left(\sum_{i=1}^{2g} \alpha_i^{r+1} \right) T^r. \end{aligned}$$

Hence

$$|N_r - (1-q^r)| \leq \left|\sum_{i=1}^{2g} \alpha_i^r\right| \leq 2gq^{r/2}. \qquad \square$$

As a consequence, we know

$$A(q) = \limsup_{g\to\infty} \frac{N_q(g)}{g} \leq 2\sqrt{q}.$$

We will see improvements on this bound in later sections of this article.

We now note for the interested reader that there is a large amount of literature which discusses the question of how one can efficiently compute the zeta function of a curve X given a set of equations which define X in an ambient projective space. We refer the reader to [27] and the references therein. We can always recover the zeta function in very simple cases by counting points over a sufficient number of extensions of the ground field, as indicated in the following example.

Example 4: Consider again the curve $X : x^3y + x^3z + y^3z = 0$ over $\mathbb{F}_2$. We write

$$Z(T) = \frac{P(T)}{(1-T)(1-2T)}$$

where $P(T) \in 1 + T\mathbb{Z}[T]$ is a polynomial of degree $2g = 6$ by the Weil conjectures. We count that X has number of points $N_i = 3, 5, 24, 17, 33, 38$ for $i = 1, \dots, 6$, hence

$$(1-T)(1-2T)\exp\left(3T + \frac{5}{2}T^2 + \dots + \frac{38}{6}T^6 + \dots\right) = P(T)$$

and by matching coefficients we find that $P(T) = 1 + 5T^3 + 8T^6$, and therefore

$$Z(T) = \frac{1 + 5T^3 + 8T^6}{(1-T)(1-2T)}.$$

From this we can calculate the reciprocal roots α_i of $P(T)$; among them are $(1 \pm \sqrt{-7})/2$ which indeed have absolute value $\sqrt{2}$.

3.2. *Serre bound*

For many years after Weil first proved these conjectures, there was essentially no investigation into how close the bound was to being sharp. In many cases, the bound can be improved greatly, and in the next few sections we will discuss the varied techniques which have been used to obtain improvements in certain cases. The first is due to Serre.

Proposition 5: (Serre, [18, 19]) *We have*

$$N_q(g) \le q + 1 + g\lfloor 2\sqrt{q} \rfloor.$$

Proof: Let $\beta_i = \lfloor 2\sqrt{q} + 1 \rfloor + \alpha_i + \overline{\alpha_i}$. Then $\beta_i \in \mathbb{R}$ are algebraic integers stable under $Gal(\overline{\mathbb{Q}}/\mathbb{Q})$ with $\beta_i > 0$ (since $|\alpha_i| = \sqrt{q}$). Therefore $\prod_{i=1}^{g} \beta_i > 0$ is an integer, and by the arithmetic-geometric mean

$$\frac{1}{g}\sum_{i=1}^{g} \beta_i \ge \left(\prod_{i=1}^{g} \beta_i\right)^{1/g} \ge 1$$

we deduce

$$g \leq \sum_{i=1}^{g} \beta_i = g\left(\lfloor 2\sqrt{q} \rfloor + 1\right) + \sum_{i=1}^{g} (\alpha_i + \overline{\alpha_i})$$

which gives the result. □

As a consequence, we immediately obtain $A(q) \leq \lfloor 2\sqrt{q} \rfloor$.

Example 6: Serre's modification already gives an essential improvement. For $g = 3$, $q = 8$, the Weil bound gives $N_8(3) \leq 25$, whereas the improved bound gives $N_8(3) \leq 24$; but this is exactly the number of points on the Klein quartic (Example 24). We conclude that $N_8(3) = 24$.

3.3. *Ihara bound*

When the genus is large in comparison to the size of the field, a significant improvement can be made, due to Ihara.

Theorem 7: (Ihara [11]) We have

$$N_q(g) \leq \frac{1}{2}\left(\sqrt{(8q+1)g^2 + (4q^2 - 4q)g} - (g - 2q - 2)\right)$$

and asymptotically

$$A(q) \leq \frac{1}{2}\left(\sqrt{8q+1} - 1\right).$$

Proof: Let X be a curve of genus g over $\mathbb{F}_q$. Then by Theorem 2,

$$\#X(\mathbb{F}_{q^r}) = q + 1 - \sum_{i=1}^{g} (\alpha_i^r + \overline{\alpha_i}^r).$$

If we write $a_i = \alpha_i + \overline{\alpha_i}$, then

$$q + 1 - \sum_{i=1}^{g} a_i = \#X(\mathbb{F}_q) \leq \#X(\mathbb{F}_{q^2}) = q^2 + 1 + 2qg - \sum_{i=1}^{g} a_i^2$$

using the fact that $\alpha_i \overline{\alpha_i} = q$.

Let $N = \#X(\mathbb{F}_q)$. Then by the Cauchy-Schwarz inequality, we have

$$g\left(\sum_{i=1}^{g} a_i^2\right) \leq \left(\sum_{i=1}^{g} a_i\right)^2,$$

hence

$$N = \#X(\mathbb{F}_q) \leq q^2 + 1 + 2qg - \frac{1}{g}\left(\sum_{i=1}^{g} a_i\right)^2 = q^2 + 1 + 2qg - \frac{1}{g}(q + 1 - N)^2.$$

Simplifying, we have

$$N^2 + (g - 2 - 2q)N + (q+1)^2 - (q^2+1)g - 2qg^2 \leq 0;$$

solving for N we obtain

$$N \leq \frac{1}{2}\left(\sqrt{(8q+1)g^2 + (4q^2 - 4q)g} - g + (2q-2)\right)$$

as claimed. Finally,

$$\frac{N}{g} \leq \frac{1}{2}\left(\sqrt{(8q+1) + \frac{4q^2 - 4q}{g}} - 1 + \frac{2q-2}{g}\right) \to \frac{1}{2}\left(\sqrt{8q+1} - 1\right)$$

as $g \to \infty$. □

Remark 8: This bound is better than the Hasse-Weil bound when

$$\begin{aligned} 2\,(q + 1 + 2g\sqrt{q}) &> \sqrt{(8q+1)g^2 + (4q^2 - 4q)g} - (g - 2q - 2) \\ g^2(1 + 4\sqrt{q})^2 &> (8q+1)g^2 + (4q^2 - 4q)g \\ g &> \frac{4q^2 - 4q}{(1+4\sqrt{q})^2 - 8q - 1} = \frac{4q(q-1)}{8\sqrt{q}(\sqrt{q}+1)} = \frac{\sqrt{q}(\sqrt{q}-1)}{2}, \end{aligned}$$

i.e. when the genus is large in comparison to the size of the field. The bound is already better for $q = 4$ and $g > 1$.

Example 9: For example, the Weil bound tells us that $N_2(100) \leq 285$, but the Ihara bound tells us that $N_2(100) \leq 159$.

3.4. *Drinfel'd-Vladut bound*

Generalizing Ihara's argument, Drinfel'd-Vladut obtained the following.

Theorem 10: (Drinfel'd-Vladut bound [26]) We have

$$A(q) \leq \sqrt{q} - 1.$$

Proof: For every positive integer k, by the Weil conjectures we have

$$\#X(\mathbb{F}_{q^k}) = q^k + 1 - \sum_{i=1}^{g} (\alpha_i^k + \overline{\alpha_i}^k)$$

with $\alpha_i = \sqrt{q}\omega_i$ where $|\omega_i| = 1$. Then

$$0 \leq \sum_{i=1}^{g} |1 + \omega_i + \cdots + \omega_i^k|^2 = \sum_{i=1}^{g} \left(\sum_{r=0}^{k} \omega_i^r\right)\left(\sum_{r=0}^{k} \overline{\omega_i}^r\right).$$

But

$$(1+\omega_i+\cdots+\omega_i^k)(1+\overline{\omega_i}+\cdots+\overline{\omega_i}^k)=\sum_{r=0}^{k}(k+1-r)(\omega_i^r+\overline{\omega_i}^r)+(k+1),$$

hence

$$\begin{aligned}0 &\leq g(k+1)+\sum_{r=1}^{k}(k+1-r)\sum_{i=1}^{g}(\omega_i^r+\overline{\omega_i}^r)\\ &= g(k+1)+\sum_{r=1}^{k}(k+1-r)\frac{q^r+1-\#X(\mathbb{F}_{q^r})}{(\sqrt{q})^r}\\ &\leq g(k+1)+\sum_{r=1}^{k}(k+1-r)\frac{q^r+1-N}{q^{r/2}}.\end{aligned}$$

where $N=\#X(\mathbb{F}_q)$. Solving for N gives

$$N\sum_{r=1}^{k}(k+1-r)\frac{1}{q^{r/2}}\leq g(k+1)+\sum_{r=1}^{k}(k+1-r)q^{r/2}+\sum_{r=1}^{k}(k+1-r)\frac{1}{q^{r/2}}$$

$$N\leq 1+\frac{g(k+1)+\sum_{r=1}^{k}(k+1-r)q^{r/2}}{\sum_{r=1}^{k}(k+1-r)q^{-r/2}}.$$

Therefore as $g\to\infty$ we have

$$\begin{aligned}A(q)=\limsup_{g\to\infty}\frac{N_q(g)}{g} &\leq \frac{k+1}{\sum_{r=1}^{k}(k+1-r)q^{-r/2}}\to\frac{1}{\sum_{r=1}^{\infty}q^{-r/2}}\\ &= \left(\frac{q^{-1/2}}{1-q^{-1/2}}\right)^{-1}=\sqrt{q}-1\end{aligned}$$

as claimed. □

Example 11: The argument of Drinfel'd-Vladut also gives upper bounds for $N_q(g)$ when g is much larger than q. For example, choosing $k=8$ in the above proof we find that $N_2(100)\leq 77$.

3.5. *Oesterlé bound*

We would like to modify the preceding argument to find the best possible minimal bound obtainable by such methods.

As an overview to motivate the method, suppose we start with a set of complex numbers α_i that might feasibly arise as *Frobenius eigenvalues* of a curve X, i.e. reciprocal roots of the numerator of the zeta function of X.

Suppose we forget almost all properties of these numbers α_i, remembering only that they occur in complex conjugate pairs and have complex absolute value $|\alpha_i| = \sqrt{q}$. We form a formal collection of these numbers by allowing them to have multiplicities which are rational numbers. Then from any such formal collection, if we assume that they in fact arise from a curve X, we can write down the zeta function of the corresponding curve, and hence the genus and number of points. The Oesterlé bound we will exhibit below is the optimal upper bound on $N_q(g)$ for this general type of object.

We carry out this program following Serre [18]. If we let $\omega_j = e^{i\theta_j}$ for $\theta_j \in \mathbb{R}$ in the above notation, then

$$N_r = \#X(\mathbb{F}_{q^r}) = q^r + 1 + \sum_{j=1}^{g}(\alpha_j^r + \overline{\alpha_j}^r) = q^r + 1 - q^{r/2}\sum_{j=1}^{g}(e^{ir\theta_j} + e^{-ir\theta_j})$$
$$= q^r + 1 - 2q^{r/2}\sum_{j=1}^{g}\cos(r\theta_j).$$

Therefore if we are given real numbers $c_r \in \mathbb{R}$, we can multiply the above equation by c_r and divide by $q^{r/2}$ to obtain

$$2\sum_{j=1}^{g} c_r \cos(r\theta_j) + c_r N_r q^{-r/2} = c_r q^{r/2} + c_r q^{-r/2}.$$

Define

$$f(\theta) = 1 + 2\sum_{n=1}^{\infty} c_r \cos(r\theta), \qquad \Psi_d(t) = \sum_{r=1}^{\infty} c_{rd} t^{rd},$$

so after summing over r we obtain

$$\sum_{j=1}^{g} 2\sum_{r=1}^{\infty} c_r \cos(r\theta_j) + \sum_{r=1}^{\infty} c_r N_r q^{-r/2} = \sum_{r=1}^{\infty} c_r q^{r/2} + \sum_{r=1}^{\infty} c_r q^{-r/2}$$
$$\sum_{j=1}^{g} f(\theta_j) + \sum_{r=1}^{\infty} c_r N_r q^{-r/2} = g + \Psi_1(q^{1/2}) + \Psi_1(q^{-1/2}).$$

Now from $N_r = \sum_{d|r} d a_d$ where a_d is the number of closed points of X of degree d, we have

$$\sum_{r=1}^{\infty} c_r N_r q^{-r/2} = \sum_{r=1}^{\infty} c_r \left(\sum_{d|r} d a_d\right) q^{-r/2}$$
$$= \sum_{d=1}^{\infty} d a_d \left(\sum_{m=1}^{\infty} c_{md} q^{-md/2}\right) = \sum_{d=1}^{\infty} d a_d \Psi_d(q^{-1/2}).$$

We have shown the following result.

Proposition 12: (Weil's explicit formulas) *With the notation above, we have*

$$\sum_{j=1}^{g} f(\theta_j) + \sum_{d=1}^{\infty} d a_d \Psi_d(1/\sqrt{q}) = g + \Psi_1(\sqrt{q}) + \Psi_1(1/\sqrt{q}).$$

Serre [18] used these formulas to obtain bounds on N_r as follows.

Corollary 13: *If the c_r are chosen such that $c_r \geq 0$ for all r and $f(\theta) \geq 0$ for all $\theta \in \mathbb{R}$, and only finitely many c_r are nonzero, then*

$$N_q(g) \leq \frac{g}{\Psi(1/\sqrt{q})} + 1 + \frac{\Psi(\sqrt{q})}{\Psi(1/\sqrt{q})},$$

where $\Psi = \Psi_1$.

Proof: We have

$$\begin{aligned} N\Psi_1(1/\sqrt{q}) &= g + \Psi_1(\sqrt{q}) + \Psi_1(1/\sqrt{q}) - \sum_{d=2}^{\infty} d a_d \Psi_d(1/\sqrt{q}) - \sum_{j=1}^{g} f(\theta_j) \\ &\leq g + \Psi_1(\sqrt{q}) + \Psi_1(1/\sqrt{q}) \end{aligned}$$

since by assumption $f(\theta_j) \geq 0$ and $c_r \geq 0$. □

For every choice of c_r satisfying these conditions, we obtain an upper bound. In particular, the Weil bound corresponds to the choice $f(\theta) = 1 + \cos\theta$.

Example 14: (Serre [18]) We can, for example, take f of the form

$$f(\theta) = \frac{\left(1 + 2\sum_{r=1}^{\infty} u_r \cos r\theta\right)^2}{1 + 2\sum_{r=1}^{\infty} u_r^2}.$$

It has constant term 1 (just check the Taylor expansion), and using trigonometric identities we can write it in the desired form if we take $u_r \geq 0$. If we take $q = 2$, then the choice $u_1 = 1$, $u_2 = 3/4$, $u_3 = 1/2$ gives

$$f(\theta) = 1 + \frac{31}{17}\cos\theta + \frac{24}{17}\cos 2\theta + \frac{16}{17}\cos 3\theta + \frac{1}{2}\cos 4\theta + \frac{3}{17}\cos 5\theta + \frac{1}{34}\cos 6\theta$$

and thus

$$N_2(g) \leq \frac{544}{318\sqrt{2} + 227} g + \frac{16(75\sqrt{2} + 86)}{318\sqrt{2} + 227} \leq 0.8038g + 4.514.$$

We find $N_2(g) \leq 7, 8, 9$ for $g = 3, 4, 5$. Compare this to the Weil bound,

$$N_2(g) \leq 3 + 2\sqrt{2}g \leq 2.828g + 3,$$

which gives $N_2(g) \leq 11, 14, 17$.

In this case, these upper bounds are in fact best possible. For example, the curve

$$X : x^3y + y^3z + xz^3 + x^2y^2 + x^2z^2 + y^2z^2 + x^2yz + xy^2z = 0$$

passes through each of the 7 points of the projective plane $\mathbb{P}^2_{\mathbb{F}_2}$, and X is nonsingular and hence has genus 3. We will briefly mention some methods of constructing curves with many points below (§4).

Corollary 15: (Serre [20]) *We have*

$$N_q(g) \leq \begin{cases} q^2 + 1, & \text{if } g \leq \sqrt{2q}(q-1)/2; \\ q^3 + 1, & \text{if } g \leq \sqrt{q}\left(\sqrt{3}(q+1) + \sqrt{q}\right)(q-1)/2. \end{cases}$$

Proof: If we take

$$f(\theta) = \frac{1}{2}(1 + \sqrt{2}\cos\theta)^2 = 1 + \sqrt{2}\cos\theta + \frac{1}{2}\cos 2\theta$$

then

$$\Psi(1/\sqrt{q}) = \frac{1}{4q}(2\sqrt{2q} + 1), \quad \Psi(\sqrt{q}) = \frac{\sqrt{q}}{4}(2\sqrt{2} + \sqrt{q})$$

and hence

$$\begin{aligned} N &\leq \frac{4gq}{2\sqrt{2q}+1} + 1 + q\sqrt{q}\left(\frac{2\sqrt{2}+\sqrt{q}}{2\sqrt{2q}+1}\right) \\ &\leq \frac{2q\sqrt{2q}(q-1) + 2q\sqrt{2q} + q^2}{2\sqrt{2q}+1} + 1 = q^2 + 1. \end{aligned}$$

The second statement follows from applying the same argument to

$$\begin{aligned} f(\theta) &= \frac{1}{3}\cos^2\theta\left(\sqrt{3} + 2\cos\theta\right)^2 \\ &= 1 + \sqrt{3}\cos\theta + \frac{7}{6}\cos 2\theta + \frac{\sqrt{3}}{3}\cos 3\theta + \frac{1}{6}\cos 4\theta. \end{aligned}$$ □

As can be seen from the above, the next problem is to find an optimal choice of the c_r which will minimize this upper bound. Equivalently, given a fixed number of points N and field q, by (13) we have

$$g \geq (N-1)\Psi(1/\sqrt{q}) - \Psi(\sqrt{q}) = (N-1)\sum_{r=1}^{\infty} c_r q^{-r/2} - \sum_{r=1}^{\infty} c_r q^{r/2}$$

and thus we want the c_r that give the best possible lower bound on g. Oesterlé was able to give an explicit recipe for finding the c_r using linear programming. It is really something of a miracle that one can give a formula for every q and g as follows.

Theorem 16: (Oesterlé) Let $\lambda = N - 1$ and m be the integer such that $q^{m/2} < \lambda \leq q^{(m+1)/2}$. Let

$$u = \frac{q^{(m+1)/2} - \lambda}{\lambda\sqrt{q} - q^{m/2}}.$$

Then there is a unique solution $\vartheta \in [\pi/(m+1), \pi/m)$ to

$$\cos(\vartheta(m+1)) + u\cos(\vartheta(m-1)/2) = 0$$

and if we let

$$a_r = (m-r)\cos(r\vartheta)\sin\vartheta + \sin((m-r)\vartheta)$$

and $c_r = a_r/a_0$, then

$$g \geq \sum_{r=1}^{m-1} c_r(\lambda q^{-r/2} - q^{r/2}) = \frac{(\lambda-1)\sqrt{q}\cos(\vartheta) + q - \lambda}{q - 2\sqrt{q}\vartheta + 1}.$$

Proof: (Ingredients of proof) The proof is given in [10], following Serre in an exposition of Oesterlé's argument in unpublished notes from a course at Harvard in 1985.

There exists a measure μ on $\mathbb{S}^1 = \{z : \mathbb{C} : |z| = 1\}$ such that

$$\frac{1}{2}\int_{\mathbb{S}^1} d\mu = (N-1)\sum_{r=1}^{\infty} c_r q^{-r/2} - \sum_{r=1}^{\infty} c_r q^{r/2}$$

with the property that equality occurs exactly when the c_r are optimal (if $q \geq 3$; one can treat the case $q = 2$ by different methods). This measure is given by $\mu = \sum_{t\in T} \nu_t \delta_t$ where δ_t is the Dirac measure ($\delta_t(x) = 1$ or 0 as $x = t$ or $x \neq t$), the ν_t are the zeros of the function $f(t) = 1 + \sum_{n=1}^{\infty} c_r(t^r + t^{-r})$ on $\mathbb{S}^1$, and T is a symmetric set with $\#T = m - 1$. One then verifies that ϑ exists and gives rise to the solutions t in the above formula directly. □

Example 17: For $q = 2$, $N = 71$, we find $m = 12$,

$$u = \frac{64\sqrt{2} - 70}{70\sqrt{2} - 64},$$

and $\vartheta = 0.2562$, and hence

$$c_1 = 0.9698,\ c_2 = 0.8868,\ \ldots,\ c_{11} = 0.004227$$

hence

$$g \geq \left\lceil \frac{66.75\sqrt{2} - 68}{3 - 1.935\sqrt{2}} \right\rceil = 101.$$

Hence a curve with genus $g \leq 100$ has $N_3(g) \leq 70 \leq 77$, which is an improvement on the Drinfel'd-Vladut bound.

In sum, when the genus is small in comparison to the size of the field, the Serre-Weil bound is minimal, and as the genus becomes larger, one ascends from the Ihara bound to the Oesterlé bound (which is best possible in the sense that it minimizes the bound coming from Weil's explicit formula).

4. Lower bounds

One is also interested in knowing in general a lower bound on $N_q(g)$. There are two essentially different cases.

4.1. *q a square*

In the case that q is a square, we can get quite precise asymptotic information.

Proposition 18: (Ihara [11], Tsfasman-Vladut-Zink [24]) *If q is a square, then*

$$A(q) \geq \sqrt{q} - 1.$$

Proof: [Sketch of proof] The curves that give this bound are Shimura curves considered over $\mathbb{F}_q$. In the case $q = p^2$, the Shimura curves are the modular curves $X_0(\ell)$ and the proof runs as follows. We can consider the curves $X_0(\ell)$ for $\ell \neq p$, $\ell \equiv 11 \pmod{12}$ over $\mathbb{F}_{p^2}$, parametrizing elliptic curves equipped with an isogeny of degree ℓ.

The genus of this curve is $(\ell + 1)/12$ (this can be computed using the Riemann-Hurwitz formula using the quotient map $X_0(\ell) \to X_0(1)$, since ramification happens only at i, ρ, and ∞). There are roughly $(p - 1)/12$ supersingular j-invariants in characteristic p all defined over $\mathbb{F}_{p^2}$ (with a slight error due to the cusps and the j-values 0 and 1728). Since there are $\ell + 1$ distinct cyclic subgroups of order ℓ of any elliptic curve defined over $\overline{\mathbb{F}}_p$, and when $j \neq 0, 1728$ the automorphism group of E is $\langle \pm 1 \rangle$, these give rise to distinct isogenies. Therefore there are approximately $(\ell+1)(p-1)/12$ rational points on $X_0(\ell)$, so the ratio to its genus is roughly $p - 1$. One then verifies that it actually approaches $p - 1$ as $\ell \to \infty$. □

For a different proof that counts points using Hecke operators, see [16, §5.6]. These curves exceed the Varshamov-Gilbert bound—see [24].

Theorem 19: If q is a square, $A(q) = \sqrt{q} - 1$.

This follows from the theorem and the upper bound (Theorem 10) above.

Garcia and Stichtenoth constructed an explicit tower X_i over $\mathbb{F}_{q^2}$ which meets this bound.

Theorem 20: (Garcia and Stichtenoth [5]**)** Start with $X_1 = \mathbb{P}^1_{\mathbb{F}_q^2}$ with coordinate x_1 and define Artin-Schreier covers X_i by

$$y_{i+1}^q + y_{i+1} = x_i^{q+1}, \quad x_{i+1} = y_{i+1}/x_i$$

for $i \geq 1$. Then

$$\#X_i(\mathbb{F}_{q^2})/g(X_i) \to q - 1 \quad (g \to \infty).$$

Proof: (Ingredients of proof) One uses the ramification theory of Artin-Schreier extensions (see [22, Proposition VI.4.1]) and the Hurwitz genus formula. The hard steps are determining precisely the places that ramify in each relative extension $K(X_i)/K(X_{i-1})$ and calculating the genus of X_i, which is found to be approximately $q^{i-1}(q-1)$. Then one calculates that

$$\#X_i(\mathbb{F}_{q^2}) \geq (q^2 - 1)q^{i-1} + 2q$$

by counting places that split completely and those that ramify. This implies the result. □

Such an infinite sequence of distinct curves X_i over $\mathbb{F}_q$ satisfying

$$\frac{\#X_i(\mathbb{F}_q)}{g(X_i)} \to \sqrt{q} - 1$$

as $i \to \infty$ is called an *optimal tower*. Remarkably, at the time of this writing, all such optimal towers have been shown to be *modular*, meaning they arise as parameter spaces for objects such as elliptic curves, Shimura curves, or Drinfel'd modules for sufficiently large i. We refer the reader to [3] for a proof that the tower above is modular and to [2] for more on this fascinating subject.

4.2. *q not a square*

In the much more difficult case when q is not a square, Serre has given the following asymptotic result.

Proposition 21: (Serre [18]) *There exists a constant $c > 0$ such that*

$$A(q) \geq c \log q$$

for all q.

The proof uses an infinite class field tower of the function field of a hyperelliptic curve over $\mathbb{F}_q$. We also have the following lower bound.

Theorem 22: (Elkies-Kresch-Poonen-Wetherell-Zieve [28]) For all q and g,

$$N_q(g) > g a_q$$

where $a_q > 0$ is a constant that depends only on q (if q is a square, we can take $a_q = (\sqrt{q} - 1)/6$).

Proof: (Ingredients of proof) The proof follows from two claims. The first is due to Ihara-Serre: for any q, there exist curves X_i of strictly increasing genera g_i such that for all i, $\#X_i(\mathbb{F}_q) \geq b_q g_i$ where $b_q > 0$ and $\lim_{i\to\infty} g_{i+1}/g_i < \infty$. The second statement which requires proof is the following: for any $X/\mathbb{F}_q$ and any $h > 4g_X$, there exists a curve $X'/\mathbb{F}_q$ such that $g_{X'} = h$ and $\#X'(\mathbb{F}_q) \geq \#X(\mathbb{F}_q)$. It is clear that these imply the theorem. To prove the second statement, one proves that there exists a degree two cover $X' \to X$ with $h = g_{X'}$ for any $h > 4g_X$. □

5. Determination of $N_q(g)$

Actual values of $N_q(g)$ for small values of q and g are interesting. Lower bounds are given by the best curve we know, and with luck, this attains one of the upper bounds above; otherwise we obtain an interval in which $N_q(g)$ lies.

5.1. *Maximal curves*

First, we seek out curves that attain the Weil bound. A curve X is called *maximal* if it attains the Weil bound (3), i.e.

$$\#X(\mathbb{F}_q) = q + 1 + 2g\sqrt{q}.$$

It follows that if X is maximal, then q is a square (so that the right-hand side is an integer) and $g \le (q - \sqrt{q})/2$ by Remark 8.

Example 23: The *Hermitian curve*

$$X : x^{q+1} + y^{q+1} = z^{q+1}$$

has genus $q(q-1)/2$ and has q^3+1 points over $\mathbb{F}_{q^2}$. (See [22, Lemma VI.4.4] for the properties of the function field of X.) This curve is also birational to $X : y^q z + y z^q = x^{q+1}$, which clarifies many of the properties of the curve; see [22, Example VI.4.3].

We may count points as follows: for any $a \in \mathbb{F}_{q^2}^*$, we have $a^{q+1} = N(a) \in \mathbb{F}_q$; the norm map $N : \mathbb{F}_{q^2}^* \to \mathbb{F}_q^*$ is surjective and hence

$$\# \ker N = (q^2 - 1)/(q - 1) = q + 1.$$

Therefore for a choice of $a \neq 0$, there are $q + 1$ choices for b such that $N(a) + N(b) = a^{q+1} + b^{q+1} = 0$, and there is the unique value $c = 0$ such that $(a : b : c) \in X$. A similar count holds when $a = 0$. Accounting for scalar multiples, and we find

$$\#X(\mathbb{F}_{q^2}) = \frac{2q^2(q+1) + q^2\left(q^2 - (q+1)\right)(q+1)}{q^2 - 1} = q^3 + 1.$$

Since

$$q^2 + 1 + 2g\sqrt{q^2} = q^2 + 1 + q^2(q-1) = q^3 + 1,$$

this curve attains the Hasse-Weil bound.

It can be shown that maximal curves only occur for specific genera.

Theorem 24: (Fuhrmann-Garcia-Torres [4]**)** If X is a maximal curve of genus g over $\mathbb{F}_q$ (where q is a square) then $g = (q - \sqrt{q})/2$ or $g \le (\sqrt{q} - 1)^2/4$. Moreover, if q is odd and

$$(\sqrt{q} - 1)(\sqrt{q} - 2)/4 < g \le (\sqrt{q} - 1)^2/4$$

then $g = (\sqrt{q} - 1)^2/4$.

Proof: (Ingredients of proof) If X is a maximal curve and X' a curve dominated by X then X' is also a maximal curve, since the Jacobian of X' is an isogeny factor of the Jacobian of X. For a maximal curve X the action of Frobenius on the Jacobian is by $-\sqrt{q}$. Choose a $\mathbb{F}_{q^2}$-rational point

P_0 on X and map X to the Jacobian $Jac(X)$ by $P \mapsto [P - P_0]$. If $F(P)$ is the Frobenius image of P on X then

$$-\sqrt{q}[P - P_0] = [F(P) - P_0]$$

in the Jacobian, so for any point P the divisor $\sqrt{q}P + F(P)$ is linearly equivalent to $(\sqrt{q}+1)P_0$ and this gives a canonically defined linear system on such curves. The result then follows by looking at this linear system, as it gives a bound on g. □

We refer the reader [23] for an introduction to Stöhr-Voloch theory which concerns the geometric consequences of the arithmetic property of having many rational points.

5.2. *Fixed g*

Serre [18] began the systematic study of $N_q(g)$ for small q and g. For $g = 0$ (the projective line), we have $N_q(0) = q + 1$.

For $g = 1$ (elliptic curves), we have the following classical result.

Proposition 25: (Hasse-Deuring) *We have*

$$N_q(1) = \begin{cases} q + \lfloor 2\sqrt{q} \rfloor, & \text{if } q = p^n,\ n \geq 3,\ n \text{ odd, and } p \mid \lfloor 2\sqrt{q} \rfloor; \\ q + 1 + \lfloor 2\sqrt{q} \rfloor, & \text{otherwise.} \end{cases}$$

The smallest exceptional case is $q = 128 = 2^7$; we have $N_{128}(1) = 150$, $\lfloor 2\sqrt{q} \rfloor = 22$; other examples are $q = 2^{11}, 2^{15}, 3^7, 5^9$.

For curves of genus $g = 2$, Serre proved the following result.

Theorem 26: (Serre [19]) If q is a square and $q \neq 4, 9$ then

$$N_q(2) = q + 1 + 4\sqrt{q}.$$

Furthermore, we have $N_4(2) = 10$ and $N_9(2) = 20$.

If q is not a square, then

$$N_q(2) = \begin{cases} q + 1 + 2\lfloor 2\sqrt{q} \rfloor, & \text{if } q \text{ is nonspecial;} \\ q + 2\lfloor 2\sqrt{q} \rfloor, & \text{if } q \text{ is special and, } 2\sqrt{q} - \lfloor 2\sqrt{q} \rfloor > (\sqrt{5}-1)/2; \\ q - 1 + 2\lfloor 2\sqrt{q} \rfloor, & \text{otherwise.} \end{cases}$$

We deem q to be *special* if $q = p^e$ and $p \mid \lfloor 2\sqrt{q} \rfloor$ or if there exist solutions to one of the equations

$$q = x^2 + 1,\ q = x^2 + x + 1,\ q = x^2 + x + 2$$

with $x \in \mathbb{Z}$. For example, if $e = 1$, i.e. $q = p$, then p is special iff $p = x^2 + 1$ or $p = x^2 + x + 1$. When $e \geq 3$, it can be shown that the Diophantine condition is satisfied only for $q = 7^3$ or $q = 2^3, 2^5, 2^{13}$.

Proof: (Ingredients of proof, see [20]) We endeavor to construct a curve of genus 2 having many rational points. One begins with an elliptic curve E having $q+1+\lfloor 2\sqrt{q} \rfloor$ points with $End(E) = R = \mathbb{Z}[\pi]$ with π the Frobenius map. The ring R is an order in a quadratic imaginary field of discriminant $D = \lfloor 2\sqrt{q} \rfloor^2 - 4q < 0$. Find a nondegenerate Hermitian form P over R, projective of rank 2, positive definite, and indecomposable (which exists when $D < -7$). Then $A = P \otimes_R E$ is an abelian variety of dimension 2, isogeneous to $E \times E$, equipped with an indecomposable, principal polarization by P. There is a correspondence between curves of genus 2 and abelian varieties of dimension 2 equipped with such a polarization, and therefore A is the Jacobian of a curve of genus 2 with $\#X(\mathbb{F}_q) = q + 1 + 2\lfloor 2\sqrt{q} \rfloor$.

The exceptional cases occur when the form is decomposable, which translates into an occurence of p satisfying one of the three given Diophantine conditions. □

5.3. *Other examples*

For small genera, certain values of $N_q(g)$ can be eliminated by listing all of the possibilities of the zeta function and showing that zeta functions in this list imply a decomposition of the Jacobian as a product of principally polarized abelian varieties, which contradicts the irreducibility of the theta divisor of the curve. Sometimes one can rule out that $N_q(g)$ meets the Serre bound by arguments like Galois descent, which works for $q = 27$, $g = 3$ and $q = 8$, $g = 4$. (For a complete list of references, see [25].)

Tables of the best lower and upper bounds for $g \leq 50$ and $q = 2^m, 3^m$ for small m have been collected by van der Geer and van der Vlugt [25]. The most updated tables can be found at

`http://www.science.uva.nl/~geer/.`

Once one has good upper bounds, one would like to show that these bounds are met by some curve, therefore one needs good methods of constructing curves; these methods include using class field theory, quotients and covers of classical curves (like the Hermitian, Fermat, and modular curves) and other curves with many rational points, fibre products of Artin-Schreier curves, rank 1 Drinfel'd modules, Kummer covers, exhaustive computer search, and many others. We refer the reader to [25] for an overview.

References

1. Ian F. Blake, *Curves with many points and their applications*, Applied alebgra, algebraic algorithms and error correcting codes: 13th International Symposium, AAEECC-13 (Honolulu, Hawaii), Lect. Notes in Comp. Sci., vol. 1719, 1999, 55–64.
2. Noam D. Elkies, *Explicit modular towers*, Proceedings of the Thirty-Fifth Annual Allerton Conference on Communication, Control and Computing (1997), T. Basar, A. Vardy, eds., 1998, 23–32, also available at `arXiV:math.NT/0103107`.
3. Noam D. Elkies, *Explicit towers of Drinfeld modular curves*, Progress in Math. **202** (2001), 189–198.
4. Rainer Fuhrmann, Arnaldo Garcia, and Fernando Torres, *On maximal curves*, J. Num. Theory **67** (1997), 29–51.
5. Arnolda Garcia and Henning Stichtenoth, *A tower of Artin-Schreier extensions of function fields attaining the Drinfeld-Vladut bound*, Invent. Math. **121** (1995), 211–222.
6. V.D. Goppa, *Codes on algebraic curves*, Sov. Math. Dokl. **24** (1981), 170–172.
7. V.D. Goppa, *Geometry and codes*, Mathematics and its applications, Soviet series, vol. 24, Kluwer Academic Publishers: Dordrecht, Netherlands, 1988,
8. Robin Hartshorne, *Algebraic geometry*, Springer-Verlag, New York, 1977.
9. Johan P. Hansen, *Deligne-Lusztig varieties and group codes*, Coding theory and algebraic geometry, Springer: Berlin, 1992, 63–81.
10. S. H. Hansen, *Rational points on curves over finite fields*, Lect. Notes Ser., Aarhus Univ. Mat. Institute, 1995.
11. Yasutaka Ihara, *Some remarks on the number of rational points of algebraic curves over finite fields*, J. Fac. Sci. Univ. Tokyo Sect. IA Math. **28** (1981), no. 3, 721–724.
12. Kristin Lauter, *Deligne-Lusztig curves as ray class fields*, preprint available via search engine at `http://www.mpim-bonn.mpg.de/cgi-bin/preprint/preprint\_search.pl`.
13. Kristin Lauter and René Schoof, Using class field theory to construct curves with many points, Lectures given at Arizona Winter School: Arithmetic of function fields, February 2000.
14. F.J. MacWilliams and N.J.A. Sloane, *Theory of error-correcting codes*, Amsterdam, North-Holland, 1977.
15. Yu. I. Manin, *What is the maximum number of points on a curve over* $\mathbb{F}_2$*?*, J. Fac. Sci. Univ. Tokyo Sect. IA Math. **28** (1981), no. 3, 715–720.
16. Carlos Moreno, *Algebraic curves over finite fields*, Cambridge tracts in mathematics, vol. 97, Cambridge University Press: Cambridge, 1991.
17. R. Schoof, *Algebraic curves and coding theory*, UTM 336, Univ. of Trento, 1990.
18. Jean-Pierre Serre, *Sur le nombre des points rationnels d'une courbe algébrique sur un corps fini*, C. R. Acad. Sci. Paris, I **296** (1983), 397–402. (*Oeuvres*, vol. III, no.128, 658–663.)
19. Jean-Pierre Serre, *Nombres de points des courbes algébriques sur* $\mathbb{F}_q$, Sém. de

Théorie des Nombres de Bordeaux (1982/1983) exp. no. 22. (*Oeuvres*, vol. III, no. 129, 664–668.)
20. Jean-Pierre Serre, Résumé des cours de 1983–1984, Annuaire du Collège de France (1984), 79–83. (*Oeuvres*, vol. III, no. 132, 701–705.)
21. J.H. Silverman, *The arithmetic of elliptic curves*, Berlin: Springer, 1994.
22. Henning Stichtenoth, *Algebraic function fields and codes*, Springer: Berlin, 1993.
23. Fernando Torres, *The approach of Stöhr-Voloch to the Hasse-Weil bound with applications to optimal curves and plane arcs*, preprint 2000, `arXiV:math.AG/0011091`.
24. M.A. Tsfasman, S.G. Vladut, and Th. Zink, *Modular curves, Shimura curves, and Goppa codes, better than Varshamov-Gilbert bound*, Math. Nachr. **109** (1982), 21–28.
25. Gerard van der Geer and Marcel van der Vlugt, *Tables of curves with many points*, Math. Comp. **69** (2000), 797–810.
26. S.G. Vladut and V.G. Drinfel'd, *Number of points of an algebraic curve*, Functional Anal. Appl. **17** (1983), no. 1, 53–54.
27. Daqing Wan, *Computing zeta functions over finite fields*, Contemporary Math. **225** (1999), 131–141.
28. Michael Zieve, Curves of large genus with many points, Lectures given at Arizona Winter School: Arithmetic of function fields, February 2000.

Hyperelliptic curves of genus 3 and 4 in characteristic 2

Yasin Demirbas*

Department of Mathematics,
Boston University,
111 Cummingston Street,
Boston, MA, 02215
demirbas@math.bu.edu

In this paper we compute the automorphism groups of hyperelliptic curves of genus 3 and 4 in characteristic 2 and display the generators and the automorphisms explicitly.

1. Introduction

As a generalization of group of points of elliptic curves, the Jacobians of hyperelliptic curves are used in cryptography. The attack by Duursma-Gaudry-Morain [2] is feasible if the Jacobian has an automorphism of large order, and to protect against the attack by Gaudry [11], the genus g of the hyperelliptic curve must be smaller than 5. So hyperelliptic curves of genus 2, 3, and 4 are the practical ones for security purposes and one needs to know the automorphism groups as well.

The aim of this paper is to determine the automorphism groups of hyperelliptic curves of genus 3 and 4 over an algebraically closed field K of characteristic 2. In sections §3.1 and §4.1, using methods similar to those used in Igusa [15], we determine the normal forms for these curves. Using the normal forms we determine the reduced automorphism group and the full automorphism group. We give the list of automorphisms and their orders for $g = 4$ in the Appendix.

Our main result is:

*I would like to express my special thanks to my advisor, Prof. Previato for her constant help, support, and encouragement and Prof. Shaska for his help, and suggestions.

Theorem 1: Let C be a genus 3 (respectively 4) hyperelliptic curve defined over an algebraically closed field K of characteristic 2. Then the automorphism group of C is one of the following: C_2, C_4, V_4, $C_2 \times C_2 \times C_2$, C_6, C_{14}, D_{12} (respectively C_2, V_4, C_4, $C_2 \times C_2 \times C_2$, C_6, C_{18}, D_{20}).

In the case of $g = 3$, the automorphism groups have been reported independently by Nart-Sadornil [5] with emphasis on fields of definition. Their results are obtained using methods similar to those used in Cardona-Nart-Pujolas [1].

Notations: Throughout this paper C will denote a hyperelliptic curve of genus 3 or 4 and K an algebraically closed field of characteristic 2. With $\bar{G}$ we will denote the reduced group of automorphisms and G the automorphism group of C regarded as an algebraic variety, C_n, D_n, and V_4 the cyclic group of order n, dihedral group of order n, and the Klein 4-group, respectively. As usual $\mathbf{P}^1$ will denote the projective line over K.

2. Preliminaries

Let C be a hyperelliptic curve of genus g over an algebraically closed field K of characteristic 2. We use an Artin-Schreier generation $y^2+y = R'(x)$. As a consequence of Hasse's theory we can find a rational function $Q(x) \in K(x)$ such that the rational function $R'(x) + Q(x) + Q(x)^2$ has no poles of even order. Let $R(x) = R'(x) + Q(x) + Q(x)^2$ and use the normalized form $y^2+y = R(x)$. y is unique up to transformations of the form $y \longmapsto y+B(x)$, where $B(x)$ is a rational function of x.

Let $\Sigma n_a(a)$ be the polar divisor of $R(x)$ on the projective line, $\mathbf{P}^1$. C is ramified at each a and if P_a is the unique point of C over a then the curve $y^2 + y = R(x)$ has the Different

$$Diff(C/\mathbf{P}^1) = \Sigma(n_a + 1)P_a$$

where the n_a are odd ([6], Prop III.7.8)

$$2g-2 = -2[F : K(x)] + deg(Diff(C/\mathbf{P}^1)) \Longrightarrow deg(Diff(C/\mathbf{P}^1)) = 2g+2$$

Take two hyperelliptic curves, $C : y^2 + y = R(x)$ and $C' : y^2 + y = Q(x)$. Then there are finite morphisms $f_1 : C \longmapsto \mathbf{P}^1$, and $f_2 : C' \longmapsto \mathbf{P}^1$ of degree 2, and there exists a unique automorphism σ of $\mathbf{P}^1$ such that $f_2 = \sigma \circ f_1$. Any isomorphism between these curves has the form

$$(x, y) \longmapsto (\frac{ax + b}{cx + d}, y + B(x))$$

for some $B(x) \in K(x)$. Hence, these curves are isomorphic if and only if

$$Q(x) = R(\frac{ax+b}{cx+d}) + S(x) + S(x)^2$$

The ramification types determine the isomorphism classes of the hyperelliptic curves. The solutions of the equation $\Sigma(n_a+1) = 2g+2$ in the unknown odd positive integers give us the following ramification types:

$$\begin{array}{l} (1,1,1,1),(3,1,1),(3,3),(5,1),(7) \quad \textit{for genus } 3 \\ (1,1,1,1,1),(3,1,1,1),(3,3,1),(5,1,1),(5,3),(7,1),(9) \quad \textit{for genus } 4 \end{array} \tag{1}$$

Therefore we get the following normal forms for genus 3 and 4 respectively.

$$y^2+y = \begin{cases} \alpha_1 x + \alpha_2 x^{-1} + \alpha_3(x-1)^{-1} + \alpha_4(x-\lambda)^{-1} \\ x^3 + \alpha x + \beta x^{-1} + \gamma(x-1)^{-1} \\ x^3 + \alpha x + \beta x^{-3} + \gamma x^{-1} \\ x^5 + \alpha x^3 + \beta x^{-1} \\ x^7 + \alpha x^5 + \beta x^3 \end{cases}$$

$$y^2+y = \begin{cases} \alpha_1 x + \alpha_2 x^{-1} + \alpha_3(x-1)^{-1} + \alpha_4(x-\lambda)^{-1} + \alpha_5(x-\mu)^{-1} \\ x^3 + \alpha x + \beta_1 x^{-1} + \beta_2(x-1)^{-1} + \beta_3(x-\lambda)^{-1} \\ x^3 + \alpha x + \beta x^{-3} + \gamma x^{-1} + \sigma(x-1)^{-1} \\ x^5 + \alpha x^3 + \beta x^{-1} + \gamma(x-1)^{-1} \\ x^5 + \alpha x^3 + \beta x^{-3} + \gamma x^{-1} \\ x^7 + \alpha x^5 + \beta x^3 + \gamma x^{-1} \\ x^9 + \alpha_1 x^7 + \alpha_2 x^5 + \alpha_3 x^3 \end{cases} \tag{2}$$

These are plane curves given in inhomogeneous form, birational to the given nonsingular curves(i.e. the function fields are isomorphic). We will use the above normal forms to determine $\bar{G}$, the reduced group of automorphisms, namely the quotient of the group of automorphisms, G by $\langle \iota \rangle$ which is contained in the center of G. And then we will compute G.

3. Hyperelliptic curves of genus 3

3.1. *Normal forms*

Let C be a genus 3 hyperelliptic curve defined over K. The degree of the different satisfies $\Sigma(n_a+1) = 8$ with n_a odd. And this equation has the following solutions:
(A) (1, 1, 1, 1),
(B) (3, 1, 1),

(C) (3, 3),
(D) (5, 1),
(E) (7)
These are the ramification types in the case of $g = 3$, and therefore we get the following normal forms in each case:

(A) $(1,1,1,1)$: C transforms birationally into

$$y^2 + y = \alpha_1 x + \alpha_2 x^{-1} + \alpha_3 (x-1)^{-1} + \alpha_4 (x-\lambda)^{-1}$$

for some $\alpha_i \neq 0$.

(B) $(3,1,1)$: First transform C birationally into

$$y^2 + y = x^3 + \alpha' x^2 + \beta' x + \gamma' x^{-1} + \sigma' (x-1)^{-1}$$

then use $y \longmapsto y + kx$ where $k^2 = \alpha'$ to transform it into

$$y^2 + y = x^3 + \alpha x + \beta x^{-1} + \gamma (x-1)^{-1}$$

for some β, $\gamma \neq 0$.

(C) $(3,3)$: First transform C birationally into

$$y^2 + y = x^3 + \alpha' x^2 + \beta' x + \gamma' x^{-3} + \mu' x^{-2} + \sigma' x^{-1}$$

then use $y \longmapsto y + mx + nx^{-1}$, where $m^2 = \alpha', n^2 = \mu'$, to transform it into

$$y^2 + y = x^3 + \alpha x + \beta x^{-3} + \gamma x^{-1}$$

for some $\beta \neq 0$.

(D) $(5,1)$: First transform C birationally into

$$y^2 + y = x^5 + \alpha' x^4 + \beta' x^3 + \gamma' x^2 + \sigma' x + \nu' x^{-1}$$

and then use $y \longmapsto y + mx^2 + nx$ to transform it into

$$y^2 + y = x^5 + \alpha x^3 + \gamma x + \beta x^{-1}$$

and finally transform it into

$$y^2 + y = x^5 + \alpha x^3 + \beta x^{-1}$$

for some $\beta \neq 0$.

(E) (7): First transform C birationally into

$$y^2 + y = x^7 + \alpha' x^6 + \beta' x^5 + \gamma' x^4 + \sigma' x^3 + \mu' x^2 + \nu'$$

and then similar to the above case, transform it into

$$y^2 + y = x^7 + \alpha x^5 + \beta x^3 + \gamma x$$

Then, replace x with $\zeta x + b$, and $y \longmapsto y + c_3 x^3 + c_2 x^2 + c_1 x + c_0$, and choose c_0, c_1, c_2, c_3 and b such that

$$\begin{array}{ll} {c_3}^2 = \zeta^6 b, & {c_1}^2 + c_2 = \zeta^2 b^5 + \beta\zeta^2 b, \\ {c_2}^2 = \zeta^4 b^3 + \alpha\zeta^4 b, & c_1 = \zeta b^6 + \alpha\zeta b^4 + \beta\zeta b^2 + \gamma\zeta, \\ c_3 = \zeta^3 b^4 + \beta\zeta^3, & {c_0}^2 + c_0 = b^7 + \alpha b^5 + \beta b^3 + \gamma b \end{array}$$

3.2. *Group of automorphisms*

Using the above normal forms we will compute the reduced and the full group of automorphisms.

(A) (1, 1, 1, 1): $y^2 + y = \alpha_1 x + \alpha_2 x^{-1} + \alpha_3 (x-1)^{-1} + \alpha_4 (x-\lambda)^{-1}$

Proposition 2: *(a) If λ is not a third root of unity, then we get 4 linear fractional transformations, $\sigma_1, \sigma_4, \sigma_9$, and σ_{12}, permuting the 4 ramification points ∞, 0, 1, λ.*

(b) If λ is a primitive third root of unity, then we get 12 transformations permuting ∞, 0, 1, λ.

The transformations, conditions on the coefficients in order the transformations to lift to automorphisms, and explicit automorphisms $\phi_1, \phi_2, \dots \phi_{12}$, are:

$\sigma_1 : x \longmapsto x$ *and* $\phi_1(x, y) = (x, y)$

$\sigma_2 : x \longmapsto \lambda^2 x + 1$ *gives* $\lambda = 1$, *and it does not lift to an automorphism.*

$\sigma_3 : x \longmapsto \lambda x + \lambda$ *gives* $\lambda = 1$, *and it does not lift to an automorphism.*

$\sigma_4 : x \longmapsto \frac{\lambda}{x}$, *and* $\alpha_1 = \lambda^{-1}\alpha_2$, $\alpha_3 = \lambda^{-1}\alpha_4$, *and* $\phi_4(x, y) = (\frac{\lambda}{x}, y)$

$\sigma_5 : x \longmapsto \frac{x+1}{x}$ *gives* $\lambda = 1$, *and it does not lift to an automorphism.*

$\sigma_6 : x \longmapsto \frac{\lambda x + \lambda^2}{x}$ *gives* $\lambda = 1$, *and it does not lift to an automorphism.*

$\sigma_7 : x \longmapsto \frac{1}{x+1}$ *gives* $\lambda = 1$, *and it does not lift to an automorphism.*

$\sigma_8 : x \longmapsto \frac{\lambda x}{x+1}$ *gives* $\lambda = 1$, *and it does not lift to an automorphism.*

$\sigma_9 : x \longmapsto \frac{x+\lambda}{x+1}$, *and* $\alpha_1 = \lambda\alpha_3$, $\alpha_2 = \lambda\alpha_4$, *and* $\phi_9(x, y) = (\frac{x+\lambda}{x+1}, y)$

$\sigma_{10} : x \longmapsto \frac{\lambda^2}{x+\lambda}$ *gives* $\lambda = 1$, *and it does not lift to an automorphism.*
$\sigma_{11} : x \longmapsto \frac{x}{x+\lambda}$ *gives* $\lambda = 1$, *and it does not lift to an automorphism.*
$\sigma_{12} : x \longmapsto \frac{\lambda x+\lambda}{x+\lambda}$, *and* $\alpha_1 = \alpha_4$, $\alpha_2 = \lambda^2\alpha_3$, *and* $\phi_{12}(x, y) = (\frac{\lambda x+\lambda}{x+\lambda}, y)$.

Proof: Out of 24 possible linear fractional transformations, determining those permuting the 4 ramification points $\infty, 0, 1, \lambda$ is a direct computation. But the details are lengthy and therefore the proof is omitted. □

The transformations $\sigma_1, \sigma_4, \sigma_9$, and σ_{12} lift to automorphisms ϕ_1, ϕ_4, ϕ_9, and ϕ_{12} respectively for both of the cases (a), and (b). Hence, both of these cases give us the same automorphism groups. Since ${\phi_4}^2 = {\phi_9}^2 = {\phi_{12}}^2 = id$, we get:

(i) If $\alpha_1 = \alpha_2\lambda^{-1}$, $\alpha_3 = \alpha_4\lambda^{-1}$, and $\alpha_1 \neq \alpha_3\lambda$ then $\bar{G} = \{\phi_1, \phi_4\} \cong C_2$, and $G = \langle \iota, \phi_4 \rangle \cong V_4$.

(ii) If $\alpha_1 = \alpha_3\lambda$, $\alpha_2 = \alpha_4\lambda$, and $\alpha_1 \neq \alpha_2\lambda^{-1}$ then $\bar{G} = \{\phi_1, \phi_9\} \cong C_2$, and $G = \langle \iota, \phi_9 \rangle \cong V_4$.

(iii) If $\alpha_1 = \alpha_4$, $\alpha_2 = \alpha_3\lambda^2$, and $\alpha_1 \neq \alpha_2\lambda^{-1}$ then $\bar{G} = \{\phi_1, \phi_{12}\} \cong C_2$, and $G = \langle \iota, \phi_{12} \rangle \cong V_4$.

(iv) If $\alpha_1 = \alpha_2\lambda^{-1} = \alpha_3\lambda = \alpha_4$ then
$\bar{G} = \{\phi_1, \phi_4, \phi_9, \phi_{12}\} \cong C_2 \times C_2$, and $G = \langle \iota, \phi_4, \phi_9 \rangle \cong C_2 \times C_2 \times C_2$.

(B) $(3, 1, 1)$: $y^2 + y = x^3 + \alpha x + \beta x^{-1} + \gamma(x-1)^{-1}$

∞ is fixed and $0 \longmapsto 1 \implies x \longmapsto x + 1$. We have the following cases

(i) If $\alpha \neq 0$ or $\beta \neq \gamma$, then $\bar{G}$ is the identity, and therefore $G = \langle \iota \rangle \cong C_2$.

(ii) If $\alpha = 0$ and $\beta = \gamma$, then $\bar{G} = \{id, \phi\} \cong C_2$
where $\phi(x, y) = (x + 1, y + x + w)$ with $w^2 + w = 1$
$\phi^2(x, y) = (x, y + 1) \Rightarrow \phi^2 = \iota$. Therefore $G = \langle \phi \rangle \cong C_4$.

(C) $(3, 3)$: $y^2 + y = x^3 + \alpha x + \beta x^{-3} + \gamma x^{-1}$

In this case the transformations, the conditions on the coefficients in order the transformations to lift to automorphisms, and automorphisms $\phi_1, \phi_2, \ldots \phi_6$, are:

$\sigma_1 : x \longmapsto x$ and $\phi_1(x, y) = (x, y)$
$\sigma_2 : x \longmapsto \zeta x$ and $\alpha = \gamma = 0$ and $\phi_2(x, y) = (\zeta x, y)$
$\sigma_3 : x \longmapsto \zeta^2 x$ and $\alpha = \gamma = 0$ and $\phi_3(x, y) = (\zeta^2 x, y)$
$\sigma_4 : x \longmapsto \frac{1}{x}$ and $\beta = 1$, $\alpha = \gamma$ and $\phi_4(x, y) = (\frac{1}{x}, y)$
$\sigma_5 : x \longmapsto \frac{1}{\zeta x}$ and $\beta = 1$, $\alpha = \gamma\zeta$ and $\phi_5(x, y) = (\frac{1}{\zeta x}, y)$
$\sigma_6 : x \longmapsto \frac{1}{\zeta^2 x}$ and $\beta = 1$, $\alpha = \gamma\zeta^{-1}$ and $\phi_6(x, y) = (\frac{1}{\zeta^2 x}, y)$
where $\zeta^3 = 1$. Since ${\phi_2}^3 = {\phi_3}^3 = {\phi_4}^2 = {\phi_5}^2 = {\phi_6}^2 = id$, we get the following:

(i) If $\beta \neq 1$, and $\alpha = \gamma = 0$, then $\bar{G} = \{\phi_1, \phi_2, \phi_3\} \cong C_3$, and therefore $G = \langle \iota\phi_2 \rangle \cong C_6$.

(ii) If $\beta = 1$, and $\alpha = \gamma \neq 0$, then $\bar{G} = \{\phi_1, \phi_4\} \cong C_2$, and therefore $G = \langle \iota, \phi_4 \rangle \cong V_4$.

(iii) If $\beta = 1$, and $\alpha = \gamma\zeta \neq 0$, then $\bar{G} = \{\phi_1, \phi_5\} \cong C_2$, and therefore $G = \langle \iota, \phi_5 \rangle \cong V_4$.

(iv) If $\beta = 1$, and $\alpha = \gamma\zeta^2 \neq 0$, then $\bar{G} = \{\phi_1, \phi_6\} \cong C_2$, and therefore $G = \langle \iota, \phi_6 \rangle \cong V_4$.

(v) If $\beta = 1$, and $\alpha = \gamma = 0$, then $\bar{G} = \{\phi_1, \phi_2, \phi_3, \phi_4, \phi_5, \phi_6\} = \langle \phi_2, \phi_4 \rangle \cong D_6$, and $G = \langle \iota\phi_2, \phi_4 \rangle \cong D_{12}$.

Otherwise $\bar{G}$ is the identity and $G = \langle \iota \rangle \cong C_2$.

(D) $(5, 1)$: $y^2 + y = x^5 + \alpha x^3 + \beta x^{-1}$

∞, and 0 are fixed $\Longrightarrow x \longmapsto \zeta x$, where ζ is a fifth root of unity. But then we get $\beta\zeta^{-1} = \beta$. Since β is nonzero we get $\zeta = 1$. Thus, $\bar{G}$ is the identity and $G = \langle \iota \rangle \cong C_2$.

(E) (7): $y^2 + y = x^7 + \alpha x^5 + \beta x^3$

We replace x with $ax + b$, and y with $y + c_3x^3 + c_2x^2 + c_1x + c_0$ and require that the new relation be $y^2 + y = x^7 + \alpha x^5 + \beta x^3$, then

$$
\begin{aligned}
&a^7 = 1, && a^3b^4 + \beta a^3 + c_3 = \beta,\\
&{c_3}^2 = a^6 b, && a^2b^5 + \beta a^2 b = {c_1}^2 + c_2.\\
&a^5b^2 + \alpha a^5 = \alpha, && ab^6 + \alpha ab^4 + \beta ab^2 = c_1,\\
&a^4b^3 + \alpha a^4 b = {c_2}^2, && b^7 + \alpha b^5 + \beta b^3 = {c_0}^2 + c_0
\end{aligned}
$$

From these we get $a = \zeta^3$ where $\zeta^7 = 1$, and ${c_3}^2 = a^6 b \Longrightarrow b = \zeta^3 {c_3}^2$ Thus, $x \longmapsto \zeta^3(x + {c_3}^2)$, $\quad y \longmapsto y + c_3x^3 + c_2x^2 + c_1x + c_0$, and

$$
\begin{aligned}
&\alpha = \alpha\zeta + {c_3}^4, && {c_2}^2 = {c_3}^2\alpha,\\
&\beta = c_3 + {c_3}^8 + \beta\zeta^2, && {c_1}^2 + c_2 = {c_3}^3 + {c_3}^2\beta,\\
&c_1 = {c_3}^4(\alpha {c_3}^4 + \beta\zeta^2), && {c_0}^2 + c_0 = {c_3}^2 c_1
\end{aligned}
$$

And we want $\alpha = \alpha\zeta + {c_3}^4 = \alpha \Longrightarrow \alpha(\zeta + 1) = {c_3}^4$, and $\beta = c_3 + {c_3}^8 + \beta\zeta^2 \Longrightarrow \beta(\zeta^2 + 1) = {c_3}^8 + c_3$. Once ζ is fixed c_3 is determined uniquely as $c_3 = \beta(\zeta^2 + 1) + \alpha^2(\zeta^2 + 1)$. We have two cases:

(i) $\alpha = 0 \Longrightarrow c_3 = 0$.

If $\beta = 0$ then ζ is arbitrary. $\bar{G} = \langle \phi \rangle \cong C_7$, where $\phi(x, y) = (\zeta^3 x, y)$. And $G = \langle \iota\phi \rangle \cong C_{14}$.

If $\beta \neq 0$ then $\zeta = 1$. Hence $\bar{G}$ is the identity and therefore $G = \langle \iota \rangle \cong C_2$.

(ii) $\alpha \neq 0$

In this case if $c_3 = 0$ then $\zeta = 1$ and $\bar{G}$ is the identity, $G = \langle \iota \rangle \cong C_2$. Otherwise $\bar{G} = \langle \phi \rangle \cong C_7$, where $\phi(x, y) = (\zeta^3(x + {c_3}^2), y + c_3x^3 + c_2x^2 + c_1x + c_0)$. And $G = \langle \iota\phi \rangle \cong C_{14}$.

All of the above are summarized in the following theorem and table:

Ramification Type	Curve	Condition	G
$(1,1,1,1)$	$y^2+y=\alpha_1 x+\alpha_2 x^{-1}$ $+\alpha_3(x-1)^{-1}+\alpha_4(x-\lambda)^{-1}$	$\alpha_1=\alpha_2\lambda^{-1}, \alpha_3=\alpha_4\lambda^{-1}, \alpha_1\neq\alpha_3\lambda$	V_4
		$\alpha_1=\alpha_3\lambda,\ \alpha_2=\alpha_4\lambda,\ \alpha_1\neq\alpha_2\lambda^{-1}$	V_4
		$\alpha_1=\alpha_4,\ \alpha_2=\alpha_3\lambda^2,\ \alpha_1\neq\alpha_2\lambda^{-1}$	V_4
		$\alpha_1=\alpha_2\lambda^{-1}, \alpha_3=\alpha_4\lambda^{-1}, \alpha_1=\alpha_3\lambda$	${C_2}^3$
$(3,1,1)$	$y^2+y=x^3+\alpha x$ $+\beta x^{-1}+\gamma(x-1)^{-1}$	$\alpha\neq 0$, or $\beta\neq\gamma$	C_2
		$\alpha=0$, and $\beta=\gamma$	V_4
$(3,3)$	$y^2+y=x^3+\alpha x$ $+x^{-3}+\beta x^{-1}$	none	C_2
		$\beta\neq 1, \alpha=\gamma=0$	C_6
		$\beta=1, \alpha=\gamma\neq 0$	V_4
		$\beta=1, \alpha=\gamma\zeta\neq 0$	V_4
		$\beta=1, \alpha=\gamma\zeta^2\neq 0$	V_4
		$\beta=1, \alpha=\gamma=0$	D_{12}
$(5,1)$	$y^2+y=x^5+\alpha x^3+\beta x^{-1}$	none	C_2
(7)	$y^2+y=x^7+\alpha x^5+\beta x^3$	$\alpha=\beta=0$	C_{14}
		$\alpha=0, \beta\neq 0$	C_2
		$\alpha\neq 0, \beta=c_3=0$	C_2
		$\alpha\neq 0, \beta=0, c_3\neq 0$	C_{14}
		$\alpha\neq 0, \beta\neq 0, c_3=0$	C_2
		$\alpha\neq 0, \beta\neq 0, c_3\neq 0$	C_{14}

Theorem 3: Let C be a genus 3 hyperelliptic curve defined over an algebraically closed field K of characteristic 2. Then the automorphism group of C is one of the following: C_2, C_4, V_4, $C_2 \times C_2 \times C_2$, C_6, C_{14}, D_{12}.

4. Hyperelliptic curves of genus 4

4.1. *Normal forms*

Let C be a genus 4 hyperelliptic curve defined over K. The degree of the different satisfies

$$\Sigma(n_a + 1) = 10$$

with n_a odd. And this equation has the following solutions:
(A) (1, 1, 1, 1, 1),
(B) (3, 1, 1, 1),
(C) (3, 3, 1),
(D) (5, 1, 1),
(E) (5, 3),
(F) (7, 1),
(G) (9)

These are the ramification types in the case of $g = 4$. Except for case (G), we will use the normal forms in section 3.1 to get the following normal forms for $g = 4$:

(A) $(1, 1, 1, 1, 1)$: C transforms birationally into

$$y^2 + y = \alpha_1 x + \alpha_2 x^{-1} + \alpha_3 (x-1)^{-1} + \alpha_4 (x-\lambda)^{-1} + \alpha_5 (x-\mu)^{-1}$$

for some $\alpha_i \neq 0$.

(B) $(3, 1, 1, 1)$: C transforms birationally into

$$y^2 + y = x^3 + \alpha x + \beta_1 x^{-1} + \beta_2 (x-1)^{-1} + \beta_3 (x-\lambda)^{-1}$$

for some $\beta_i \neq 0$.

(C) $(3, 3, 1)$: C transforms birationally into

$$y^2 + y = x^3 + \alpha x + \beta x^{-3} + \gamma x^{-1} + \sigma (x-1)^{-1}$$

for some β, $\sigma \neq 0$.

(D) $(5, 1, 1)$: C transforms birationally into

$$y^2 + y = x^5 + \alpha x^3 + \beta x^{-1} + \gamma (x-1)^{-1}$$

for some β, $\gamma \neq 0$.

(E) (5, 3): C transforms birationally into

$$y^2 + y = x^5 + \alpha x^3 + \beta x^{-3} + \gamma x^{-1}$$

for some $\beta \neq 0$.

(F) (7, 1): C transforms birationally into

$$y^2 + y = x^7 + \alpha x^5 + \beta x^3 + \gamma x^{-1}$$

for some $\gamma \neq 0$.

(G) (9): First transform C birationally into

$$y^2 + y = x^9 + \alpha_1' x^8 + \alpha_2' x^7 + \alpha_3' x^6 + \alpha_4' x^5 + \alpha_5' x^4 + \alpha_6' x^3 + \alpha_7' x^2 + \alpha_8' x + \alpha_9'$$

and then similar to the case (7) in section 4.1, transform into

$$y^2 + y = x^9 + \alpha_1 x^7 + \alpha_2 x^5 + \alpha_3 x^3 + \alpha_4 x$$

Then, use $x \longmapsto \zeta x + b$, $y \longmapsto y + c_4 x^4 + c_3 x^3 + c_2 x^2 + c_1 x + c_0$, to transform C into

$$y^2 + y = x^9 + \alpha_1 x^7 + \alpha_2 x^5 + \alpha_3 x^3$$

4.2. *Group of automorphisms*

(A) (1, 1, 1, 1, 1): $y^2 + y = \alpha_1 x + \alpha_2 x^{-1} + \alpha_3 (x-1)^{-1} + \alpha_4 (x - \lambda_1)^{-1} + \alpha_5 (x - \lambda_2)^{-1}$

Proposition 4: *We have the following linear fractional transformations permuting the 5 ramification points ∞, 0, 1, λ, and μ*

(a) If $\mu = \lambda + 1$, then we get 4 transformations σ_1, σ_2', σ_3', σ_4'. When λ, and μ are the two primitive third roots of unity σ_2', σ_3', σ_4' become σ_2, σ_3, σ_4 respectively.

$\sigma_1 : x \longmapsto x$
$\sigma_2' : x \longmapsto x + 1$
$\sigma_3' : x \longmapsto x + \lambda$
$\sigma_4' : x \longmapsto x + \lambda + 1$

(b) If $\mu = \frac{\lambda}{\lambda+1}$, then we get 4 transformations σ_1, σ_{25}', σ_{43}', σ_{55}'. When λ, and μ are the two primitive third roots of unity σ_{25}', σ_{43}', σ_{55}' become σ_{25}, σ_{43}, σ_{55} respectively.

$\sigma_1 : x \longmapsto x$
$\sigma_{25}' : x \longmapsto \frac{x}{x+1}$
$\sigma_{43}' : x \longmapsto \frac{\lambda x}{x+\lambda}$
$\sigma_{55}' : x \longmapsto \frac{\mu x}{x+\mu}$

(c) If $\mu = \lambda^{-1}$, *then we get* 4 *transformations* σ_1, σ'_{23}, σ'_{42}, σ'_{56}. *When* λ, *and* μ *are the two primitive third roots of unity* σ'_{23}, σ'_{42}, σ'_{56} *become* σ_{23}, σ_{42}, σ_{56} *respectively.*

$\sigma_1 : x \longmapsto x$
$\sigma'_{23} : x \longmapsto \frac{1}{x}$
$\sigma'_{42} : x \longmapsto \frac{\lambda x+1}{x+\lambda}$
$\sigma'_{56} x \longmapsto \frac{\mu x+1}{x+\mu}$

(d) If $\mu = \lambda^2$, *then we get* 4 *transformations* σ_1, σ'_{24}, σ'_{28}, σ'_{57}. *When* λ, *and* μ *are the two primitive third roots of unity* σ'_{24}, σ'_{28}, σ'_{57} *become* σ_{24}, σ_{28}, σ_{57} *respectively.*

$\sigma_1 : x \longmapsto x$
$\sigma'_{24} : x \longmapsto \frac{\lambda^2}{x}$
$\sigma'_{28} : x \longmapsto \frac{x+\mu}{x+1}$
$\sigma'_{57} : x \longmapsto \frac{\mu x+\mu}{x+\mu}$

(e) If $\mu^2 = \lambda$, *then we get* 4 *transformations* σ_1, σ'_{13}, σ'_{26}, σ'_{45}. *When* λ, *and* μ *are the two primitive third roots of unity* σ'_{13}, σ'_{26}, σ'_{45} *become* σ_{13}, σ_{26}, σ_{45} *respectively.*

$\sigma_1 : x \longmapsto x$
$\sigma'_{13} : x \longmapsto \frac{\mu^2}{x}$
$\sigma'_{26} : x \longmapsto \frac{x+\lambda}{x+1}$
$\sigma'_{45} : x \longmapsto \frac{\lambda x+\lambda}{x+\lambda}$

(f) If λ, *and* μ *are the two distinct primitive third roots of unity, then we get* 60 *transformations* σ_1, σ_2, $\cdots$, σ_{60} *permuting.*

Proof: The proof is by a direct computation. The details are lengthy and therefore the proof is omitted. See the Appendix for the list of all the transformations. □

We get 60 transformations permuting the ramification points ∞, 0, 1, λ, and μ. But only 40 of them lift to automorphisms of the curve. Although it suffices to determine the generators and the order of lifts to determine the automorphism groups, we want to display the 60 linear fractional transformations, and the explicit automorphisms labeled as ϕ_1, ϕ_2, $\cdots$, ϕ_{60}. See the Appendix. Now, we will compute the reduced and full group of automorphisms for each case of Prop 3.1.

We will consider 2 cases.

Case 1: If λ, and μ are not third roots of 1 then each set of automorphisms of parts (a) - (e) of Proposition 4 give the same groups, and only the conditions on the coefficients α_i change. We will give the automorphism groups for part (a) only. The rest is treated in the same way.

(i) If $\alpha_2 = \alpha_3$, $\alpha_4 = \alpha_5$, and $\alpha_2 \neq \alpha_4$, then $\bar{G} = \{\phi_1, \phi_2\} \cong C_2$. And $G = \langle \iota, \phi_2 \rangle \cong V_4$.

(ii) If $\alpha_2 = \alpha_4$, $\alpha_3 = \alpha_5$, and $\alpha_2 \neq \alpha_3$, then $\bar{G} = \{\phi_1, \phi_3\} \cong C_2$. And $G = \langle \iota, \phi_3 \rangle \cong V_4$.

(iii) If $\alpha_2 = \alpha_5$, $\alpha_3 = \alpha_4$, and $\alpha_2 \neq \alpha_3$, then $\bar{G} = \{\phi_1, \phi_4\} \cong C_2$. And $G = \langle \iota, \phi_4 \rangle \cong V_4$.

(iv) If $\alpha_2 = \alpha_3 = \alpha_4 = \alpha_5$, then $\bar{G} = \{\phi_1, \phi_2, \phi_3, \phi_4\} \cong V_4$. And $G = \langle \iota, \phi_2, \phi_3 \rangle \cong C_2 \times C_2 \times C_2$.

Case 2: If $\lambda^3 = \mu^3 = 1$ then we have 40 automorphisms. And, automorphisms of order 5 form 6 disjoint sets under the following set of conditions respectively:

$\{\alpha_1 = \alpha_2,\ \alpha_2 = \alpha_3\lambda\}$ $\{\alpha_1 = \alpha_2,\ \alpha_2 = \alpha_3\lambda^{-1}\}$
$\{\alpha_1 = \alpha_2\lambda,\ \alpha_2 = \alpha_3\}$ $\{\alpha_1 = \alpha_2\lambda,\ \alpha_2 = \alpha_3\lambda^{-1}\}$
$\{\alpha_1 = \alpha_2\lambda^{-1},\ \alpha_2 = \alpha_3\}$ $\{\alpha_1 = \alpha_2\lambda^{-1},\ \alpha_2 = \alpha_3\lambda\}$

And under these conditions, automorphisms of order 2 can become elements of some of these sets as we can see in (i) - (vi) below.

(i) If $\alpha_1 = \alpha_2$, $\alpha_2 = \alpha_3\lambda$, $\alpha_3 = \alpha_4$, and $\alpha_4 = \alpha_5\lambda^{-1}$, then $\bar{G} = \{\phi_1, \phi_4, \phi_{20}, \phi_{23}, \phi_{26}, \phi_{27}, \phi_{38}, \phi_{43}, \phi_{57}, \phi_{58}\} = \langle \phi_{20}, \phi_{23} \rangle \cong D_{10}$. And $G = \langle \iota\phi_{20}, \phi_{23} \rangle \cong D_{20}$.

(ii) If $\alpha_1 = \alpha_2$, $\alpha_2 = \alpha_3\lambda^{-1}$, $\alpha_3 = \alpha_4\lambda$, and $\alpha_4 = \alpha_5\lambda^{-1}$, then $\bar{G} = \{\phi_1, \phi_3, \phi_{16}, \phi_{28}, \phi_{32}, \phi_{39}, \phi_{45}, \phi_{50}, \phi_{55}\} = \langle \phi_{16}, \phi_{23} \rangle \cong D_{10}$. And $G = \langle \iota\phi_{16}, \phi_{23} \rangle \cong D_{20}$.

(iii) If $\alpha_1 = \alpha_2\lambda$, $\alpha_2 = \alpha_3$, $\alpha_3 = \alpha_4\lambda$, and $\alpha_4 = \alpha_5$, then $\bar{G} = \{\phi_1, \phi_2, \phi_{17}, \phi_{24}, \phi_{26}, \phi_{36}, \phi_{42}, \phi_{48}, \phi_{54}, \phi_{55}\} = \langle \phi_{17}, \phi_{24} \rangle \cong D_{10}$. And $G = \langle \iota\phi_{17}, \phi_{24} \rangle \cong D_{20}$.

(iv) If $\alpha_1 = \alpha_2\lambda$, $\alpha_2 = \alpha_3\lambda^{-1}$, $\alpha_3 = \alpha_4$, and $\alpha_4 = \alpha_5\lambda$, then $\bar{G} = \{\phi_1, \phi_4, \phi_{22}, \phi_{24}, \phi_{25}, \phi_{30}, \phi_{44}, \phi_{45}, \phi_{56}, \phi_{60}\} = \langle \phi_{22}, \phi_{24} \rangle \cong D_{10}$. And $G = \langle \iota\phi_{22}, \phi_{24} \rangle \cong D_{20}$.

(v) If $\alpha_1 = \alpha_2\lambda^{-1}$, $\alpha_2 = \alpha_3$, $\alpha_3 = \alpha_4\lambda^{-1}$, and $\alpha_4 = \alpha_5$, then $\bar{G} = \{\phi_1, \phi_2, \phi_{13}, \phi_{15}, \phi_{28}, \phi_{35}, \phi_{43}, \phi_{47}, \phi_{53}, \phi_{56}\} = \langle \phi_{13}, \phi_{15} \rangle \cong D_{10}$. And $G = \langle \iota\phi_{15}, \phi_{13} \rangle \cong D_{20}$.

(vi) If $\alpha_1 = \alpha_2\lambda^{-1}$, $\alpha_2 = \alpha_3\lambda$, $\alpha_3 = \alpha_4\lambda^{-1}$, $\alpha_4 = \alpha_5\lambda$, then $\bar{G} = \{\phi_1, \phi_3, \phi_{13}, \phi_{18}, \phi_{25}, \phi_{33}, \phi_{40}, \phi_{42}, \phi_{51}, \phi_{57}\} = \langle \phi_{13}, \phi_{18} \rangle \cong D_{10}$. And $G = \langle \iota\phi_{18}, \phi_{13} \rangle \cong D_{20}$.

Now it remains to group the automorphisms of order 2 under certain conditions. Again we will only display the list under the condition $\alpha_1 = \alpha_2 = \alpha_3$ for the sake of simplicity. The other cases are treated in a similar way.

(i) If $\alpha_1 = \alpha_2 = \alpha_3 = \alpha_4 = \alpha_5$, then $\bar{G} = \{\phi_1, \phi_2, \phi_3, \phi_4\} \cong V_4$. And $G = \langle \iota, \phi_2, \phi_3 \rangle \cong C_2 \times C_2 \times C_2$.

(ii) If $\alpha_1 = \alpha_2 = \alpha_3 = \alpha_4 = \alpha_5\lambda$, then $\bar{G} = \{\phi_1, \phi_{25}\} \cong C_2$. And $G = \langle \iota, \phi_{25} \rangle \cong V_4$.

(iii) If $\alpha_1 = \alpha_2 = \alpha_3 = \alpha_4 = \alpha_5\lambda^{-1}$, then $\bar{G} = \{\phi_1, \phi_{23}\} \cong C_2$. And $G = \langle \iota, \phi_{23} \rangle \cong V_4$.

(iv) If $\alpha_1 = \alpha_2 = \alpha_3 = \alpha_4\lambda = \alpha_5$, then $\bar{G} = \{\phi_1, \phi_{23}\} \cong C_2$. And $G = \langle \iota, \phi_{23} \rangle \cong V_4$.

(v) If $\alpha_1 = \alpha_2 = \alpha_3 = \alpha_4\lambda = \alpha_5\lambda$, then $\bar{G} = \{\phi_1, \phi_2\} \cong C_2$. And $G = \langle \iota, \phi_2 \rangle \cong V_4$.

(vi) If $\alpha_1 = \alpha_2 = \alpha_3 = \alpha_4\lambda = \alpha_5\lambda^{-1}$, then $\bar{G} = \{\phi_1, \phi_{25}, \phi_{43}, \phi_{55}\} \cong V_4$. And $G = \langle \iota, \phi_{25}, \phi_{43} \rangle \cong C_2 \times C_2 \times C_2$.

(vii) If $\alpha_1 = \alpha_2 = \alpha_3 = \alpha_4\lambda^{-1} = \alpha_5$, then $\bar{G} = \{\phi_1, \phi_{25}\} \cong C_2$. And $G = \langle \iota, \phi_{25} \rangle \cong V_4$.

(viii) If $\alpha_1 = \alpha_2 = \alpha_3 = \alpha_4\lambda^{-1} = \alpha_5\lambda$, then $\bar{G} = \{\phi_1, \phi_{23}, \phi_{42}, \phi_{56}\} \cong V_4$. And $G = \langle \iota, \phi_{23}, \phi_{42} \rangle \cong C_2 \times C_2 \times C_2$.

(ix) If $\alpha_1 = \alpha_2 = \alpha_3 = \alpha_4\lambda^{-1} = \alpha_5\lambda^{-1}$, then $\bar{G} = \{\phi_1, \phi_2\} \cong C_2$. And $G = \langle \iota, \phi_2 \rangle \cong V_4$.

(x) If $\alpha_1 = \alpha_2 = \alpha_3$, $\alpha_3 \neq \alpha_4\beta$, where $\beta = 1$, λ, or λ^{-1}, and $\alpha_4 = \alpha_5$, then $\bar{G} = \{\phi_1, \phi_2\} \cong C_2$. And $G = \langle \iota, \phi_2 \rangle \cong V_4$.

(xi) If $\alpha_1 = \alpha_2 = \alpha_3$, $\alpha_3 \neq \alpha_4\beta$, where $\beta = 1$, λ, or λ^{-1}, and $\alpha_4 = \alpha_5\lambda$, then $\bar{G} = \{\phi_1, \phi_{25}\} \cong C_2$. And $G = \langle \iota, \phi_{25} \rangle \cong V_4$.

(xii) If $\alpha_1 = \alpha_2 = \alpha_3$, $\alpha_3 \neq \alpha_4\beta$, where $\beta = 1$, λ, or λ^{-1}, and $\alpha_4 = \alpha_5\lambda^{-1}$, then $\bar{G} = \{\phi_1, \phi_{23}\} \cong C_2$. And $G = \langle \iota, \phi_{23} \rangle \cong V_4$.

(B) $(3, 1, 1, 1)$: $y^2 + y = x^3 + \alpha x + \beta_1 x^{-1} + \beta_2(x-1)^{-1} + \beta_3(x-\lambda)^{-1}$

∞ is fixed, so from $(1, 1, 1, 1)$ case of $g = 3$ we get:

$\phi_1(x, y) = (\lambda^2 x + 1, y + \lambda^2 x + \lambda)$, with $\alpha = 0$, $\beta = \gamma\lambda$, $\gamma = \sigma\lambda$, $\sigma = \beta\lambda$

$\phi_2(x, y) = (\lambda x + \lambda, y + x + \lambda)$, with $\alpha = 0$, $\beta = \sigma\lambda^2$, $\gamma = \beta\lambda^2$, $\sigma = \gamma\lambda^2$

where $\phi_1 = {\phi_2}^2$, ${\phi_2}^3 = \iota$, and λ is a third root of unity. Therefore, if $\alpha = 0$, $\beta = \gamma\lambda$, $\gamma = \sigma\lambda$, and $\sigma = \beta\lambda$ then $\bar{G} = \{id, \phi_2, {\phi_2}^2\} \cong C_3$, and $G = \langle \phi_2 \rangle \cong C_6$.

Otherwise, $\bar{G}$ is the identity, and therefore $G = \langle \iota \rangle \cong C_2$.

(C) $(3, 3, 1)$: $y^2 + y = x^3 + \alpha x + \beta x^{-3} + \gamma x^{-1} + \sigma(x-1)^{-1}$

Since 1 is fixed, from $(3, 3)$ of $g = 3$ we get $x \longmapsto \frac{1}{x}$. $\bar{G} = \{id, \phi\} \cong C_2$, where $\phi(x, y) = (\frac{1}{x}, y + w)$ with $w^2 + w = \sigma$, $\beta = 1$, and $\alpha = \gamma$.

Since $\phi^2 = id$, $G = \langle \iota, \phi \rangle \cong V_4$.

(D) $(5, 1, 1)$: $y^2 + y = x^5 + \alpha x^3 + \beta x^{-1} + \gamma(x-1)^{-1}$

In this case ∞ is fixed, $0 \longmapsto 1$, and $1 \longmapsto 0$, so $x \longmapsto x + 1$. If $\alpha = 0$, or 1, then $\bar{G} = \{id, \phi\} \cong C_2$, where $\phi(x, y) = (x + 1, y + x^2 + (\alpha + 1)x + w)$

with $w^2 + w = \alpha + 1$. Otherwise it is the identity. Hence,

(i) If $\alpha = 0$, then $\phi^2 = id$ and therefore $G = \langle \iota, \phi \rangle \cong V_4$.

(ii) If $\alpha = 1$, then $\phi^2 = \iota$ and therefore $G = \langle \phi \rangle \cong C_4$.

Otherwise $G = \langle \iota \rangle \cong C_2$.

(E) $(5, 3)$: $y^2 + y = x^5 + \alpha x^3 + \beta x^{-3} + \gamma x^{-1}$

∞, and 0 are fixed, so $x \longmapsto \zeta x$ where ζ is a third root of unity. But then we get $\alpha\zeta^3 = \alpha$, and since $\alpha \neq 0$ we get $\zeta = 1$. Thus $\bar{G}$ is the identity, and $G = \langle \iota \rangle \cong C_2$.

(F) $(7, 1)$: $y^2 + y = x^7 + \alpha x^5 + \beta x^3 + \gamma x^{-1}$

∞, and 0 are fixed, so $x \longmapsto \zeta x$ where ζ is a seventh root of unity. But then we get $\alpha\zeta^5 = \alpha$, and since $\alpha \neq 0$ we get $\zeta = 1$. Therefore $\bar{G}$ is the identity, and $G = \langle \iota \rangle \cong C_2$.

(G) (9): Similar to the case (E) in section 4.1, replacing x with $ax + b$, and y with $y + c_4x^4 + c_3x^3 + c_2x^2 + c_1x^1 + c_0$, and requiring that the new relation be $y^2 + y = x^9 + \alpha_1 x^7 + \alpha_2 + x^5 \alpha_3 x^3$, we get the following relations:

$a = \zeta^4$ where ζ is a 9th root of unity, $\quad {c_3}^2 = \alpha_1 \zeta {c_4}^2$,

${c_0}^2 + c_0 = \alpha_1 \zeta {c_4}^{14} + \alpha_2 \zeta^2 {c_4}^{10} + \alpha_3 \zeta^3 {c_4}^6$, $\quad {c_4}^2 = \zeta^5 b$,

$c_1 = \alpha_1 \zeta {c_4}^{12} + \alpha_2 \zeta^2 {c_4}^8 + \alpha_3 \zeta^3 {c_4}^4$, $\quad \alpha_1 = \alpha_1 \zeta$,

${c_1}^2 + c_2 = \alpha_1 \zeta {c_4}^{10} + \alpha_3 \zeta^3 {c_4}^2$, $\quad \alpha_2 = \zeta^2(\alpha_2 + \alpha_1 \zeta^8 {c_4}^4)$,

${c_2}^2 + c_4 = \zeta^7(\alpha_1 \zeta^5 {c_4}^6 + \alpha_2 \zeta^4 {c_4}^2)$, $\quad \alpha_3 = c_3 = \alpha_1 \zeta {c_4}^8 + \zeta^3 \alpha_3$.

From the above relations we get $x \longmapsto \zeta^4(x + {c_4}^2)$, where c_4 is any root of the following polynomial which has distinct roots

${\alpha_1}^4 \zeta^4 {c_4}^{48} + {\alpha_2}^4 \zeta^8 {c_4}^{32} + {\alpha_1}^2 \zeta^2 {c_4}^{20} + {\alpha_3}^4 \zeta^3 {c_4}^{16} + \alpha_1 \zeta^3 {c_4}^6 + {\alpha_3}^2 \zeta^6 {c_4}^4 + \alpha_2 \zeta^2 {c_4}^2 + c_4 = 0$.

We also want the following to be satisfied:

$$\alpha_1 \zeta = \alpha_1, \quad \alpha_2(\zeta^2 + 1) + \alpha_1 {c_4}^4 = 0, \quad \alpha_1 {c_4}^8 + \alpha_3(\zeta^3 + 1) = 0.$$

(i): $\alpha_1 = 0 \implies \zeta$ is an arbitrary root of 1, and therefore $\alpha_2 = \alpha_3 = 0$. And from the polynomial we get $c_4 = 0$. Thus in this case the $\bar{G} = \langle \phi \rangle \cong C_9$, where $\phi(x, y) = (\zeta^4 x, y)$, and $G = \langle \iota\phi \rangle \cong C_{18}$.

(ii): $\alpha_1 \neq 0 \implies \zeta = 1$, and $c_4 = 0$. So $\bar{G}$ is the identity, and $G = \langle \iota \rangle \cong C_2$.

All these will be summarized in the following theorem and table. However, for the sake of simplicity most of the cases of $(1, 1, 1, 1, 1)$ will be omitted from the table.

Theorem 5: Let C be a genus 4 hyperelliptic curve defined over an algebraically closed field K of characteristic 2. Then the automorphism group of C is one of the following: C_2, V_4, C_4, $C_2 \times C_2 \times C_2$, C_6, C_{18}, D_{20}.

Type	Curve	Condition	Aut
$(1,1,1,1,1)$	$y^2+y=\alpha_1 x+\alpha_2 x^{-1}+\alpha_3(x-1)^{-1}$ $+\alpha_4(x-\lambda_1)^{-1}+\alpha_5(x-\lambda_2)^{-1}$	$\alpha_2=\alpha_3, \alpha_4=\alpha_5, \alpha_2\neq\alpha_4$	V_4
		$\alpha_2=\alpha_4, \alpha_3=\alpha_5, \alpha_2\neq\alpha_3$	V_4
		$\alpha_2=\alpha_5, \alpha_3=\alpha_4, \alpha_2\neq\alpha_3$	V_4
		$\alpha_2=\alpha_3=\alpha_4=\alpha_5$	${C_2}^3$
		$\alpha_1=\alpha_2=\alpha_3\lambda=\alpha_4\lambda=\alpha_5$	D_{20}
		$\alpha_1=\alpha_2=\alpha_3\lambda^{-1}=\alpha_4=\alpha_5\lambda^{-1}$	D_{20}
		$\cdots$	$\cdots$
		$\cdots$	$\cdots$
$(3,1,1,1)$	$y^2+y=x^3+\alpha x+\beta_1 x^{-1}$ $+\beta_2(x-1)^{-1}+\beta_3(x-\lambda)^{-1}$	$\alpha\neq 0$	C_2
		$\alpha=0, \beta=\gamma\lambda, \gamma=\sigma\lambda, \sigma=\beta\lambda$	C_6
$(3,3,1)$	$y^2+y=x^3+\alpha x$ $+x^{-3}+\beta x^{-1}+\gamma(x-1)^{-1}$	none	C_2
		$\beta=1, \alpha=\gamma$	V_4
$(5,1,1)$	$y^2+y=x^5+\alpha x^3$ $+\beta x^{-1}+\gamma(x-1)^{-1}$	$\alpha\neq 0, or 1$	C_2
		$\alpha=0$	V_4
		$\alpha=1$	C_4
$(5,3)$	$y^2+y=x^5+\alpha x^3+x^{-3}+\beta x^{-1}$	none	C_2
$(7,1)$	$y^2+y=x^7+\alpha x^5+\beta x^3+\gamma x^{-1}$	none	C_2
(9)	$y^2+y=x^9+\alpha_1 x^7+\alpha_2 x^5+\alpha_3 x^3$	$\alpha_1\neq 0$	C_2
		$\alpha_1=0$	C_{18}

5. APPENDIX

Transformation	Automorphism	Order(ϕ_i)
$\sigma_1 : x \longmapsto x$	$\phi_1(x,y) = (x,y)$	1
$\sigma_2 : x \longmapsto x+1$	$\phi_2(x,y) = (x+1, y+w)$ where $w^2+w=\alpha_1$	2
$\sigma_3 : x \longmapsto x+\lambda$	$\phi_3(x,y) = (x+\lambda, y+w)$ where $w^2+w=\alpha_1\lambda$	2
$\sigma_4 : x \longmapsto x+\lambda^{-1}$	$\phi_4(x,y) = (x+\lambda^{-1}, y+w)$ where $w^2+w=\alpha_1\lambda^{-1}$	2
$\sigma_5 : x \longmapsto \lambda x$	does not lift to an automorphism	
$\sigma_6 : x \longmapsto \lambda x+1$	does not lift to an automorphism	
$\sigma_7 : x \longmapsto \lambda x+\lambda$	does not lift to an automorphism	
$\sigma_8 : x \longmapsto \lambda x+\lambda^{-1}$	does not lift to an automorphism	
$\sigma_9 : x \longmapsto \lambda^{-1} x$	does not lift to an automorphism	
$\sigma_{10} : x \longmapsto \lambda^{-1} x+1$	does not lift to an automorphism	
$\sigma_{11} : x \longmapsto \lambda^{-1} x+\lambda$	does not lift to an automorphism	
$\sigma_{12} : x \longmapsto \lambda^{-1} x+\lambda^{-1}$	does not lift to an automorphism	
$\sigma_{13} : x \longmapsto \frac{\lambda}{x}$	$\phi_{13}(x,y) = (\frac{\lambda}{x}, y+w)$ where $w^2+w=\alpha_5\lambda$	2
$\sigma_{14} : x \longmapsto \frac{x+1}{x}$	does not lift to an automorphism	
$\sigma_{15} : x \longmapsto \frac{x+\lambda}{x}$	$\phi_{15}(x,y) = (\frac{x+\lambda}{x}, y)$	5
$\sigma_{16} : x \longmapsto \frac{\lambda x+1}{x}$	$\phi_{16}(x,y) = (\frac{\lambda x+1}{x}, y)$	5
$\sigma_{17} : x \longmapsto \frac{x+\lambda^{-1}}{x}$	$\phi_{17}(x,y) = (\frac{x+\lambda^{-1}}{x}, y)$	5
$\sigma_{18} : x \longmapsto \frac{\lambda x+\lambda}{x}$	$\phi_{18}(x,y) = (\frac{\lambda x+\lambda}{x}, y)$	5
$\sigma_{19} : x \longmapsto \frac{\lambda x+\lambda^{-1}}{x}$	does not lift to an automorphism	
$\sigma_{20} : x \longmapsto \frac{\lambda^{-1}x+1}{x}$	$\phi_{20}(x,y) = (\frac{\lambda^{-1}x+1}{x}, y)$	5
$\sigma_{21} : x \longmapsto \frac{\lambda^{-1}x+\lambda}{x}$	does not lift to an automorphism	
$\sigma_{22} : x \longmapsto \frac{\lambda^{-1}x+\lambda^{-1}}{x}$	$\phi_{22}(x,y) = (\frac{\lambda^{-1}x+\lambda^{-1}}{x}, y)$	5
$\sigma_{23} : x \longmapsto \frac{1}{x}$	$\phi_{23}(x,y) = (\frac{1}{x}, y+w)$ where $w^2+w=\alpha_3$	2
$\sigma_{24} : x \longmapsto \frac{\lambda^{-1}}{x}$	$\phi_{24}(x,y) = (\frac{\lambda^{-1}}{x}, y+w)$ where $w^2+w=\alpha_4\lambda^{-1}$	2
$\sigma_{25} : x \longmapsto \frac{x}{x+1}$	$\phi_{25}(x,y) = (\frac{x}{x+1}, y+w)$ where $w^2+w=\alpha_2$	2
$\sigma_{26} : x \longmapsto \frac{x+\lambda}{x+1}$	$\phi_{26}(x,y) = (\frac{x+\lambda}{x+1}, y+w)$ here $w^2+w=\alpha_5\lambda^{-1}$	2
$\sigma_{27} : x \longmapsto \frac{\lambda x+1}{x+1}$	$\phi_{27}(x,y) = (\frac{\lambda x+1}{x+1}, y)$	5
$\sigma_{28} : x \longmapsto \frac{x+\lambda^{-1}}{x+1}$	$\phi_{28}(x,y) = (\frac{x+\lambda^{-1}}{x+1}, y+w)$ where $w^2+w=\alpha_4\lambda$	2

$\sigma_{29} : x \longmapsto \frac{\lambda x}{x+1}$	does not lift to an automorphism	
$\sigma_{30} : x \longmapsto \frac{\lambda x+\lambda^{-1}}{x+1}$	$\phi_{30}(x,y) = (\frac{\lambda x+\lambda^{-1}}{x+1}, y)$	5
$\sigma_{31} : x \longmapsto \frac{\lambda^{-1}x}{x+1}$	does not lift to an automorphism	
$\sigma_{32} : x \longmapsto \frac{\lambda^{-1}x+1}{x+1}$	$\phi_{32}(x,y) = (\frac{\lambda^{-1}x+1}{x+1}, y)$	5
$\sigma_{33} : x \longmapsto \frac{\lambda^{-1}x+\lambda}{x+1}$	$\phi_{33}(x,y) = (\frac{\lambda^{-1}x+\lambda}{x+1}, y)$	5
$\sigma_{34} : x \longmapsto \frac{1}{x+1}$	does not lift to an automorphism	
$\sigma_{35} : x \longmapsto \frac{\lambda}{x+1}$	$\phi_{35}(x,y) = (\frac{\lambda}{x+1}, y)$	5
$\sigma_{36} : x \longmapsto \frac{\lambda^{-1}}{x+1}$	$\phi_{36}(x,y) = (\frac{\lambda^{-1}}{x+1}, y)$	5
$\sigma_{37} : x \longmapsto \frac{x}{x+\lambda}$	does not lift to an automorphism	
$\sigma_{38} : x \longmapsto \frac{x+1}{x+\lambda}$	$\phi_{38}(x,y) = (\frac{x+1}{x+\lambda}, y)$	5
$\sigma_{39} : x \longmapsto \frac{1}{x+\lambda}$	$\phi_{39}(x,y) = (\frac{1}{x+\lambda}, y)$	5
$\sigma_{40} : x \longmapsto \frac{\lambda}{x+\lambda}$	$\phi_{40}(x,y) = (\frac{\lambda}{x+\lambda}, y)$	5
$\sigma_{41} : x \longmapsto \frac{\lambda^{-1}}{x+\lambda}$	does not lift to an automorphism	
$\sigma_{42} : x \longmapsto \frac{\lambda x+1}{x+\lambda}$	$\phi_{42}(x,y) = (\frac{\lambda x+1}{x+\lambda}, y+w)$ where $w^2+w = \alpha_3\lambda$	2
$\sigma_{43} : x \longmapsto \frac{\lambda x}{x+\lambda}$	$\phi_{43}(x,y) = (\frac{\lambda x}{x+\lambda}, y+w)$ where $w^2+w = \alpha_2\lambda^{-1}$	2
$\sigma_{44} : x \longmapsto \frac{x+\lambda^{-1}}{x+\lambda}$	$\phi_{44}(x,y) = (\frac{x+\lambda^{-1}}{x+\lambda}, y)$	5
$\sigma_{45} : x \longmapsto \frac{\lambda x+\lambda}{x+\lambda}$	$\phi_{45}(x,y) = (\frac{\lambda x+\lambda}{x+\lambda}, y+w)$ where $w^2+w = \alpha_5$	2
$\sigma_{46} : x \longmapsto \frac{\lambda^{-1}x}{x+\lambda}$	does not lift to an automorphism	
$\sigma_{47} : x \longmapsto \frac{\lambda^{-1}x+\lambda}{x+\lambda}$	$\phi_{47}(x,y) = (\frac{\lambda^{-1}x+\lambda}{x+\lambda}, y)$	5
$\sigma_{48} : x \longmapsto \frac{\lambda^{-1}x+\lambda^{-1}}{x+\lambda}$	$\phi_{48}(x,y) = (\frac{\lambda^{-1}x+\lambda^{-1}}{x+\lambda}, y)$	5
$\sigma_{49} : x \longmapsto \frac{x}{x+\lambda^{-1}}$	does not lift to an automorphism	
$\sigma_{50} : x \longmapsto \frac{x+1}{x+\lambda^{-1}}$	$\phi_{50}(x,y) = (\frac{x+1}{x+\lambda^{-1}}, y)$	5
$\sigma_{51} : x \longmapsto \frac{x+\lambda}{x+\lambda^{-1}}$	$\phi_{51}(x,y) = (\frac{x+\lambda}{x+\lambda^{-1}}, y)$	5
$\sigma_{52} : x \longmapsto \frac{\lambda x}{x+\lambda^{-1}}$	does not lift to an automorphism	
$\sigma_{53} : x \longmapsto \frac{\lambda x+\lambda}{x+\lambda^{-1}}$	$\phi_{53}(x,y) = (\frac{\lambda x+\lambda}{x+\lambda^{-1}}, y)$	5
$\sigma_{54} : x \longmapsto \frac{\lambda x+\lambda^{-1}}{x+\lambda^{-1}}$	$\phi_{54}(x,y) = (\frac{\lambda x+\lambda^{-1}}{x+\lambda^{-1}}, y)$	5
$\sigma_{55} : x \longmapsto \frac{\lambda^{-1}x}{x+\lambda^{-1}}$	$\phi_{55}(x,y) = (\frac{\lambda^{-1}x}{x+\lambda^{-1}}, y+w)$ where $w^2+w = \alpha_2\lambda$	2
$\sigma_{56} : x \longmapsto \frac{\lambda^{-1}x+1}{x+\lambda^{-1}}$	$\phi_{56}(x,y) = (\frac{\lambda^{-1}x+1}{x+\lambda^{-1}}, y+w)$ $w^2+w = \alpha_3\lambda^{-1}$	2
$\sigma_{57} : x \longmapsto \frac{\lambda^{-1}x+\lambda^{-1}}{x+\lambda^{-1}}$	$\phi_{57}(x,y) = (\frac{\lambda^{-1}x+\lambda^{-1}}{x+\lambda^{-1}}, y+w)$ $w^2+w = \alpha_4$	2
$\sigma_{58} : x \longmapsto \frac{1}{x+\lambda^{-1}}$	$\phi_{58}(x,y) = (\frac{1}{x+\lambda^{-1}}, y)$	5
$\sigma_{59} : x \longmapsto \frac{\lambda}{x+\lambda^{-1}}$	does not lift to an automorphism	
$\sigma_{60} : x \longmapsto \frac{\lambda^{-1}}{x+\lambda^{-1}}$	$\phi_{60}(x,y) = (\frac{\lambda^{-1}}{x+\lambda^{-1}}, y)$	5

References

1. G. Cardona, E. Nart, J. Pujolas *Curves of genus two over fields of even characteristic*, http://www.arxiv.org/math.NT/0210105.
2. I. M. Duursma, P. Gaudry, F. Morain *Speeding Up The Discrete Log Computation On Curves With Automorphisms*, Advances in Cryptology - Asia Cryp '99, Springer-Verlag LNCS **1716**, 103 - 121.
3. P. Gaudry, *An Algorithm for Solving the Discrete Log Problems on Hyperelliptic Curves*, Advances in Cryptology-Eurocrypt 2000, Lecture Notes in Computer Science **1807**, Springer, Berlin (2000), 19 - 34.
4. J-I. Igusa, *Arithmetic variety of moduli for genus two*, Ann. of Math. **72** (1960), 612 - 649.
5. E. Nart, D. Sadornil *Hyperelliptic curves of genus three over finite fields of even characteristic*, Finite Fields Appl., **10** (2004), 198 - 220.
6. H. Stichtenoth, *Algebraic Function Fields and Codes*, Springer-Verlag, Berlin Heidelberg (1970).

Modular representations on some Riemann-Roch spaces of modular curves $X(N)$

David Joyner* and Amy Ksir

Mathematics Dept.
U.S. Naval Academy
Annapolis, MD 21402
wdj@usna.edu, ksir@usna.edu

We compute the $PSL(2, N)$-module structure of the Riemann-Roch space $L(D)$, where D is an invariant non-special divisor on the modular curve $X(N)$, with $N \geq 7$ prime. This depends on a computation of the ramification module, which we give explicitly. These results hold for characteristic p if $X(N)$ has good reduction mod p and p does not divide the order of $PSL(2, N)$. We give as examples the cases $N = 7, 11$, which were also computed using GAP [4]. Applications to AG codes associated to this curve are considered, and specific examples are computed using GAP [4] and MAGMA [6].

1. Introduction

The modular curve $X(N)$ has a natural action by the finite group $G = PSL(2, N)$, with quotient $X(1)$. If D is a $PSL(2, N)$-invariant divisor on $X(N)$, then there is a natural representation of G on the Riemann-Roch space $L(D)$. In this paper, we give some results about the $PSL(2, N)$-module structure of the Riemann-Roch space $L(D)$, in the case where N is prime and $N \geq 7$. If D is non-special then a formula in Borne [5] gives

$$[L(D)] = (1 - g_{X(1)})[k[G]] + [deg_{eq}(D)] - [\tilde{\Gamma}_G]. \qquad (1)$$

Here $g_{X(1)}$ is the genus of $X(1)$ (which is zero), square brackets denote the equivalence class of a representation of G, $deg_{eq}(D)$ is the equivariant

*I'd like to thank Tony Shaska for his hospitality during the Computational Aspects of Algebraic Curves conference in May, 2005.

degree of D, and $\tilde{\Gamma}_G$ is the ramification module (these will be defined in sections 5 and 6).

Our main result (see Theorem 9 in section 5) gives an explicit computation of the G-module structure of the ramification module $\tilde{\Gamma}_G$. We show it can be explicitly decomposed into irreducible G-modules where the multiplicity of an irreducible G-module π in $\tilde{\Gamma}_G$ only depends on π and the residue class of N modulo 24. We then go on in section 6 to compute the equivariant degrees of G-invariant divisors on $X(N)$, and use these and Borne's formula to compute the G-module structure of the corresponding Riemann-Roch spaces.

If $K = \mathbb{Q}(G)$ denotes the abelian extension of $\mathbb{Q}$ generated by the character values of G then $\mathcal{G} = Gal(K/\mathbb{Q})$ acts on the set of irreducible representations of G. We call this the *Galois action.* In the case where $\tilde{\Gamma}_G$ is invariant under the Galois action, the authors have given a somewhat simpler formula for $\tilde{\Gamma}_G$ [11]. In section 5 (see Theorem 15), we also prove that the ramification module is $\mathcal{G}$-invariant if and only if $N \equiv 1 \pmod 4$.

As a corollary, it is an easy exercise now to compute explicitly the decomposition

$$H^1(X(N), k) = H^0(X(N), \Omega^1) \oplus \overline{H^0(X(N), \Omega^1)} = L(K) \oplus \overline{L(K)},$$

into irreducible G-modules, where K is a canonical divisor. This was discussed in Khare-Prasad [13] (over $k = \mathbb{C}$) and Schoen [20] (over the finite field $k = GF(N)$). Indeed, Schoen observes that the multiplicities of the irreducible representations occurring in $H^1(X(N), k)$ can be interpreted in terms of dimension of cusp forms and number of cusps on $X(N)$.

In section 7 we look at the examples $N = 7, 11$, using GAP [4] to do many of the computations. In the last section, applications to AG codes associated to this curve are considered (MAGMA [6] was used to do some of these computations).

Notation: Throughout this paper, $N > 5$ is a prime, $\mathbb{F} = GF(N)$ is the finite field with N elements, and $G = PSL(2, N)$.

Acknowledgements: We thank D. Prasad and R. Guralnick for enlightening correspondence, and in particular for the references Khare-Prasad [13] and Janusz [10].

2. Modular curves

Let H denote the complex upper half-plane, let $H^* = H \cup \mathbb{Q} \cup \{\infty\}$, and recall that $SL(2, \mathbb{Q})$ acts on H^* by fractional linear transformations. Let

$X(N)$ denote the modular curve defined over $\mathbb{Q}$ whose complex points are given by $\Gamma(N)\backslash H^*$, where

$$\Gamma(N) = \{\begin{pmatrix} a & b \\ c & d \end{pmatrix} \in SL_2(\mathbb{Z}) \mid a-1 \equiv d-1 \equiv b \equiv c \equiv 0 \pmod{N}\}.$$

Throughout this paper, we will assume that N is prime and $N > 6$. In this case, the genus of $X(N)$ is given by the formula

$$g = 1 + \frac{(N-6)(N^2-1)}{24}.$$

For example, $X(7)$ is genus 3 and $X(11)$ is genus 26.

Let N be a prime and, for $j \in \mathbb{Z}/N\mathbb{Z}$, let y_j be variables satisfying

$$y_j + y_{-j} = 0, \quad y_{a+b}y_{a-b}y_{c+d}y_{c-d} + y_{a+c}y_{a-c}y_{d+b}y_{d-b} + y_{a+d}y_{a-d}y_{b+c}y_{b-c} = 0, \tag{2}$$

for all $a, b, c, d \in \mathbb{Z}/N\mathbb{Z}$. These are Klein's equations for $X(N)$ (see Adler [1] or Ritzenthaler [17]).

Example 6: When $N = 7$, this reduces to $y_1^3 y_2 - y_2^3 y_3 - y_3^3 y_1 = 0$, the famous Klein quartic.

When $N = 11$, the 20 equations which arise reduce to the 10 equations

$$\begin{aligned}
-y_1^2 y_2 y_3 + y_2 y_4 y_5^2 + y_3^2 y_4 y_5 &= 0, \\
-y_1^3 y_4 + y_2 y_4^3 - y_3^3 y_5 &= 0, \\
-y_1 y_3^3 - y_1^3 y_5 + y_2^3 y_4 &= 0, \\
-y_1^2 y_3 y_4 + y_1 y_3 y_5^2 + y_2^2 y_4 y_5 &= 0, \\
-y_1^2 y_2 y_5 + y_1 y_3 y_4^2 - y_2^2 y_3 y_5 &= 0, \\
y_1^3 y_2 - y_3 y_5^3 - y_4^3 y_5 &= 0, \\
y_1 y_5^3 - y_2^3 y_3 + y_3^3 y_4 &= 0, \\
-y_1 y_2^2 y_4 + y_1 y_4^2 y_5 + y_2 y_3^2 y_5 &= 0, \\
y_1 y_2^3 + y_2 y_5^3 - y_3 y_4^3 &= 0, \\
y_1 y_2 y_3^2 + y_1 y_4 y_5^2 - y_2 y_3 y_4^2 &= 0.
\end{aligned}$$

The curve $X(N)$ over a field k parametrizes pairs of an elliptic curve over k and a subgroup of order N of the group structure on the curve. This can be extended to fields of positive characteristic, if $X(N)$ has good reduction. Since Klein's equations have integer coefficients they can also be extended to an arbitrary field k. However, Velu [10] (see also Ritzenthaler [19]) has shown that $X(N)$ has good reduction over fields of characteristic p where p does not divide N (in our case, $p \neq N$, since N is itself assumed to be a prime).

Let

$$G = PSL_2(\mathbb{Z}/N\mathbb{Z}) \cong \overline{\Gamma(1)}/\overline{\Gamma(N)},$$

where the overline denotes the image in $PSL_2(\mathbb{Z})$. This group acts on $X(N)$. (In characteristic 0, see Shimura [3], in characteristic $\ell > 0$, see Ritzenthaler [17].) When $N > 2$ is prime, $|G| = N(N^2 - 1)/2$. The quotient of $X(N)$ by the action of $PSL(2, N)$ is $X(1) \cong \mathbb{P}^1$.

Definition 7: When X has good reduction to a finite field k and, in addition, the characteristic ℓ of k does not divide $|G|$, we say that ℓ is **good**.

If k is a field of good characteristic, the automorphism group of $X(N)$ is known to be $PSL(2, N)$ Bending-Camina-Guralnick [3].

The action of $G = SL_2(\mathbb{Z}/N\mathbb{Z})$ on the set of points of the projective curve defined by Klein's equations is described in Ritzenthaler [17] (see also Adler [2], Ritzenthaler [18]). The element $g = \begin{pmatrix} a & b \\ c & d \end{pmatrix} \in G$ sends $(y_j)_{j\in\mathbb{Z}/N\mathbb{Z}} \in X(N)$ to $(\rho(g)y_j)_{j\in\mathbb{Z}/N\mathbb{Z}} \in X(N)$, where

$$\rho(g)(y_j) = \sum_{t\in\mathbb{Z}/N\mathbb{Z}} \zeta^{b(aj^2+2jtc)+t^2cd} y_{aj+tc},$$

where ζ denotes a primitive N-th root of unity in k.

Remark 8: When the formulas for the special cases $\rho\begin{pmatrix} 0 & 1 \\ -1 & 0 \end{pmatrix}$, $\rho\begin{pmatrix} 1 & 1 \\ 0 & 1 \end{pmatrix}$, and $\rho\begin{pmatrix} a & 0 \\ 0 & a^{-1} \end{pmatrix}$, are written down separately, the similarity with the Weil representation for $SL_2(\mathbb{Z}/N\mathbb{Z})$ is striking (see also Adler [2]).

2.1. *Ramification data*

We will now consider the ramification data of the maps

$$\phi_N : H^* \to X(N)$$

and

$$\psi_N : X(N) \to X(1) \cong \mathbb{P}^1. \tag{3}$$

We will first study the characteristic zero case. We recall some facts from Shimura [3].

We call a matrix $g = \begin{pmatrix} a & b \\ c & d \end{pmatrix} \in SL(2, \mathbb{Z})$ *elliptic*, *parabolic*, or *hyperbolic* if $|\mathrm{tr}(g)|$ is < 2, $= 2$ and $g \neq I$, or > 2, respectively. We call a point $z \in H^*$

elliptic, *parabolic*, or *hyperbolic* it it is fixed by such a matrix. Parabolic points are also called *cuspidal points* or *cusps.* Note that $g = \begin{pmatrix} a & b \\ c & d \end{pmatrix}$, $c \neq 0$, fixes $z \in H$ if and only if

$$z = \frac{a - d \pm \sqrt{(a+d)^2 - 4}}{2c}. \tag{4}$$

(If $c = 0$ then g has no fixed points in H.)

If a group G acts on a set X then we denote by $G_x = \{g \in G \mid g(x) = x\}$ the stabilizer in G of x.

Lemma 9:

- *If $z \in H^*$ is elliptic then z belongs to the $SL_2(\mathbb{Z})$-orbit of $z_1 = i$ or $z_2 = \frac{1+\sqrt{3}i}{2}$.*
- *The stabilizer in $SL_2(\mathbb{Z})$ of z_1 is*

  $$SL_2(\mathbb{Z})_{z_1} = \langle \begin{pmatrix} 0 & 1 \\ -1 & 0 \end{pmatrix} \rangle, \quad \Gamma(N)_{z_1} = \{I\},$$

  and of z_2 is

  $$SL_2(\mathbb{Z})_{z_2} = \langle \begin{pmatrix} 0 & 1 \\ -1 & 1 \end{pmatrix} \rangle, \quad \Gamma(N)_{z_2} = \{I\}.$$

- *If $z \in H^*$ is parabolic then z belongs to the $SL_2(\mathbb{Z})$-orbit of ∞. We have*

  $$SL_2(\mathbb{Z})_\infty = \langle \begin{pmatrix} 1 & 1 \\ 0 & 1 \end{pmatrix} \rangle, \quad \Gamma(N)_\infty = \langle \begin{pmatrix} 1 & N \\ 0 & 1 \end{pmatrix} \rangle.$$

- *There are no hyperbolic points in H^*,*

The last part follows from (4), the others are proven in Shimura [3]. The following is well-known.

Lemma 10: *The ramification index of ψ_N at $P = \phi_N(z)$ is $[\overline{\Gamma(1)}_z : \overline{\Gamma(N)}_z]$. The stabilizer at P is*

$$G_P = \overline{\Gamma(1)}_z / \overline{\Gamma(N)}_z.$$

In particular, we have the following.

- *If $z = z_1$ then*

  $$G_P = H_1 = \langle \begin{pmatrix} 0 & 1 \\ -1 & 0 \end{pmatrix} \rangle$$

  has order 2.

- *If* $z = z_2$ *then*

$$G_P = H_2 = \langle \begin{pmatrix} 0 & 1 \\ -1 & 1 \end{pmatrix} \rangle$$

has order 3.

- *If* $z = \infty$ *then*

$$G_P = H_3 = \langle \begin{pmatrix} 1 & 1 \\ 0 & 1 \end{pmatrix} \rangle \cong \mathbb{Z}/N\mathbb{Z}$$

has order N.

When $k = \mathbb{C}$, no other cyclic subgroups of G occur as stabilizers.

Proposition 11: *Assume k is an algebraically closed field of good characteristic in the sense of Definition 7. If H is a cyclic subgroup of $G = Aut_k(X(N)) = PSL_2(\mathbb{Z}/N\mathbb{Z})$ which is not of order 2, 3, or N then H is not the stabilizer of any point on $X(N)$. Furthermore, there is only one orbit of points with stabilizer of order 2, one orbit of points with stabilizer of order 3, and one orbit of points with stabilizer of order N.*

Proof: In the proof of Ritzenthaler [17, Proposition 1.3] it is stated that ϕ_N is ramified over three points of indices 2, 3 and N. The Hurwitz(-Riemann-Zeuthen) formula says that (see for example Hartshorne [9, Corollary IV.2.4]):

$$2 \cdot g(X(N)) - 2 = -2d + \deg(R),$$

where R is the ramification divisor

$$R = \sum_{P \in X(N)} length(\Omega_{X(N)/X(1)})_P \cdot P.$$

(Indeed, there are more explicit formulas for $\deg(R)$, since we are in the tamely ramified case.) Since the genus of $X(N)$ and of $X(1)$ do not depend on the characteristic, so we can conclude that the ramification divisor of $X(N)/\mathbb{C}$ has the same degree as the ramification divisor of $X(N)/k$. In other words, "$\deg(R)$ over $\mathbb{C}$" equals "$\deg(R)$ over k". Since all the summands in $\deg(R)$ are ≥ 0, this is enough to show that no other cyclic subgroups of G arise as stabilizers over k. □

We call a cyclic subgroup of G whose order is not 1, 2, 3, or N **spurious**. Spurious subgroups are not uncommon. Here is a general construction. If N is prime, so $\mathbb{F} = \mathbb{Z}/N\mathbb{Z}$ is a field, let $\mathbb{E}/\mathbb{F}$ denote a quadratic field with norm $nm : \mathbb{E} \to \mathbb{F}$. The kernel of the norm map, $T = ker(nm)$ embeds into $SL_2(\mathbb{F})$. Moreover, since $\mathbb{E}^\times$ is cyclic, so is T, as well as all of its subgroups. It's known that T has order $N+1$ since the norm map is surjective in this case.

Definition 12: At each point $P \in X(N)$, the **ramification character** θ_P is the character of the action of G_P on the cotangent space to $X(N)$ at P.

3. Representation theory of $PSL(2, N)$

We first consider the representation theory of G over $\mathbb{C}$, following the treatment in Fulton-Harris [7].

The group $PSL(2, N)$ has $3+(N-1)/2$ conjugacy classes. Let $\varepsilon \in \mathbb{F}$ be a generator for the cyclic group $\mathbb{F}^\times$. Then each class will have a representative of one of the following forms:

$$\begin{pmatrix} 1 & 0 \\ 0 & 1 \end{pmatrix}, \begin{pmatrix} x & 0 \\ 0 & x^{-1} \end{pmatrix}, \begin{pmatrix} 1 & 1 \\ 0 & 1 \end{pmatrix}, \begin{pmatrix} 1 & \varepsilon \\ 0 & 1 \end{pmatrix}, \begin{pmatrix} x & \varepsilon y \\ y & x \end{pmatrix}.$$

The irreducible representations of $PSL(2, N)$ include the trivial representation **1** and one irreducible V of dimension N. All but two of the others fall into two types: representations W_α of dimension $N+1$ ("principal series"), and X_β of dimension $N-1$ ("discrete series"). The first type, W_α, is labeled by a homomorphism $\alpha : \mathbb{F}^\times \to \mathbb{C}^\times$. The second type is indexed by a homomorphism $\beta : T \to \mathbb{C}^\times$, where T is a cyclic subgroup of order $N+1$ of $\mathbb{F}(\sqrt{\varepsilon})^\times$. The characters of these are as follows:

	$\begin{pmatrix} 1 & 0 \\ 0 & 1 \end{pmatrix}$	$\begin{pmatrix} x & 0 \\ 0 & x^{-1} \end{pmatrix}$	$\begin{pmatrix} 1 & 1 \\ 0 & 1 \end{pmatrix}$	$\begin{pmatrix} 1 & \varepsilon \\ 0 & 1 \end{pmatrix}$	$\begin{pmatrix} x & \varepsilon y \\ y & x \end{pmatrix}$
1	1	1	1	1	1
X_β	$N-1$	0	-1	-1	$-\beta(x+\sqrt{\varepsilon}y) - \beta(x-\sqrt{\varepsilon}y)$
V	N	1	0	0	-1
W_α	$N+1$	$\alpha(x)+\alpha(x^{-1})$	1	1	0

Let τ denote a generator of T. Let ζ be a primitive Nth root of unity in $\mathbb{C}$. Let q and q' be defined by

$$q = \sum_{(\frac{a}{N})=1} \zeta^a \text{ and } q' = \sum_{(\frac{a}{N})=-1} \zeta^a, \tag{5}$$

where the sums are over the quadratic residues and nonresidues $\pmod N$, respectively. If $N \equiv 1 \bmod 4$, then the "principal series" representation

W_{α_0} corresponding to

$$\begin{aligned} \alpha_0 : \mathbb{F}^\times &\to \mathbb{C}^\times \\ \varepsilon &\mapsto -1 \end{aligned}$$

is not irreducible, but splits into two irreducibles W' and W'', each of dimension $(N+1)/2$. Their characters satisfy:

	$\begin{pmatrix} 1 & 0 \\ 0 & 1 \end{pmatrix}$	$\begin{pmatrix} x & 0 \\ 0 & x^{-1} \end{pmatrix}$	$\begin{pmatrix} 1 & 1 \\ 0 & 1 \end{pmatrix}$	$\begin{pmatrix} 1 & \varepsilon \\ 0 & 1 \end{pmatrix}$	$\begin{pmatrix} x & \varepsilon y \\ y & x \end{pmatrix}$
W'	$\frac{N+1}{2}$	$\alpha_0(x)$	$1+q$	$1+q'$	0
W''	$\frac{N+1}{2}$	$\alpha_0(x)$	$1+q'$	$1+q$	0

Similarly, if $N \equiv 3 \bmod 4$, then the "discrete series" representation X_β corresponding to

$$\begin{aligned} \beta_0 : T &\to \mathbb{C}^\times \\ \tau &\mapsto -1 \end{aligned}$$

sending the generator τ to -1, splits into two irreducibles X' and X'', each of dimension $(N-1)/2$. Their characters satisfy:

	$\begin{pmatrix} 1 & 0 \\ 0 & 1 \end{pmatrix}$	$\begin{pmatrix} x & 0 \\ 0 & x^{-1} \end{pmatrix}$	$\begin{pmatrix} 1 & 1 \\ 0 & 1 \end{pmatrix}$	$\begin{pmatrix} 1 & \varepsilon \\ 0 & 1 \end{pmatrix}$	$\begin{pmatrix} x & \varepsilon y \\ y & x \end{pmatrix}$
X'	$\frac{N-1}{2}$	0	q	q'	$-\beta_0(x+y\sqrt{\varepsilon})$
X''	$\frac{N-1}{2}$	0	q'	q	$-\beta_0(x+y\sqrt{\varepsilon})$

According to Janusz [10], the Schur index of each irreducible representation of G is 1.

The action of the Galois group $\mathcal{G}$ on the irreducible representations of G is as follows. The character values lie in $\mathbb{Q}(\mu)$, where μ is a primitive m^{th} root of unity and $m = N(N^2-1)/4$. For each integer j relatively prime to m, there is an element σ_j of the Galois group $\mathcal{G} = Gal(\mathbb{Q}(\mu)/\mathbb{Q})$ taking μ to μ^j. This Galois group element will act on representations by taking a representation with character values $(a_1, \ldots, a_n)$ to a representation with character values $(\sigma_j(a_1), \ldots, \sigma_j(a_n))$. Representations with rational character values will be fixed under this action.

Therefore the Galois group $\mathcal{G}$ will fix the trivial representation and the N-dimensional representation V. Its action preserves the set of $N-1$-dimensional "principal series" representations X_β, and the set of $N+1$-dimensional "discrete series" representations W_α. In the case $N \equiv 1 \pmod 4$, the Galois group will exchange the two $(N+1)/2$-dimensional representations W' and W''; if $N \equiv 3 \pmod 4$, the Galois group will exchange the two $(N-1)/2$-dimensional representations X' and X''.

3.1. *Brauer characters of some induced representations*

Let k be a field of characteristic ℓ, assume k contains a primitive $|G|$-th root of unity[†], and assume that ℓ is good in the sense of (7). Because ℓ does not divide the order of G, the characters of $\mathbb{C}$-representations of G are the same as the Brauer characters of k-representations of G. Some general remarks on the Brauer characters of induced modular representations follow.

Let G be a finite group, let H be a subgroup. Let $\sigma : H \to GL_m(k)$ be an m-dimensional representation. Let

$$\sigma^o(g) = \begin{cases} \sigma(g), & g \in H, \\ 0, & g \in G - H. \end{cases}$$

If $G = g_1H \cup g_2H \cup ... \cup g_rH$ is a disjoint union into cosets (where $g_i \in G$ and $g_1 = 1$) then let

$$\pi = Ind_H^G \, \sigma(g) = (\sigma^o(g_i^{-1}gg_j))_{1\leq i,j\leq r}.$$

This $rm \times rm$ matrix is the induced representation of σ to G. The trace of this representation is given by the k-valued class function

$$\mathrm{tr}\,\pi(g) = \sum_{x\in G/H} \mathrm{tr}\,\sigma^o(x^{-1}gx), \tag{6}$$

where we identify G/H with $\{g_1, ..., g_r\}$.

Fix an embedding $k^\times \hookrightarrow \mathbb{C}^\times$ (recall k contains all the eigenvalues of each $\pi(g)$). Let χ_σ denote the Brauer character of σ and let χ_π denote the Brauer character of π.

Proposition 13: *If H has no elements of order $\ell = \mathrm{char}(k)$ then*

$$\chi_\pi(g) = \sum_{x\in G/H} \chi_\sigma^o(x^{-1}gx),$$

where χ_σ^o is the function χ_σ extended by 0 of H.

This is well-known and the proof of this is straightforward, using only (6) and the definitions, so omitted.

In the next few sections, we will use induced characters to compute the $deg_{eq}(D)$ and $\tilde{\Gamma}_G$. By the above proposition, such character computations will hold over $\mathbb{C}$ and in good positive characteristic. In particular, (10) and all the examples given in §7 also hold in this case.

[†]We really only need for k to contain the character values of the irreducible $\overline{k}$-representations of G, where $\overline{k}$ denotes an algebraic closure of k.

4. Induced characters

To compute the $PSL(2,N)$-module structure of the Riemann-Roch space, we are interested in induced representations from the non-spurious cyclic subgroups

$$H_1 = \langle \begin{pmatrix} 0 & 1 \\ -1 & 0 \end{pmatrix} \rangle, \quad H_2 = \langle \begin{pmatrix} 0 & 1 \\ -1 & 1 \end{pmatrix} \rangle, \quad H_3 = \langle \begin{pmatrix} 1 & 1 \\ 0 & 1 \end{pmatrix} \rangle$$

of orders 2, 3, and N, respectively. We compute these by computing the restrictions of the irreducibles to these subgroups and using Frobenius reciprocity.

4.1. *Induced characters from H_1*

To compute the restrictions, we must find the conjugacy classes of $PSL(2,N)$ containing elements of these cyclic groups. For H_1 and H_2, these depend on N.

Define the number i as follows. When $N \equiv 1 \pmod 4$, let i denote an element in $\mathbb{F}^\times$ whose square is -1 (one can take $i = \varepsilon^{(N-1)/4}$, where ε is a generator of $\mathbb{F}^\times$). Then

$$\begin{pmatrix} 0 & 1 \\ -1 & 0 \end{pmatrix} \text{ is conjugate to } \begin{pmatrix} i & 0 \\ 0 & i^{-1} \end{pmatrix}.$$

When $N \equiv 3 \pmod 4$, there is no square root of -1 in $\mathbb{F}$, so we pass to the quadratic extension and let $i = x + \sqrt{\varepsilon} y$ denote a square root of -1 in $\mathbb{F}(\sqrt{\varepsilon})^\times$. (If τ is a generator of the cyclic subgroup T of order $N+1$ in $\mathbb{F}(\sqrt{\varepsilon})^\times$, we can take $i = \tau^{(N+1)/4}$.) Then

$$\begin{pmatrix} 0 & 1 \\ -1 & 0 \end{pmatrix} \text{ is conjugate to } \begin{pmatrix} x & \varepsilon y \\ y & x \end{pmatrix}.$$

Now let us compute the restrictions of the irreducible representations of $PSL(2,N)$ to H_1. We do this by examining the relevant columns of the character table of $PSL(2,N)$ from section 3, and comparing them to the character table of the two element group H_1.

If $N \equiv 1 \pmod 4$, the relevant columns read

	$\begin{pmatrix} 1 & 0 \\ 0 & 1 \end{pmatrix}$	$\begin{pmatrix} i & 0 \\ 0 & i^{-1} \end{pmatrix}$
$\mathbf{1}$	1	1
W'	$\frac{N+1}{2}$	$\alpha_0(i)$
W''	$\frac{N+1}{2}$	$\alpha_0(i)$
X_β	$N-1$	0
V	N	1
W_α	$N+1$	$\alpha(i)+\alpha(i^{-1})$

Let θ_1 be the nontrivial character of H_1. From these columns we see that the restricted representations are:

$$Res^G_{H_1}\mathbf{1} = \mathbf{1}_{H_1}$$

$$Res^G_{H_1}W' = \frac{N-1}{4}\mathbf{1}_{H_1} + \frac{N-1}{4}\theta_1 + \begin{cases} \mathbf{1}_{H_1} & \text{if } \alpha_0(i)=1 \\ \theta_1 & \text{if } \alpha_0(i)=-1 \end{cases}$$

$$Res^G_{H_1}W'' = \text{same}$$

$$Res^G_{H_1}X_\beta = \frac{N-1}{2}\mathbf{1}_{H_1} + \frac{N-1}{2}\theta_1$$

$$Res^G_{H_1}V = \frac{N+1}{2}\mathbf{1}_{H_1} + \frac{N-1}{2}\theta_1$$

$$Res^G_{H_1}W_\alpha = \frac{N-1}{2}\mathbf{1}_{H_1} + \frac{N-1}{2}\theta_1 + \begin{cases} 2\mathbf{1}_{H_1} & \text{if } \alpha(i)=1 \\ 2\theta_1 & \text{if } \alpha(i)=-1 \end{cases}$$

The sign of $\alpha_0(i)$ depends on the equivalence class of $N \pmod 8$. Since $\alpha_0(\varepsilon)=-1$ and $i=\varepsilon^{(N-1)/4}$, we will have $\alpha_0(i)=1$ if $(N-1)/4$ is even ($N\equiv 1 \pmod 8$) and $\alpha_0(i)=-1$ if $(N-1)/4$ is odd ($N \equiv 5 \pmod 8$). For the induced representation of the nontrivial character, we have, for $N\equiv 1 \pmod 8$,

$$Ind^G_{H_1}\theta_1 = \frac{N-1}{4}(W'+W'') + \frac{N-1}{2}\sum_\beta X_\beta + \frac{N-1}{2}V + \frac{N-1}{2}\sum_{\alpha(i)=1} W_\alpha + \frac{N+3}{2}\sum_{\alpha(i)=-1} W_\alpha,$$

and for $N\equiv 5 \pmod 8$,

$$Ind^G_{H_1}\theta_1 = \frac{N+3}{4}(W'+W'') + \frac{N-1}{2}\sum_\beta X_\beta + \frac{N-1}{2}V + \frac{N-1}{2}\sum_{\alpha(i)=1} W_\alpha + \frac{N+3}{2}\sum_{\alpha(i)=-1} W_\alpha.$$

Now let us compute the restricted representations for $N \equiv 3 \pmod 4$. The relevant columns of the character table of $PSL(2,N)$ read

	$\begin{pmatrix} 1 & 0 \\ 0 & 1 \end{pmatrix}$	$\begin{pmatrix} x & \varepsilon y \\ y & x \end{pmatrix}$
$\mathbf{1}$	1	1
X'	$\frac{N-1}{2}$	$-\beta_0(i)$
X''	$\frac{N-1}{2}$	$-\beta_0(i)$
X_β	$N-1$	$-\beta(i)-\beta(i^{-1})$
V	N	-1
W_α	$N+1$	0

Therefore, the restricted representations are:

$$\begin{aligned}
Res_{H_1}^G \mathbf{1} &= \mathbf{1}_{H_1} \\
Res_{H_1}^G X' &= \frac{N-3}{4}\mathbf{1}_{H_1} + \frac{N-3}{4}\theta_1 + \begin{cases} \mathbf{1}_{H_1} & \text{if } \beta_0(i) = -1 \\ \theta_1 & \text{if } \beta_0(i) = 1 \end{cases} \\
Res_{H_1}^G X'' &= \text{same} \\
Res_{H_1}^G X_\beta &= \frac{N-3}{2}\mathbf{1}_{H_1} + \frac{N-3}{2}\theta_1 + \begin{cases} 2\mathbf{1}_{H_1} & \text{if } \beta(i) = -1 \\ 2\theta_1 & \text{if } \beta(i) = 1 \end{cases} \\
Res_{H_1}^G V &= \frac{N-1}{2}\mathbf{1}_{H_1} + \frac{N+1}{2}\theta_1 \\
Res_{H_1}^G W_\alpha &= \frac{N+1}{2}\mathbf{1}_{H_1} + \frac{N+1}{2}\theta_1
\end{aligned}$$

As before, the sign of $\beta_0(i)$ in the representations induced from X' and X'' depends on the equivalence class of $N \pmod 8$. If $N \equiv 3 \pmod 8$, then $\beta_0(i) = -1$ and if $N \equiv 7 \pmod 8$, then $\beta_0(i) = 1$. We get the following induced representations of the nontrivial character: if $N \equiv 3 \pmod 8$,

$$\begin{aligned}
Ind_{H_1}^G \theta_1 = {} & \frac{N-3}{4}(X' + X'') + \frac{N+1}{2} \sum_{\beta(i)=1} X_\beta + \frac{N-3}{2} \sum_{\beta(i)=-1} X_\beta \\
& + \frac{N+1}{2} V + \frac{N+1}{2} \sum_\alpha W_\alpha
\end{aligned}$$

and if $N \equiv 7 \pmod 8$,

$$\begin{aligned}
Ind_{H_1}^G \theta_1 = {} & \frac{N+1}{4}(X' + X'') + \frac{N+1}{2} \sum_{\beta(i)=1} X_\beta + \frac{N-3}{2} \sum_{\beta(i)=-1} X_\beta \\
& + \frac{N+1}{2} V + \frac{N+1}{2} \sum_\alpha W_\alpha.
\end{aligned}$$

4.2. *Induced characters from H_2*

Now we consider the induced characters from the cyclic group H_2 of order 3. As in the case of H_1, the conjugacy classes of the elements of H_2 depend on N.

We define a number ω as follows. When $N \equiv 1 \pmod 3$, ω is a root of $x^2 + 1 = x$ in $\mathbb{F}^\times$ (one can take $\omega = \varepsilon^{(N-1)/6}$). Note

$$\begin{pmatrix} 0 & 1 \\ -1 & 1 \end{pmatrix} \text{ is conjugate to } \begin{pmatrix} \omega & 0 \\ 0 & \omega^{-1} \end{pmatrix},$$

in G. When $N \equiv -1 \pmod 3$, $x + \sqrt{\varepsilon} y$ is a root of $x^2 + 1 = x$ in $\mathbb{F}(\sqrt{\varepsilon})^\times$ (one can take $\omega = \tau^{(N+1)/6}$). Note

$$\begin{pmatrix} 0 & 1 \\ -1 & 0 \end{pmatrix} \text{ is conjugate to } \begin{pmatrix} x & \varepsilon y \\ y & x \end{pmatrix},$$

in G. In any case, the nontrivial elements of H_2 are G-conjugate.

Let θ_2 and θ_2^2 be the nontrivial characters of H_2. Again, to compute $Ind_{H_2}^G \theta_2$ and $Ind_{H_2}^G \theta_2^2$, we compute the restrictions of the irreducible characters of G and use Frobenius reciprocity.

If $N \equiv 1 \pmod 3$, the relevant columns of the character table of $G = PSL(2, N)$ read

	$\begin{pmatrix} 1 & 0 \\ 0 & 1 \end{pmatrix}$	$\begin{pmatrix} \omega & 0 \\ 0 & \omega^{-1} \end{pmatrix}$
$\mathbf{1}$	1	1
X'	$\frac{N-1}{2}$	0
X''	$\frac{N-1}{2}$	0
W'	$\frac{N+1}{2}$	$\alpha_0(\omega)$
W''	$\frac{N+1}{2}$	$\alpha_0(\omega)$
X_β	$N-1$	0
V	N	1
W_α	$N+1$	$\alpha(\omega) + \alpha(\omega^{-1})$

where the character table contains W' and W'' if $N \equiv 1 \pmod 4$, and X' and X'' if $N \equiv 3 \pmod 4$. If $N \equiv 1 \pmod 4$, the value of $\alpha_0(\omega)$ is $(-1)^{(N-1)/6}$ and $\frac{N-1}{6}$ is even, so $\alpha_0(\omega) = 1$. Therefore, the restricted representations are:

$$\begin{aligned}
Res^G_{H_2}\mathbf{1} &= \mathbf{1}_{H_2}\\
Res^G_{H_2}X' &= \frac{N-1}{6}(\mathbf{1}_{H_2}+\theta_2+\theta_2^2)\\
Res^G_{H_2}X'' &= \text{same}\\
Res^G_{H_2}W' &= \frac{N-1}{6}(\mathbf{1}_{H_2}+\theta_2+\theta_2^2)+\mathbf{1}_{H_2}\\
Res^G_{H_2}W'' &= \text{same}\\
Res^G_{H_2}X_\beta &= \frac{N-1}{3}(\mathbf{1}_{H_2}+\theta_2+\theta_2^2)\\
Res^G_{H_2}V &= \frac{N-1}{3}(\mathbf{1}_{H_2}+\theta_2+\theta_2^2)+\mathbf{1}_{H_2}\\
Res^G_{H_2}W_\alpha &= \frac{N-1}{3}(\mathbf{1}_{H_2}+\theta_2+\theta_2^2)+\begin{cases} 2\mathbf{1}_{H_2} & \text{if } \alpha(\omega)=1\\ \theta_2+\theta_2^2 & \text{if } \alpha(\omega)\neq 1\end{cases}\\
Ind^G_{H_2}\theta_2 = Ind^G_{H_2}\theta_2^2 &= \frac{N-1}{3}\sum_\beta X_\beta + \frac{N-1}{3}V\\
&\quad + \frac{N-1}{3}\sum_{\alpha(\omega)=1} W_\alpha + \frac{N+2}{3}\sum_{\alpha(\omega)\neq 1} W_\alpha\\
&\quad + \frac{N-1}{6}\begin{cases} W'+W'' & \text{if } N\equiv 1 \pmod{12}\\ X'+X'' & \text{if } N\equiv 7 \pmod{12}\end{cases}.
\end{aligned}$$

Assume next $N \equiv 2 \pmod 3$. Again, the character table contains W' and W'' if $N \equiv 1 \pmod 4$ (so $N \equiv 5 \pmod{12}$), and X' and X'' if $N \equiv 3 \pmod 4$ (so $N \equiv 11 \pmod{12}$). The relevant columns of the character table of G are as follows. Note that for the irreducibles X' and X'' (in the case $N \equiv 11 \pmod{12}$), the character value depends on $\beta_0(\omega)$, where β_0 is the character of the cyclic group of order $N+1$ sending a generator to -1. In this case, $\omega = \tau^{\frac{N+1}{6}}$, and $\frac{N+1}{6}$ is even, so $\beta_0(\omega) = 1$.

	$\begin{pmatrix}1&0\\0&1\end{pmatrix}$	$\begin{pmatrix}x&\varepsilon y\\y&x\end{pmatrix}$
$\mathbf{1}$	1	1
X'	$\frac{N-1}{2}$	-1
X''	$\frac{N-1}{2}$	-1
W'	$\frac{N+1}{2}$	0
W''	$\frac{N+1}{2}$	0
X_β	$N-1$	$-\beta(\omega)-\beta(\omega^{-1})$
V	N	-1
W_α	$N+1$	0

Therefore, we see that when $N \equiv 2 \pmod 3$, the restricted representations are:

$$\begin{aligned}
Res^G_{H_2}\mathbf{1} &= \mathbf{1}_{H_2} \\
Res^G_{H_2}X' &= \frac{N-5}{6}(\mathbf{1}_{H_2} + \theta_2 + \theta_2^2) + \theta_2 + \theta_2^2 \\
Res^G_{H_2}X'' &= \text{same} \\
Res^G_{H_2}W' &= \frac{N+1}{6}(\mathbf{1}_{H_2} + \theta_2 + \theta_2^2) \\
Res^G_{H_2}W'' &= \text{same} \\
Res^G_{H_2}X_\beta &= \frac{N-5}{3}(\mathbf{1}_{H_2} + \theta_2 + \theta_2^2) + \begin{cases} 2(\theta_2 + \theta_2^2) & \text{if } \beta(\omega) = 1 \\ 2\mathbf{1}_{H_2} + \theta_2 + \theta_2^2 & \text{if } \beta(\omega) \neq 1 \end{cases} \\
Res^G_{H_2}V &= \frac{N-2}{3}(\mathbf{1}_{H_2} + \theta_2 + \theta_2^2) + \theta_2 + \theta_2^2 \\
Res^G_{H_2}W_\alpha &= \frac{N+1}{3}(\mathbf{1}_{H_2} + \theta_2 + \theta_2^2)
\end{aligned}$$

And the induced representations for $N \equiv 2 \pmod 3$ are:

$$\begin{aligned}
Ind^G_{H_2}\theta_2 = Ind^G_{H_2}\theta_2^2 &= \frac{N+1}{3}\sum_\beta X_\beta + \frac{N-2}{3}\sum_\beta X_\beta + \frac{N+1}{3}\sum_\alpha W_\alpha \\
&\quad + \frac{N+1}{3}V + \frac{N+1}{6}\begin{cases} W' + W'' & \text{if } N \equiv 5 \pmod{12} \\ X' + X'' & \text{if } N \equiv 11 \pmod{12} \end{cases}
\end{aligned}$$

4.3. *Induced characters from H_3*

Now we consider the induced characters from H_3, the cyclic group of order N. For the restrictions of the irreducible characters of $PSL(2, N)$ to H_3, the relevant columns of the character table of $G = PSL(2, N)$ are as follows. As before, the representations W' and W'' appear if and only if $N \equiv 1 \pmod 4$, and X' and X'' appear if and only if $N \equiv 3 \pmod 4$.

	$\begin{pmatrix} 1 & 0 \\ 0 & 1 \end{pmatrix}$	$\begin{pmatrix} 1 & 1 \\ 0 & 1 \end{pmatrix}$	$\begin{pmatrix} 1 & \varepsilon \\ 0 & 1 \end{pmatrix}$
$\mathbf{1}$	1	1	1
X'	$\frac{N-1}{2}$	q	q'
X''	$\frac{N-1}{2}$	q'	q
W'	$\frac{N+1}{2}$	$1+q$	$1+q'$
W''	$\frac{N+1}{2}$	$1+q'$	$1+q$
X_β	$N-1$	-1	-1
V	N	0	0
W_α	$N+1$	1	1

Recall

$$q = \sum_{(\frac{a}{N})=1} \zeta^a \text{ and } q' = \sum_{(\frac{a}{N})=-1} \zeta^a.$$

Now let θ_3 be the nontrivial character of H_3 such that

$$\theta_3 \begin{pmatrix} 1 & 1 \\ 0 & 1 \end{pmatrix} = \zeta.$$

The restricted representations are:

$$\begin{aligned}
Res_{H_3}^G \mathbf{1} &= \mathbf{1}_{H_3} \\
Res_{H_3}^G X' &= \sum_{(\frac{k}{n})=1} \theta_3^k \\
Res_{H_3}^G X'' &= \sum_{(\frac{k}{n})=-1} \theta_3^k \\
Res_{H_3}^G W' &= \mathbf{1}_{H_3} + \sum_{(\frac{k}{n})=1} \theta_3^k \\
Res_{H_3}^G W'' &= \mathbf{1}_{H_3} + \sum_{(\frac{k}{n})=-1} \theta_3^k \\
Res_{H_3}^G X_\beta &= \sum_{k=1}^{N-1} \theta_3^k \\
Res_{H_3}^G V &= \mathbf{1}_{H_3} + \sum_{k=1}^{N-1} \theta_3^k \\
Res_{H_3}^G W_\alpha &= \mathbf{2}\mathbf{1}_{H_3} + \sum_{k=1}^{N-1} \theta_3^k
\end{aligned}$$

Therefore,

$$Ind_{H_3}^G \theta_3^k = \sum_\beta X_\beta + V + \sum_\alpha W_\alpha + \begin{cases} W', & N \equiv 1 \pmod 4,\ k \text{ a quad. residue mod } N, \\ X', & N \equiv 3 \pmod 4,\ k \text{ a quad. residue mod } N, \\ W'', & N \equiv 1 \pmod 4,\ k \text{ not a quad. residue mod } N, \\ X'', & N \equiv 3 \pmod 4,\ k \text{ not a quad. residue mod } N. \end{cases}$$

5. Ramification module

For the remainder of this paper, let G^* denote the set of equivalence classes of irreducible representations of G. We often abuse notation by using the same symbol for an element of G^*, a representation in its equivalence class, and the character of such a representation. Over a field k of good positive characteristic, we abuse notation by using G^* instead to represent Brauer characters of G over k.

Define the module Γ_G by

$$\Gamma_G = \sum_{P \in X(N)} Ind_{G_P}^{G} \left(\sum_{\ell=1}^{e_P - 1} \ell \theta_P^{\ell} \right),$$

where θ_P is the ramification character at a point P. Thanks to Nakajima [16], it is known that there is a G-module $\tilde{\Gamma}_G$ such that $[\Gamma_G] = |G| \cdot [\tilde{\Gamma}_G]$ (see also Borne [5]). We call $\tilde{\Gamma}_G$ the *ramification module.*

In the case of the modular curve $X(N)$, let π_{θ_P} denote the induced representation from a character θ_P. Then by definition,

$$\tilde{\Gamma}_G = \frac{\overline{R}_2[\pi_{\theta_1}] + \overline{R}_3([\pi_{\theta_2}] + 2[\pi_{\theta_2^2}]) + \overline{R}_N([\pi_{\theta_3}] + ... + (N-1)[\pi_{\theta_3^{N-1}}])}{|G|}, \tag{7}$$

where $\overline{R}_i$ is the number of $P \in X(N)$ with $G_P = H_i$. Recall from section 2.1 that

$$\overline{R}_2 = \frac{|G|}{2}, \quad \overline{R}_3 = \frac{|G|}{3}, \quad \overline{R}_N = \frac{|G|}{N},$$

so that in fact

$$\tilde{\Gamma}_G = \frac{1}{2}[\pi_{\theta_1}] + \frac{1}{3}([\pi_{\theta_2}] + 2[\pi_{\theta_2^2}]) + \frac{1}{N}([\pi_{\theta_3}] + ... + (N-1)[\pi_{\theta_3^{N-1}}]). \tag{8}$$

This can easily be computed using the induced characters from section 4. To combine the results of sections 4.1 and 4.2, we must look at the congruence class of N modulo 24. We have

$N \equiv 1 \pmod{24} \implies N \equiv 1 \pmod 4, N \equiv 1 \pmod 8, N \equiv 1 \pmod{12},$
$N \equiv 5 \pmod{24} \implies N \equiv 1 \pmod 4, N \equiv 5 \pmod 8, N \equiv 5 \pmod{12},$
$N \equiv 7 \pmod{24} \implies N \equiv 3 \pmod 4, N \equiv 7 \pmod 8, N \equiv 7 \pmod{12},$
$N \equiv 11 \pmod{24} \implies N \equiv 3 \pmod 4, N \equiv 3 \pmod 8, N \equiv 11 \pmod{12},$
$N \equiv 13 \pmod{24} \implies N \equiv 1 \pmod 4, N \equiv 5 \pmod 8, N \equiv 1 \pmod{12},$
$N \equiv 17 \pmod{24} \implies N \equiv 1 \pmod 4, N \equiv 1 \pmod 8, N \equiv 5 \pmod{12},$
$N \equiv 19 \pmod{24} \implies N \equiv 3 \pmod 4, N \equiv 3 \pmod 8, N \equiv 7 \pmod{12},$

$N \equiv 23 \pmod{24} \implies N \equiv 3 \pmod 4, N \equiv 7 \pmod 8, N \equiv 11 \pmod{12}$.

Let $\mathcal{Q}$ denote the set of quadratic residues $\pmod N$, let $\mathcal{N}$ denote the set of quadratic non-residues $\pmod N$ and let

$$S_{\mathcal{Q}} = \sum_{\ell \in \mathcal{Q}} \ell, \quad S_{\mathcal{N}} = \sum_{\ell \in \mathcal{N}} \ell = \frac{N(N-1)}{2} - S_{\mathcal{Q}}.$$

If $N \equiv 1 \pmod 4$, then

$$S_{\mathcal{Q}} = S_{\mathcal{N}} = \frac{N(N-1)}{4}. \tag{9}$$

If $N \equiv 3 \pmod 4$, then $S_{\mathcal{Q}}$ and $S_{\mathcal{N}}$ are not equal.

We prove the following result. An empty sum is by convention zero.

Theorem 14: We have the following decomposition of the ramification module:

- If $N \equiv 1 \pmod{24}$, let $m = \frac{13N-13}{12}$. Then

$$\begin{aligned}\tilde{\Gamma}_G = &\frac{m}{2}W' + \frac{m}{2}W'' + mV + m\sum_{\beta} X_\beta \\ &+m \sum_{\alpha(i)=\alpha(\omega)=1} W_\alpha + (m+1) \sum_{\alpha(i)=1,\alpha(\omega)\neq 1} W_\alpha \\ &+(m+1) \sum_{\alpha(i)\neq 1,\alpha(\omega)=1} W_\alpha + (m+2) \sum_{\alpha(i)\neq 1,\alpha(\omega)\neq 1} W_\alpha.\end{aligned}$$

- If $N \equiv 5 \pmod{24}$, let $m = \frac{13N-5}{12}$. Then

$$\begin{aligned}\tilde{\Gamma}_G = &\frac{m+1}{2}W' + \frac{m+1}{2}W'' + mV \\ &+m \sum_{\beta(\omega)=1} X_\beta + (m-1) \sum_{\beta(\omega)\neq 1} X_\beta \\ &+m \sum_{\alpha(i)=1} W_\alpha + (m+1) \sum_{\alpha(i)\neq 1} W_\alpha.\end{aligned}$$

- If $N \equiv 7 \pmod{24}$, let $m = \frac{13N-7}{12}$. Then

$$\begin{aligned}\tilde{\Gamma}_G = &+(\frac{7N-1}{24} + \frac{S_{\mathcal{Q}}}{N}))X' + (\frac{7N-1}{24} + \frac{S_{\mathcal{N}}}{N}))X'' + mV \\ &+m \sum_{\beta(i)=1} X_\beta + (m-1) \sum_{\beta(i)=-1} X_\beta \\ &+m \sum_{\alpha(\omega)=1} W_\alpha + (m+1) \sum_{\alpha(\omega)=-1} W_\alpha.\end{aligned}$$

- If $N \equiv 11 \pmod{24}$, let $m = \frac{13N+1}{12}$. Then

$$\begin{aligned}\tilde{\Gamma}_G = (\frac{7N-5}{24} + \frac{S_{\mathcal{Q}}}{N})X' + (\frac{7N-5}{24} + \frac{S_{\mathcal{N}}}{N})X'' + mV + m\sum_{\alpha} W_\alpha \\ +m \sum_{\beta(i)=\beta(\omega)=1} X_\beta + (m-1) \sum_{\beta(i)=1,\beta(\omega)\neq 1} X_\beta \\ +(m-1) \sum_{\beta(i)=-1,\beta(\omega)=1} X_\beta + (m-2) \sum_{\beta(i)=-1,\beta(\omega)\neq 1} X_\beta.\end{aligned}$$

- If $N \equiv 13 \pmod{24}$, let $m = \frac{13N-13}{12}$. Then

$$\begin{aligned}\tilde{\Gamma}_G = \frac{m}{2}W' + \frac{m}{2}W' + mV + m\sum_{\beta} X_\beta \\ +m \sum_{\alpha(i)=\alpha(\omega)=1} W_\alpha + (m+1) \sum_{\alpha(i)=1,\alpha(\omega)\neq 1} W_\alpha \\ +(m+1) \sum_{\alpha(i)\neq 1,\alpha(\omega)=1} W_\alpha + (m+2) \sum_{\alpha(i)\neq 1,\alpha(\omega)\neq 1} W_\alpha.\end{aligned}$$

- If $N \equiv 17 \pmod{24}$, let $m = \frac{13N-5}{12}$. Then

$$\begin{aligned}\tilde{\Gamma}_G = \frac{m}{2}W' + \frac{m}{2}W'' + mV \\ +m \sum_{\beta(\omega)=1} X_\beta + (m-1) \sum_{\beta(\omega)\neq 1} X_\beta \\ +m \sum_{\alpha(i)=1} W_\alpha + (m+1) \sum_{\alpha(i)\neq 1} W_\alpha.\end{aligned}$$

- If $N \equiv 19 \pmod{24}$, let $m = \frac{13N-7}{12}$. Then

$$\begin{aligned}\tilde{\Gamma}_G = (\frac{7N-13}{24} + \frac{S_{\mathcal{Q}}}{N})X' + (\frac{7N-13}{24} + \frac{S_{\mathcal{N}}}{N})X'' + mV \\ +m \sum_{\beta(i)=1} X_\beta + (m-1) \sum_{\beta(i)=-1} X_\beta \\ +m \sum_{\alpha(\omega)=1} W_\alpha + (m+1) \sum_{\alpha(\omega)=1} W_\alpha.\end{aligned}$$

- If $N \equiv 23 \pmod{24}$, let $m = \frac{13N+1}{12}$. Then

$$\tilde{\Gamma}_G = (\frac{7N+7}{24} + \frac{S_{\mathcal{Q}}}{N})X' + (\frac{7N+7}{24} + \frac{S_{\mathcal{N}}}{N})X'' + mV + m\sum_{\alpha} W_\alpha$$
$$+m \sum_{\beta(i)=\beta(\omega)=1} X_\beta + (m-1) \sum_{\beta(i)=1,\beta(\omega)=\neq 1} X_\beta$$
$$+(m-1) \sum_{\beta(i)=-1,\beta(\omega)=1} X_\beta + (m-2) \sum_{\beta(i)=-1,\beta(\omega)\neq 1} X_\beta.$$

Proof: It is straightforward to verify the theorem case-by-case using (8). As an intermediate step, we tabulate the multiplicities of the irreducible representations in the induced representations which comprise Γ_G. These tables are included in Appendix A (see §10 below). We sketch the case $N \equiv 1 \pmod{24}$.

We begin with the multiplicity of V. According to (8) and the $N \equiv 1 \pmod{24}$ table in Appendix A, the multiplicity of V in $\tilde{\Gamma}_G$ is

$$\frac{1}{2}\frac{N-1}{2} + \frac{1}{3}N - 1 + \frac{1}{N}\frac{N(N-1)}{2} = \frac{13N-13}{12} = m.$$

The multiplicity of an X_β (or a W_α with $\alpha(i) = \alpha(\omega) = 1$) in $\tilde{\Gamma}_G$ is the same, since they have the same rows in the table as V. Next consider the multiplicities of W' and W''. Since $S_{\mathcal{Q}} = S_{\mathcal{N}} = \frac{N(N-1)}{4}$, these rows in the table are exactly half of the V row, so the multiplicities of W' and W'' are both

$$\frac{1}{2}\frac{N-1}{4} + \frac{1}{3}\frac{N-1}{2} + \frac{1}{N}\frac{N(N-1)}{4} = \frac{13N-13}{24} = \frac{m}{2}.$$

Next consider a W_α with $\alpha(i) = 1$, $\alpha(\omega) \neq 1$. This row of the table differs from the V row in the second column, where it is $N+2$ instead of $N-1$. Since this entry is divided by 3 in the computation, the multiplicity in $\tilde{\Gamma}_G$ of such a W_α will be one more than the multiplicity of V, or $m+1$. More directly,

$$\frac{1}{2}\frac{N-1}{2} + \frac{1}{3}N + 2 + \frac{1}{N}\frac{N(N-1)}{2} = \frac{13N-13}{12} = m+1.$$

Similarly, consider a W_α with $\alpha(i) \neq 1$, $\alpha(\omega) = 1$. This row of the table differs from the V row in the first column, where again the net result is to add one to the multiplicity. In this case,

$$\frac{1}{2}\frac{N+3}{2} + \frac{1}{3}N - 1 + \frac{1}{N}\frac{N(N-1)}{2} = \frac{13N-13}{12} = m+1,$$

as desired. Finally, consider a W_α with $\alpha(i) \neq 1$, $\alpha(\omega) \neq 1$. We see that both of the first two columns of the table differ in this row from the V row, for a net addition of 2 to the multiplicity. In this case,

$$\frac{1}{2}\frac{N+3}{2} + \frac{1}{3}N + 2 + \frac{1}{N}\frac{N(N-1)}{2} = \frac{13N-13}{12} = m + 2.$$

This completes the verification of the theorem in the case $N \equiv 1 \pmod{24}$. A similar analysis can be used on the other cases. □

If $\tilde{\Gamma}_G$ has a $\mathbb{Q}[G]$-module structure, it may be computed more simply (see Joyner-Ksir [11]). In this case, the formula for it is

$$\tilde{\Gamma}_G = \bigoplus_{\pi \in G^*} \left[\sum_{\ell=1}^{L} (\dim \pi - \dim (\pi^{H_\ell})) \frac{R_\ell}{2} \right] \pi \tag{10}$$

where $\{H_1, ..., H_L\}$ represent the set of conjugacy classes of cyclic subgroups of G). If $\tilde{\Gamma}_G$ is not $\mathcal{G} = Gal(\mathbb{Q}(G)/\mathbb{Q})$-invariant, so there is no $\mathbb{Q}[G]$-module structure, then

$$\bigoplus_{\pi \in G^*} |\mathcal{G}| \cdot \left[\sum_{\ell=1}^{L} (\dim \pi - \dim (\pi^{H_\ell})) \frac{R_\ell}{2} \right] \pi \tag{11}$$

yields the G-module decomposition of the Galois-closure of $\tilde{\Gamma}_G$ (i.e., the smallest $\mathcal{G}$-invariant G-module containing $\tilde{\Gamma}_G$). See the remark after Corollary 6 in Joyner-Ksir [11]. Also, compare this with Lemma 3 in Khare-Prasad [13].

This motivates the following, as stated in the introduction.

Theorem 15: For $N > 5$ prime, the ramification module of $X(N)$ over $X(1)$ is $\mathcal{G}$-invariant if and only if $N \equiv 1 \pmod 4$.

Proof: As noted in §3, the action of $\mathcal{G}$ is as follows:

- *On X' and X'' or W' and W''*: The induced characters from H_1 and H_2 are invariant under this action. The induced characters from H_3 are not; the action exchanges θ^k where k is a quadratic residue $\pmod N$ with $\theta^{k'}$ where k' is not a quadratic residue $\pmod N$. However the sum appearing in the ramification module

$$Ind_{G_P}^G \left(\sum_{\ell=1}^{e_P - 1} \ell \theta_P^\ell \right) = \sum_{\ell=1}^{N-1} \ell Ind_{H_3}^G \theta^\ell$$

will include

$$\sum_{(\frac{\ell}{N})=1} \ell W' + \sum_{(\frac{\ell}{N})=-1} \ell W''$$

if $N \equiv 1 \pmod 4$, or

$$\sum_{(\frac{\ell}{N})=1} \ell X' + \sum_{(\frac{\ell}{N})=-1} \ell X''$$

if $N \equiv 3 \pmod 4$. Here $\left(\frac{*}{N}\right)$ denotes the Legendre symbol. If $N \equiv 1 \pmod 4$, then the sum of the quadratic residues is the same as the sum of the nonresidues, so the multiplicities of W' and W'' are the same. If $N \equiv 3 \pmod 4$, then the sum of the quadratic residues is not the same as the sum of the quadratic nonresidues, so these multiplicities are different. Therefore this part of the ramification module module is Galois invariant if and only if $N \equiv 1 \pmod 4$.

- *On* X_β: The multiplicity of a discrete series representation X_β depends only on $\beta(i)$ and $\beta(\omega)$. Observe that $\mathcal{G}$ fixes elements of $\mathbb{Q}$, therefore cannot exchange X_β having $\beta(i) = 1$ with an X_β having $\beta(i) = -1$, or X_β having $\beta(\omega) = 1$ with X_β having $\beta(\omega) \neq 1$. Therefore this part of the ramification module is Galois invariant.
- *On* W_α: Similarly, the multiplicity of a principal series representation W_α depends only on $\alpha(i)$ and $\alpha(\omega)$. Again, $\mathcal{G}$ will not exchange these, so this part of the module is Galois invariant.

Thus we see that $\tilde{\Gamma}_G$ is Galois-invariant if and only if $N \equiv 1 \pmod 4$□

In this case, we can use formula (10) to compute the ramification module directly from the restricted representations in section 4, and get the same result with fewer steps. If $N \equiv 3 \pmod 4$, formula (11) will yield the Galois closure of the ramification module.

6. Equivariant degree and Riemann-Roch space

Now we will define and compute the equivariant degree of a G-invariant divisor. This, together with Borne's formula (1) will allow us to compute the G-module structure of the Riemann-Roch space $L(D)$.

Fix a point $P \in X(N)$ and let D be a divisor on $X(N)$ of the form

$$D = \frac{r}{e_P} \sum_{g \in G} g(P) = r \sum_{g \in G/G_P} g(P),$$

where G_P denotes the stabilizer in G of P and $e_P = |G_P|$ denotes the ramification index at P.

When $r = 1$ such a divisor is called a *reduced orbit* by Borne [5]. When r is a multiple of e_P then D is the pull-back of a divisor on $X(1)$ via ψ_N in (3).

When D is as above, the *equivariant degree* of D is the virtual representation

$$deg_{eq}(D) = Ind_{G_P}^{G}\left(\sum_{\ell=1}^{r}\theta_P^{-\ell}\right),$$

where θ_P is the ramification character of $X(N)$ at P (a character of G_P). In general, the equivariant degree is additive on disjointly supported divisors. See for example Borne [5] for more details.

Example 16: Consider a hyperelliptic curve X for which every ramification point P has $e_P = 2$. Let D denote the sum of all the ramification points of X. Then (by definition) $deg_{eq}(D) = \Gamma_G = 2\tilde{\Gamma}_G$.

On the modular curve $X(N)$, the results of section 2.1 tell us that there are only four types of reduced orbits to consider: the stabilizer G_P of a point P in the support of D may have order 1, 2, 3, or N. Let D_0, D_1, D_2, and D_3 denote reduced orbits of each type. We compute the equivariant degree of rD_i in each case.

Case 1 : G_P is trivial. If G_P has order 1, then the ramification character is trivial. D_0 is an entire orbit, in fact the pullback of a divisor on $X(1)$. The equivariant degree of rD_0 is $r \cdot k[G]$, a multiple of the regular representation $k[G]$ of G.

Case 2 : $G_P \cong H_1$. If G_P has order 2, then the equivariant degree of D_1 is $Ind_{G_P}^{G}\theta_1$ (which was computed in section 4.1). Even multiples of D_1 will be pullbacks from $X(1)$, so $\deg_{eq}(2rD_1) = r \cdot k[G]$, and $\deg_{eq}((2r+1)D_1) = r \cdot k[G] + Ind_{G_P}^{G}\theta_1$.

Case 3 : $G_P \cong H_2$. The equivariant degree of D_2 is $Ind_{G_P}^{G}\theta_2^2$, which was computed in section 4.2. Note that $Ind_{G_P}^{G}\theta_2^2 = Ind_{G_P}^{G}\theta_2$, so $\deg_{eq}(2D_2) = 2Ind_{G_P}^{G}\theta_2$. But $3D_2$ is a pullback from $X(1)$, so $\deg_{eq}(3D_2) = k[G]$, and more generally $\deg_{eq}(3rD_2) = r \cdot k[G]$, $\deg_{eq}((3r+1)D_2) = r \cdot k[G] + Ind_{G_P}^{G}\theta_2$, and $\deg_{eq}((3r+2)D_2) = r \cdot k[G] + 2Ind_{G_P}^{G}\theta_2$.

Case 4 : $G_P \cong H_3$. This case is more complicated because $Ind_{H_3}^{G}\theta_3^k$ depends on whether k is a quadratic residue modulo N (see section 4.3). If $1 \leq r \leq N-1$, let q_r denote the number of quadratic residues modulo

N in the set $N-r, \ldots, N-1$, and let n_r denote the number of quadratic nonresidues modulo N in the same set. Then the equivariant degree $\deg_{eq}(rD_3)$ will be

$$\deg_{eq}(rD_3) = r\left(\sum_{\beta} X_\beta + V + \sum_{\alpha} W_\alpha\right) + \begin{cases} q_r W' + n_r W'', & N \equiv 1 \pmod 4 \\ q_r X' + n_r X'', & N \equiv 3 \pmod 4). \end{cases}$$

Now we would like to compute the G-module structure of the Riemann-Roch space $L(D)$ for a non-special G-invariant divisor D. First, let us consider which G-invariant divisors are non-special. To be non-special, it is sufficient to have $degD > 2g-2$, where

$$g = 1 + \frac{(N-6)(N^2-1)}{24}$$

is the genus of $X(N)$, so

$$2g-2 = \frac{(N-6)(N^2-1)}{12} = \frac{(N-6)}{6N}|G|.$$

The reduced orbits D_0, D_1 and D_2 have degrees $|G|$, $|G|/2$ and $|G|/3$, respectively, which are greater than $2g-2$ for any N, so these are always non-special. However, the reduced orbit D_3 has degree $|G|/N$, which is greater than $2g-2$ only if $N < 12$. A multiple of this reduced orbit, rD_3 will be non-special if $r > (N-6)/6$.

Once the equivariant degree of a non-special divisor has been computed, it is simple to combine this with the ramification module, using Borne's formula, to find the G-module structure of the Riemann-Roch space $L(D)$. The number of cases becomes quite cumbersome, but in each case there will be a number M such that all but the two smallest nontrivial irreducibles (either W' and W'' or X' and X'') will have multiplicity either M, $M+1$, $M+2$, or $M+3$, and the multiplicities of the two smallest irreducibles will be half of one of these numbers.

There is one special case worth mentioning: if D is a pull-back of an effective divisor $\overline{D}$ on $X(1)$ via ψ_N, then it will be non-special; moreover in this case

$$[\deg_{eq}(D)] = \deg(D_0)[k[G]],$$

and we will have the following decomposition of the Riemann-Roch space

$$[L(D)] = (1 + \deg(\overline{D}))[k[G]] - [\tilde{\Gamma}_G],$$

where $\tilde{\Gamma}_G$ is as above.

7. Examples

The computer algebra system GAP [4] computes information about $PSL(2,N)$; one can use it to compute character tables, induced characters and Schur inner products. In the examples of $X(7)$ and $X(11)$ below, we use GAP to explicitly compute the G-module structure of the ramification module and some Riemann-Roch spaces, in the cases $N = 7$ and $N = 11$. Appendix B (see §11 below) gives computations of the ramification module for higher N, and the GAP programs we used.

7.1. *The case $N = 7$*

The equivalence classes of irreducible representations of $PSL(2,7)$ are $G^* = \{\pi_1, \pi_2, ..., \pi_6\}$, where

$$\dim(\pi_1) = 1, \quad \dim(\pi_2) = \dim(\pi_3) = 3, \quad \dim(\pi_4) = 6, \quad \dim(\pi_5) = 7, \quad \dim(\pi_6) = 8.$$

We can identify

$$\pi_1 = \mathbf{1}_G, \quad \pi_2 = X', \quad \pi_3 = X'', \quad \pi_4 = X_\beta, \quad \pi_5 = V, \quad \pi_5 = W_\alpha.$$

Let $\zeta = e^{\frac{2\pi i}{7}}$ and let $\mathbb{Q}(q)$ denote the (quadratic) extension of $\mathbb{Q}$ by $q = \zeta + \zeta^2 + \zeta^4$. Let $\mathcal{G}$ denote the Galois group of $\mathbb{Q}(q)/\mathbb{Q}$. Then $\mathcal{G}$ acts on the irreducible representations G^* by swapping the two 3-dimensional representations (i.e., $\pi_2 = X'$ and $\pi_3 = X''$) and fixing the others.

There are 4 conjugacy classes of non-trivial cyclic subgroups of G, whose representatives are denoted by H_1 (order 2), H_2 (order 3), H_3 (order 7), H_4 (order 4). Thus H_4 is spurious. We use GAP to compute the induced characters:

- If $\theta_1 \in H_1^*$ then $\pi_{\theta_1} = Ind_{H_1}^G \theta_1$ is 84-dimensional. Moreover,

$$\pi_{\theta_1} \cong \begin{cases} 2\pi_2 \oplus 2\pi_3 \oplus 2\pi_4 \oplus 4\pi_5 \oplus 4\pi_6, & \theta_1 \neq 1, \\ \pi_1 \oplus \pi_2 \oplus \pi_3 \oplus 4\pi_4 \oplus 3\pi_5 \oplus 4\pi_6, & \theta_1 = 1. \end{cases}$$

- If $\theta_2 \in H_2^*$ then $\pi_{\theta_2} = Ind_{H_2}^G \theta_2$ is 56-dimensional. Moreover,

$$\pi_{\theta_2} \cong \begin{cases} \pi_2 \oplus \pi_3 \oplus 2\pi_4 \oplus 2\pi_5 \oplus 3\pi_6, & \theta_2 \neq 1, \\ \pi_1 \oplus \pi_2 \oplus \pi_3 \oplus 2\pi_4 \oplus 3\pi_5 \oplus 2\pi_6, & \theta_2 = 1. \end{cases}$$

- If $\theta_3 \in H_3^*$ is a fixed non-trivial character then $\pi_{\theta_3} = Ind_{H_3}^G \theta_3$ is 24-dimensional. Moreover,

$$\pi_{\theta_3^k} \cong \begin{cases} \pi_3 \oplus \pi_4 \oplus \pi_5 \oplus \pi_6 & k \text{ quad. non-res.} \pmod 7, \\ \pi_2 \oplus \pi_4 \oplus \pi_5 \oplus \pi_6, & k \text{ quad. res.} \pmod 7, \\ \pi_1 \oplus \pi_5 \oplus 2\pi_6, & k \equiv 0 \pmod 7). \end{cases}$$

These computations agree with the computations in section 4. This data allows us to easily compute the ramification module using equation (8):

$$[\tilde{\Gamma}_G] = [3\pi_2 \oplus 4\pi_3 \oplus 6\pi_4 \oplus 7\pi_5 \oplus 8\pi_6]. \qquad (12)$$

Note that this is not Galois-invariant, because π_2 and π_3 have different multiplicities. A naïve computation of the ramification module, using (10), yields the following. For brevity, we represent $m_1[\pi_1] + ... + m_6[\pi_6]$ as $(m_1, ..., m_6)$. We compute, using GAP, the quantities

$$(\dim \pi - \dim (\pi^{H_1}))_{i=1..6} = (1,3,3,6,7,8) - (1,1,1,4,3,4) = (0,2,2,2,4,4),$$

$$(\dim \pi - \dim (\pi^{H_2}))_{i=1..6} = (1,3,3,6,7,8) - (1,1,1,2,3,2) = (0,2,2,4,4,6),$$

$$(\dim \pi - \dim (\pi^{H_3}))_{i=1..6} = (1,3,3,6,7,8) - (1,0,0,0,1,2) = (0,3,3,6,6,6).$$

$$(\dim \pi - \dim (\pi^{H_4}))_{i=1..6} = (1,3,3,6,7,8) - (1,1,1,2,1,2) = (0,2,2,4,6,6).$$

Combining this with $R_1 = R_2 = R_3 = 1$ and $R_4 = 0$ in (8) gives

$$\begin{aligned}[\tilde{\Gamma}_G] &= [\textstyle\bigoplus_{i=1}^{6} \left[\sum_{\ell=1}^{4} (\dim \pi_i - \dim (\pi_i^{H_\ell})) \frac{R_\ell}{2}\right] \pi_i] \\ = (0,2,2,2,4,4)\tfrac{1}{2} &+ (0,2,2,4,4,6)\tfrac{1}{2} + (0,3,3,6,6,6)\tfrac{1}{2} + (0,2,2,4,6,6)\tfrac{0}{2} \\ &= (0,7/2,7/2,6,7,8) \\ &= \tfrac{7}{2}[\pi_2] + \tfrac{7}{2}[\pi_3] + 6[\pi_4] + 7[\pi_5] + 8[\pi_6].\end{aligned}$$

This is impossible, and therefore we see that $\tilde{\Gamma}_G$ does not have a $\mathbb{Q}[G]$-module structure in this case. However, the result does agree with (11): the multiplicity given for π_2 and π_3 is the average of the two actual multiplicities.

Now we will use GAP to compute the equivariant degree and Riemann-Roch module for some example divisors. As noted in section 6, any effective G-invariant divisor on $X(7)$ will be non-special. Since $X(1)$ is genus zero, for $N = 7$, Borne's formula (1) becomes

$$\begin{aligned}[L(D)] &= [\pi_1 \oplus 3\pi_2 \oplus 3\pi_3 \oplus 6\pi_4 \oplus 7\pi_5 \oplus 8\pi_6] + [deg_{eq}(D)] - [\tilde{\Gamma}_G] \\ &= [\pi_1 \oplus 3\pi_2 \oplus 3\pi_3 \oplus 6\pi_4 \oplus 7\pi_5 \oplus 8\pi_6] + [deg_{eq}(D)] \\ &\quad -3[\pi_2] - 4[\pi_3] - 6[\pi_4] - 7[\pi_5] - 8[\pi_6] \\ &= [\pi_1] - [\pi_3] + [deg_{eq}(D)].\end{aligned}$$

If D_1 is the reduced orbit of a point with stabilizer H_1, then

$$[deg_{eq}(D_1)] = [\pi_{\theta_1}] = [2\pi_2 \oplus 2\pi_3 \oplus 2\pi_4 \oplus 4\pi_5 \oplus 4\pi_6]$$

and

$$[L(D_1)] = [\pi_1 \oplus 2\pi_2 \oplus \pi_3 \oplus 2\pi_4 \oplus 4\pi_5 \oplus 4\pi_6].$$

If D_2 is the reduced orbit of a point with stabilizer H_2, then

$$[deg_{eq}(D_2)] = [\pi_{\theta_2}] = [\pi_2 \oplus \pi_3 \oplus 2\pi_4 \oplus 2\pi_5 \oplus 3\pi_6],$$

$$[deg_{eq}(2D_2)] = [2\pi_{\theta_2}] = [2\pi_2 \oplus 2\pi_3 \oplus 4\pi_4 \oplus 4\pi_5 \oplus 6\pi_6],$$

and

$$[L(D_2)] = [\pi_1 \oplus \pi_2 \oplus 2\pi_4 \oplus 2\pi_5 \oplus 3\pi_6],$$

$$[L(2D_2)] = [\pi_1 \oplus 2\pi_2 \oplus \pi_3 \oplus 4\pi_4 \oplus 4\pi_5 \oplus 6\pi_6].$$

If D_3 is the reduced orbit of a point with stabilizer H_3, then

$$[deg_{eq}(D_3)] = [\pi_{\theta_3^{N-1}}] = [\pi_3 \oplus \pi_4 \oplus \pi_5 \oplus \pi_6],$$

$$[deg_{eq}(2D_3)] = 2[\pi_3 \oplus \pi_4 \oplus \pi_5 \oplus \pi_6],$$

$$[deg_{eq}(3D_3)] = [\pi_2 + 2\pi_3] + 3[\pi_4 \oplus \pi_5 \oplus \pi_6],$$

$$[deg_{eq}(4D_3)] = [\pi_2 + 3\pi_3] + 4[\pi_4 \oplus \pi_5 \oplus \pi_6],$$

$$[deg_{eq}(5D_3)] = [2\pi_2 + 3\pi_3] + 5[\pi_4 \oplus \pi_5 \oplus \pi_6],$$

$$[deg_{eq}(6D_3)] = [3\pi_2 + 3\pi_3] + 6[\pi_4 \oplus \pi_5 \oplus \pi_6].$$

It follows that

$$L(D_3) = [\pi_1] - [\pi_3] + [\pi_3 \oplus \pi_4 \oplus \pi_5 \oplus \pi_6] = [\pi_1 \oplus \pi_4 \oplus \pi_5 \oplus \pi_6],$$

which is dimension 22, and

$$L(2D_3) = [\pi_1] - [\pi_3] + 2[\pi_3 \oplus \pi_4 \oplus \pi_5 \oplus \pi_6] = [\pi_1 \oplus \pi_3] + 2[\pi_4 \oplus \pi_5 \oplus \pi_6],$$

which is dimension 46.

7.2. *The case* $N = 11$

We next consider the case of $N = 11$. Up to equivalence, the irreducible representations of $G = PSL(2, 11)$ are $G^* = \{\pi_1, \pi_2, ..., \pi_8\}$, where

$$\dim(\pi_1) = 1, \quad \dim(\pi_2) = \dim(\pi_3) = 5, \quad \dim(\pi_4) = \dim(\pi_5) = 10,$$
$$\dim(\pi_6) = 11, \quad \dim(\pi_7) = \dim(\pi_8) = 12.$$

and we can identify

$$\pi_1 = \mathbf{1}_G, \quad \pi_2 = X', \quad \pi_3 = X'', \quad \pi_4 = X_{\beta_1}, \quad \pi_5 = X_{\beta_2},$$

$$\pi_6 = V, \quad \pi_7 = W_{\alpha_1}, \quad \pi_8 = W_{\alpha_2}.$$

We can use GAP to compute the character table of G and from the character values we can deduce some things about the characters $\alpha_1, \alpha_2 : \mathbb{F}^\times \to \mathbb{C}^\times$ and $\beta_1, \beta_2 : T \to \mathbb{C}^\times$. The generator ε of $\mathbb{F}^\times$ has order 10; it is sent by α_1 to $e^{4\pi i/5}$ and by α_2 to $e^{2\pi i/5}$. The generator τ of T has order 12; it is sent by β_1 to $e^{2\pi i/3}$ and by β_2 to $e^{\pi i/3}$. Because $N \equiv -1 \pmod{12}$, the numbers "i" $= \tau^3$ and "ω" $= \tau^2$ are both in T; we see from this that $\beta_1(i) = 1$, $\beta_2(i) = -1$, $\beta_1(\omega) \neq 1$, and $\beta_2(\omega) \neq 1$.

Thanks to the above discussion, it should be clear that if $K = \mathbb{Q}(G)$ denotes the abelian extension of $\mathbb{Q}$ generated by the character values of G then $\mathcal{G} = Gal(K/\mathbb{Q})$ acts by swapping the two irreducible 5-dimensionals X' and X'', and the two irreducible 12-dimensionals W_{α_1} and W_{α_2}, but not the two irreducible 10-dimensionals X_{β_1} and X_{β_2}.

There are 5 conjugacy classes of cyclic subgroups of G, whose representatives are denoted by H_1, ..., H_5. They satisfy $|H_1| = 2$, $|H_2| = 3$, $|H_3| = 11$, $|H_4| = 5$, $|H_5| = 6$. We are interested in H_1, H_2, and H_3; H_4 and H_5 are spurious. We use GAP to compute the induced characters:

- If $\theta_1 \in H_1^*$ then $\pi_{\theta_1} = Ind_{H_1}^G \theta_1$ is 330-dimensional. Moreover,

$$\pi_{\theta_1} \cong \begin{cases} 2\pi_2 \oplus 2\pi_3 \oplus 6\pi_4 \oplus 4\pi_5 \oplus 6\pi_6 \oplus 6\pi_7 \oplus 6\pi_8, & \theta_1 \neq 1, \\ \pi_1 \oplus 3\pi_2 \oplus 3\pi_3 \oplus 4\pi_4 \oplus 6\pi_5 \oplus 5\pi_6 \oplus 6\pi_7 \oplus 6\pi_8, & \theta_1 = 1. \end{cases}$$

- If $\theta_2 \in H_2^*$ then $\pi_{\theta_2} = Ind_{H_2}^G \theta_2$ is 220-dimensional. Moreover,

$$\pi_{\theta_2} = \begin{cases} 2\pi_2 \oplus 2\pi_3 \oplus 3\pi_4 \oplus 3\pi_5 \oplus 4\pi_6 \oplus 4\pi_7 \oplus 4\pi_8, & \theta_2 \neq 1, \\ \pi_1 \oplus \pi_2 \oplus \pi_3 \oplus 4\pi_4 \oplus 4\pi_5 \oplus 3\pi_6 \oplus 4\pi_7 \oplus 4\pi_8, & \theta_2 = 1. \end{cases}$$

- If $\theta_3 \in H_3^*$ is a fixed non-trivial character then $\pi_{\theta_3} = Ind_{H_3}^G \theta_3$ is 66-dimensional. Note that this is not $\mathcal{G}$-invariant. Moreover,

$$\pi_{\theta_3^k} = \begin{cases} \pi_2 \oplus \pi_4 \oplus \pi_5 \oplus \pi_6 \oplus \pi_7 \oplus \pi_8, \ k \text{ quad. non-res.} \pmod{11}, \\ \pi_3 \oplus \pi_4 \oplus \pi_5 \oplus \pi_6 \oplus \pi_7 \oplus \pi_8, \ k \text{ quad. res.} \pmod{11}, \\ \pi_1 \oplus \pi_6 \oplus 2\pi_7 \oplus 2\pi_8, \qquad k \equiv 0 \pmod{11}. \end{cases}$$

From this data, we can easily compute the ramification module, using equation (8). The equivalence class $[\Gamma_G]$ equals

$$\begin{gathered} \overline{R}_2[\pi_{\theta_1}] + \overline{R}_3([\pi_{\theta_2}] + 2[\pi_{\theta_2^2}]) \\ +\overline{R}_N([\pi_{\theta_3}] + \ldots + (N-1)[\pi_{\theta_3^{N-1}}]) \end{gathered}$$

where $N = 11$. As was mentioned above, using Shimura [3], we have $\overline{R}_2 = 660/2$, $\overline{R}_3 = 660/3$, $\overline{R}_{11} = 660/11$. Therefore,

$$\begin{aligned} [\Gamma_G] &= 660 \cdot [\tfrac{1}{2} \cdot (0,2,2,6,4,6,6,6) + (0,2,2,3,3,4,4,4) \\ &\quad + \cdot (0,3,2,5,5,5,5,5)] \\ &= 660 \cdot (0,6,5,11,10,12,12,12), \end{aligned}$$

by the formulas for the induced characters above. Therefore,
$[\tilde{\Gamma}_G] = (0,6,5,11,10,12,12,12)$.

As in the case of $X(7)$, this is not Galois invariant, because π_2 and π_3 have different multiplicities. Again we compute the Galois closure using equation (11):

$$\begin{aligned} (\dim \pi - \dim (\pi^{H_1}))_{i=1..8} &= (1,5,5,10,10,11,12,12) - (1,3,3,4,6,5,6,6) \\ &= (0,2,2,6,4,6,6,6), \end{aligned}$$

$$\begin{aligned} (\dim \pi - \dim (\pi^{H_2}))_{i=1..8} &= (1,5,5,10,10,11,12,12) - (1,1,1,4,4,3,4,4) \\ &= (0,4,4,6,6,8,8,8), \end{aligned}$$

$$\begin{aligned} (\dim \pi - \dim (\pi^{H_3}))_{i=1..8} &= (1,5,5,10,10,11,12,12) - (1,0,0,0,0,1,2,2) \\ &= (0,5,5,10,10,10,10,10), \end{aligned}$$

These calculations, combined with $R_1 = R_2 = R_3 = 1$ and $R_4 = R_5 = 0$, give

$$\begin{gathered} [\tilde{\Gamma}_G] = [\bigoplus_{i=1}^8 \left[\sum_{\ell=1}^5 (\dim \pi_i - \dim (\pi_i^{H_\ell}))\tfrac{R_\ell}{2}\right] \pi_i] \\ = (0,2,2,6,4,6,6,6)\tfrac{1}{2} + (0,2,2,3,3,4,4,4)\tfrac{1}{2} + (0,5,5,10,10,10,10,10)\tfrac{1}{2} \\ = (0,11/2,11/2,11,10,12,12,12) \\ = \tfrac{11}{2}[\pi_2] + \tfrac{11}{2}[\pi_3] + 11[\pi_4] + 10[\pi_5] + 12[\pi_6] + 12[\pi_7] + 12[\pi_8]. \end{gathered}$$

We can see directly from this calculation $\tilde{\Gamma}_G$ does not have a $\mathbb{Q}[G]$-module structure in this case. However, the result does agree with (11): we have computed the average of the multiplicities of the representations under the Galois action.

For every non-special G-equivariant divisor D on $X(11)$, the formula (1) says

$$\begin{aligned}[L(D)] &= [\pi_1 \oplus 5\pi_2 \oplus 5\pi_3 \oplus 10\pi_4 \oplus 10\pi_5 \oplus 11\pi_6 \oplus 12\pi_7 \oplus 12\pi_8] \\ &\quad +[deg_{eq}(D)] - [\tilde{\Gamma}_G] \\ &= [\pi_1] - [\pi_2] - [\pi_4] - [\pi_6] + [deg_{eq}(D)].\end{aligned}$$

If D_1 is the reduced orbit of a point with stabilizer H_1, then D_1 is non-special and

$$[deg_{eq}(D_1)] = [\pi_{\theta_1}] = [2\pi_2 \oplus 2\pi_3 \oplus 6\pi_4 \oplus 4\pi_5 \oplus 6\pi_6 \oplus 6\pi_7 \oplus 6\pi_8],$$

and

$$[L(D_1)] = [\pi_1 \oplus \pi_2 \oplus 2\pi_3 \oplus 5\pi_4 \oplus 4\pi_5 \oplus 5\pi_6 \oplus 6\pi_7 \oplus 6\pi_8].$$

If D_2 is the reduced orbit of a point with stabilizer H_2, then D_2 is non-special and

$$[deg_{eq}(D_2)] = [\pi_{\theta_2^2}] = [2\pi_2 \oplus 2\pi_3 \oplus 3\pi_4 \oplus 3\pi_5 \oplus 4\pi_6 \oplus 4\pi_7 \oplus 4\pi_8],$$

and

$$[L(D_2)] = [\pi_1 \oplus \pi_2 \oplus 2\pi_3 \oplus 2\pi_4 \oplus 3\pi_5 \oplus 3\pi_6 \oplus 4\pi_7 \oplus 4\pi_8].$$

If D_3 is the reduced orbit of a point with stabilizer H_3, then D_3 is non-special and

$$[deg_{eq}(D_3)] = [\pi_{\theta_3^{10}}] = [\pi_2 \oplus \pi_4 \oplus \pi_5 \oplus \pi_6 \oplus \pi_7 \oplus \pi_8],$$

$$[deg_{eq}(2D_3)] = [\pi_2 \oplus \pi_3 \oplus 2\pi_4 \oplus 2\pi_5 \oplus 2\pi_6 \oplus 2\pi_7 \oplus 2\pi_8],$$

and

$$[L(D_3)] = [\pi_1 \oplus \pi_5 \oplus \pi_7 \oplus \pi_8],$$

$$[L(2D_3)] = [\pi_1 \oplus \pi_3 \oplus \pi_4 \oplus 2\pi_5 \oplus \pi_6 \oplus 2\pi_7 \oplus 2\pi_8].$$

8. Application to codes

In this section we consider applications of our previous results to we discuss connections with the theory of error-correcting codes.

Assume that ℓ is a good prime. Also, assume that k contains all the character values of G and that k is finite, where k denotes the field of definition of the reduction of X mod ℓ. (The point is that we want to be able to work over a separable algebraic closure $\overline{k}$ of k but then be able to take $Gal(\overline{k}/k)$-fixed points to obtain our results.) We recall some background on AG codes following for example Joyner-Traves [12].

Let $P_1, ..., P_n \in X(k)$ be distinct points and $E = P_1 + ... + P_n \in Div(X)$ be stabilized by G. This implies that G acts on the set $supp(E)$ by permutation. Assume D is a G-equivariant divisor of $X(k)$, so G acts on the Riemann-Roch space $L(D)$. Assume these divisors have disjoint support, $\mathrm{supp}(D) \cap \mathrm{supp}(E) = \emptyset$. Let $C = C(D, E)$ denote the AG code

$$C = \{(f(P_1), ..., f(P_n)) \mid f \in L(D)\}. \tag{13}$$

This is the image of $L(D)$ under the evaluation map

$$\begin{aligned} eval_E : L(D) &\to k^n, \\ f &\longmapsto (f(P_1), ..., f(P_n)). \end{aligned} \tag{14}$$

The group G acts on C by $g \in G$ sending $c = (f(P_1), ..., f(P_n)) \in C$ to $c' = (f(g^{-1}(P_1)), ..., f(g^{-1}(P_n)))$, where $f \in L(D)$. First, we observe that this map sending $c \longmapsto c'$, denoted $\phi(g)$, is well-defined. In other words, if $eval_E$ is not injective and c is also represented by $f' \in L(D)$, so $c = (f'(P_1), ..., f'(P_n)) \in C$, then we can easily verify $(f(g^{-1}(P_1)), ..., f(g^{-1}(P_n))) = (f'(g^{-1}(P_1)), ..., f'(g^{-1}(P_n)))$. (Indeed, G acts on the set $supp(E)$ by permutation.) This map $\phi(g)$ induces a homomorphism of G into the permutation automorphism group of the code $\mathrm{Aut}(C)$, denoted

$$\phi : G \to \mathrm{Aut}(C) \tag{15}$$

For properties of this map, see Joyner-Traves [12]. In particular, the following is known.

Lemma 17: *If D and E satisfy* $\deg(D) > 2g$ *and* $\deg(E) > 2g + 2$ *then ϕ and $eval_E$ are injective.*

Proof: We say that the space $L(D)$ **separates points** if for all points $P,Q \in X$, $f(P) = f(Q)$ (for all $f \in L(D)$) implies $P = Q$ (see Hartshorne [9, chapter II, §7]). By Proposition IV.3.1 in Hartshorne [9], D very ample implies $L(D)$ separates points. In general, if $L(D)$ separates points then

$$\mathrm{Ker}(\phi) = \{g \in G \mid g(P_i) = P_i,\ 1 \le i \le n\}.$$

It is known (proof of Prop. VII3.3, [22]) that if $n = \deg(E) > 2g + 2$ then $\{g \in G \mid g(P_i) = P_i,\ 1 \le i \le n\}$ is trivial. Therefore, if $n > 2g + 2$ and $L(D)$ separates points then ϕ is injective. Since (see Corollary IV.3.2 in Hartshorne [9]) $\deg(D) > 2g$ implies D is very ample, the lemma follows.□

As an amusing application of our theory, we show how to easily recover some results of Tsafsman and Vladut on AG codes associated to modular curves.

First, we recall some notation and results from Tsafsman-Vladut [23]. Let $A_N = \mathbb{Z}[\zeta_N, 1/N]$, where $\zeta_N = e^{2\pi i/N}$, let K_N denote the quadratic subfield of $\mathbb{Q}(\zeta_N)$, and let $B_N = A_N \cap K_N$. There is a scheme $X(N)/\mathbb{Z}[1/N]$ which represents a moduli functor "parameterizing" elliptic curves E with a level N structure α_N. There is a scheme $X_P(N)/\mathbb{Z}[1/N]$ which represents a moduli functor "parameterizing" elliptic curves E with a "projective" level N structure β_N. If P is a prime ideal in the ring of integers $\mathcal{O}_{K_N}$ dividing ℓ then the reduction of the form of $X(N)$ defined over K_N, denoted $X(N)/P$, is a smooth projective absolutely irreducible curve over the residue field $k(P)$, with a $PSL_2(\mathbb{Z}/N\mathbb{Z})$-action commuting with the reduction. Similarly, with $X(N)$ replaced by $X_P(N)$. Recall from §4.1.3 of Tsafsman-Vladut [23] that

$$k(P) = \begin{cases} GF(\ell^2), & (\ell) = P, \\ GF(\ell), & (\ell) = PP', \end{cases}$$

Let $X'_N = X_P(N)/P$ and let

$$\psi'_N : X'_N \to X'_N/PSL_2(\mathbb{Z}/N\mathbb{Z}) \cong \mathbb{P}^1$$

denote the quotient map. Let D_∞ denote the reduced orbit (in the sense of Borne) of the point ∞, so $deg(D_\infty) = |G|/N$. Let $D = rD_\infty$, for $r \ge 1$. According to [23], in general, this divisor is actually defined over $GF(\ell)$, not just $k(P)$. Moreover, $\deg(D) = r \cdot (N^2 - 1)/2$. Let $E = P_1 + ... + P_n$ be the sum of all the supersingular points of X'_N and let

$$C = C(X'_N, E, D) = \{(f(P_1), ..., f(P_n)) \mid f \in L(D)\}$$

denote the AG code associated to X'_N, D, E. This is a G-module, via (15). Moreover, choosing r suitably yields a "good" family of codes with large automorphism group.

In fact, if D is "sufficiently large" (so, D is non-special and both ϕ, $eval_E$ are injective) then the Brauer-character analoguess of formulas in section 6 give not only the G-module structure of each $L(rD_\infty)$, but that of C as well.

See also Remark 4.1.66 in Tsafsman-Vladut [23].

9. AG codes associated to $X(7)$

We focus on the Klein quartic example started in Joyner-Ksir [11]. We also use Elkies [6] as a general reference.

Let $k = GF(43)$. This field contains 7th roots of unity ($\zeta_7 = 41$), cube roots of unity ($\zeta_3 = 36$), and the square root of -7 (take $\sqrt{-7} = 6$). Consider

$$\rho_1 = \begin{pmatrix} {\zeta_7}^4 & 0 & 0 \\ 0 & {\zeta_7}^2 & 0 \\ 0 & 0 & \zeta_7 \end{pmatrix} = \begin{pmatrix} 16 & 0 & 0 \\ 0 & 4 & 0 \\ 0 & 0 & 41 \end{pmatrix},$$

$$\rho_2 = \begin{pmatrix} 0 & 1 & 0 \\ 0 & 0 & 1 \\ 1 & 0 & 0 \end{pmatrix},$$

and

$$\rho_3 = \begin{pmatrix} \frac{\zeta_7 - {\zeta_7}^6}{\sqrt{-7}} & \frac{{\zeta_7}^2 - {\zeta_7}^5}{\sqrt{-7}} & \frac{{\zeta_7}^4 - {\zeta_7}^3}{\sqrt{-7}} \\ \frac{{\zeta_7}^2 - {\zeta_7}^5}{\sqrt{-7}} & \frac{{\zeta_7}^4 - {\zeta_7}^3}{\sqrt{-7}} & \frac{\zeta_7 - {\zeta_7}^6}{\sqrt{-7}} \\ \frac{{\zeta_7}^4 - {\zeta_7}^3}{\sqrt{-7}} & \frac{\zeta_7 - {\zeta_7}^6}{\sqrt{-7}} & \frac{{\zeta_7}^2 - {\zeta_7}^5}{\sqrt{-7}} \end{pmatrix} = \begin{pmatrix} 11 & 37 & 39 \\ 37 & 39 & 11 \\ 39 & 11 & 37 \end{pmatrix}.$$

It may be checked that these matrices preserve the form

$$\phi(x, y, z) = x^3 y + y^3 z + z^3 x,$$

over k. They generate the subgroup $G \cong PSL_2(7)$ of order 168 in $PGL(3, k)$. The Klein curve $x^3y + y^3z + z^3x = 0$, denoted here by X, has no other automorphisms in characteristic 43, so $G = Aut_k(X)$.

Let D_∞ denote the reduced orbit of the point ∞, so $deg(D_\infty) = |G|/N = 24$, and let $D = rD_\infty$. Let $E = P_1 + ... + P_n$ denote the sum of the remaining $k(P)$-rational points of X, so D and E have disjoint support.

If C is as in (14) then then map ϕ in (15) is injective. Since $eval_E$ is injective as well, the G-module structure of C is the same as that of $L(D)$, which is known thanks to the Brauer-character analog of the formula (10). See §7.1 above and also Example 3, Joyner-Ksir [11].

Magma tells us that the points of $X(k)$ are

$$\begin{gathered}
\{(0:1:0),(0:0:1),(1:0:0),(19:9:1),(36:9:1),(31:9:1),\\
(19:27:1),(1:38:1),(27:38:1),(15:38:1),(12:28:1),(38:28:1),\\
(36:28:1),(40:41:1),(10:25:1),(20:25:1),(13:25:1),\\
(20:32:1),(42:10:1),(35:10:1),(9:10:1),(40:30:1),\\
(13:30:1),(33:30:1),(24:4:1),(25:36:1),(12:36:1),(6:36:1),\\
(12:22:1),(14:23:1),(8:23:1),(21:23:1),(24:26:1),(37:26:1),\\
(25\ \ 26:1),(23:35:1),(15:14:1),(33:14:1),(38:14:1),(33:42:1),\\
(38:14:1),(33:42:1),(17:40:1),\\
(4:40:1),(22:40:1),(5:34:1),(15:34:1),(23:34:1),\\
(31:16:1),(40:15:1),(37:15:1),(9:15:1),(37:2:1),(11:6:1),\\
(39:6:1),(36:6:1),(31:18:1),(9:18:1),(3:18:1),(10:11:1),\\
(5:13:1),(24:13:1),(14:13:1),(5:39:1),(41:31:1),\\
(13:31:1),(32:31:1),(10:7:1),(14:7:1),(19:7:1),(6:21:1),\\
(17:17:1),(3:17:1),(23:17:1),(3:8:1),(25:24:1),\\
(16:24:1),(2:24:1),(20:29:1),(17:29:1),(6:29:1),(38:1:1)\}
\end{gathered}$$

Magma tells us that the orbit of $(1:0:0) \in X(k)$ under G is

$$\begin{gathered}
\{(1:0:0),(36:28:1),(32:31:1),(15:34:1),(9:18:1),(37:2:1),\\
(2:24:1),(3:8:1),(22:40:1),(19:27:1),(13:30:1),\\
(5:39:1),(8:23:1),(14:7:1),(25:26:1),(33:42:1),\\
(17:29:1),(42:10:1),(0:10),(0:0:1),\\
(20:32:1),(12:22:1),(27:38:1),(39:6:1)\}
\end{gathered}$$

We evaluate each element of the Riemann-Roch space $L(D)$ at a point in the complement of the above-mentioned orbit in the set of rational points $X(k)$. For concreteness, a typical basis element of $L(D)$, when $r = 1$, looks like

$$y^j x^i (y+1)^{-1}(y+3)^{-1}(y+4)^{-1}(y+5)^{-1}(y+9)^{-1}(y+11)^{-1}(y+12)^{-1}\times$$
$$\times(y+13)^{-1}(y+14)^{-1}(y+15)^{-1}(y+16)^{-1}(y+17)^{-1}(y+19)^{-1}(y+20)^{-1}\times$$
$$\times(y+21)^{-1}(y+25)^{-1}(y+33)^{-1}(y+35)^{-1}(y+36)^{-1}(y+37)^{-1}\times$$
$$\times(y+41)^{-1}(x^2y^8+34x^2y+16xy^{11}+2xy^4+17y^{14}+33y^7+17),$$

for example with $i = 2$ and $j = 6$. In fact, all such elements with $i = 1, 2$ and $0 \leq j \leq 6$ are basis elements of $L(D)$.

With $r = 1$, Magma says that $C = C(X, D, E)$ is a $[56, 22, 32]$ code‡. It has a generator matrix of in standard form $(I \mid A)$, where I is the 22×22 identity and A can be given explicitly as well. In this case, $eval_E$ is injective.

With $r = 2$, Magma says that $C = C(X, D, E)$ is a $[56, 46, 8]$ code. It has a generator matrix of in standard form $(I \mid A)$, where I is the 46×46 identity and A can be given explicitly as well. In this case, $eval_E$ is injective.

With $r = 3$, Magma says that $C = C(X, D, E)$ is a $[56, 56, 1]$ code. In this case, $eval_E$ is not injective. Indeed, $\dim L(3D_\infty) = 70$.

Remark 18: Indeed, it is known more generally, that for an AG code constructed as above from a curve of genus g that $n \leq \dim(C)+d(C)+g-1$, where $d(C)$ denotes the minimum distance (Theorem 3.1.1 in Tsafsman-Vladut [23]). Therefore, as an AG code, the codes constructed above with $r = 1, 2$ are as "long as possible". Are they best possible over $GF(43)$ for those k, d?

Remark 19: An attempt to compute the permutation automorphism group of the above $[56, 22, 32]$ code using MAGMA failed due to lack of memory (on an AMD64 linux machine with 1.5 G of RAM).

Remark 20: In general, §4.1 of [23] shows how to construct a family of 'good" codes from the curves $X = X'_N$, for $N > 5$ is a prime, with automorphism group $G = PSL(2, p)$.

10. Appendix A: Tables for Theorem 9

The following tables are an intermediate step in the proof of theorem 9. They were obtained by compiling the results in §§4.1-4.3.

‡In other words, C has length 56, dimension 22 over k, and minimum distance 32.

$N \equiv 1 \pmod{24}$	$Ind_{H_1}^G \theta_1$	$Ind_{H_2}^G(\theta_2 + 2\theta_2^2)$	$Ind_{H_3}^G(\sum_{\ell=1}^{N-1} \ell\theta_3^\ell)$
$\mathbf{1}_G$	0	0	0
W'	$\frac{N-1}{4}$	$\frac{N-1}{2}$	$S_\mathcal{Q}$
W''	$\frac{N-1}{4}$	$\frac{N-1}{2}$	$S_\mathcal{N}$
W_α such that $\alpha(i) = \alpha(\omega) = 1$	$\frac{N-1}{2}$	$N-1$	$\frac{N(N-1)}{2}$
W_α such that $\alpha(i) = 1, \alpha(\omega) \neq 1$	$\frac{N-1}{2}$	$N+2$	$\frac{N(N-1)}{2}$
W_α such that $\alpha(i) = -1, \alpha(\omega) = 1$	$\frac{N+3}{2}$	$N-1$	$\frac{N(N-1)}{2}$
W_α such that $\alpha(i) = -1, \alpha(\omega) \neq 1$	$\frac{N+3}{2}$	$N+2$	$\frac{N(N-1)}{2}$
X_β	$\frac{N-1}{2}$	$N-1$	$\frac{N(N-1)}{2}$
V	$\frac{N-1}{2}$	$N-1$	$\frac{N(N-1)}{2}$

$N \equiv 5 \pmod{24}$	$Ind_{H_1}^G \theta_1$	$Ind_{H_2}^G(\theta_2 + 2\theta_2^2)$	$Ind_{H_3}^G(\sum_{\ell=1}^{N-1} \ell\theta_3^\ell)$
$\mathbf{1}_G$	0	0	0
W'	$\frac{N+3}{4}$	$\frac{N+1}{2}$	$S_\mathcal{Q}$
W''	$\frac{N+3}{4}$	$\frac{N+1}{2}$	$S_\mathcal{N}$
W_α such that $\alpha(i) = 1$	$\frac{N-1}{2}$	$N+1$	$\frac{N(N-1)}{2}$
W_α such that $\alpha(i) = -1$	$\frac{N+3}{2}$	$N+1$	$\frac{N(N-1)}{2}$
X_β such that $\beta(\omega) = 1$	$\frac{N-1}{2}$	$N+1$	$\frac{N(N-1)}{2}$
X_β such that $\beta(\omega) \neq 1$	$\frac{N-1}{2}$	$N-2$	$\frac{N(N-1)}{2}$
V	$\frac{N-1}{2}$	$N+1$	$\frac{N(N-1)}{2}$

$N \equiv 7 \pmod{24}$	$Ind_{H_1}^G \theta_1$	$Ind_{H_2}^G(\theta_2 + 2\theta_2^2)$	$Ind_{H_3}^G(\sum_{\ell=1}^{N-1} \ell\theta_3^\ell)$
$\mathbf{1}_G$	0	0	0
W_α such that $\alpha(\omega) = 1$	$\frac{N+1}{2}$	$N-1$	$\frac{N(N-1)}{2}$
W_α such that $\alpha(\omega) \neq 1$	$\frac{N+1}{2}$	$N+2$	$\frac{N(N-1)}{2}$
X'	$\frac{N+1}{4}$	$\frac{N-1}{2}$	$S_\mathcal{Q}$
X''	$\frac{N+1}{4}$	$\frac{N-1}{2}$	$S_\mathcal{N}$
X_β such that $\beta(i) = 1$	$\frac{N+1}{2}$	$N-1$	$\frac{N(N-1)}{2}$
X_β such that $\beta(i) = -1$	$\frac{N-3}{2}$	$N-1$	$\frac{N(N-1)}{2}$
V	$\frac{N+1}{2}$	$N-1$	$\frac{N(N-1)}{2}$

$N \equiv 11 \pmod{24}$	$Ind_{H_1}^G \theta_1$	$Ind_{H_2}^G(\theta_2 + 2\theta_2^2)$	$Ind_{H_3}^G(\sum_{\ell=1}^{N-1} \ell\theta_3^\ell)$
$\mathbf{1}_G$	0	0	0
W_α	$\frac{N+1}{2}$	$N+1$	$\frac{N(N-1)}{2}$
X'	$\frac{N-3}{4}$	$\frac{N+1}{2}$	$S_\mathcal{Q}$
X''	$\frac{N-3}{4}$	$\frac{N+1}{2}$	$S_\mathcal{N}$
X_β such that $\beta(i) = \beta(\omega) = 1$	$\frac{N+1}{2}$	$N+1$	$\frac{N(N-1)}{2}$
X_β such that $\beta(i) = 1, \beta(\omega) \neq 1$	$\frac{N+1}{2}$	$N-2$	$\frac{N(N-1)}{2}$
X_β such that $\beta(i) = -1, \beta(\omega) = 1$	$\frac{N-3}{2}$	$N+1$	$\frac{N(N-1)}{2}$
X_β such that $\beta(i) = -1, \beta(\omega) \neq 1$	$\frac{N-3}{2}$	$N-2$	$\frac{N(N-1)}{2}$
V	$\frac{N+1}{2}$	$N+1$	$\frac{N(N-1)}{2}$

$N \equiv 13 \pmod{24}$	$Ind_{H_1}^G \theta_1$	$Ind_{H_2}^G(\theta_2 + 2\theta_2^2)$	$Ind_{H_3}^G(\sum_{\ell=1}^{N-1} \ell\theta_3^\ell)$
$\mathbf{1}_G$	0	0	0
W'	$\frac{N+3}{4}$	$\frac{N-1}{2}$	$S_{\mathcal{Q}}$
W''	$\frac{N+3}{4}$	$\frac{N-1}{2}$	$S_{\mathcal{N}}$
W_α such that $\alpha(i) = \alpha(\omega) = 1$	$\frac{N-1}{2}$	$N-1$	$\frac{N(N-1)}{2}$
W_α such that $\alpha(i) = 1, \alpha(\omega) \neq 1$	$\frac{N-1}{2}$	$N+2$	$\frac{N(N-1)}{2}$
W_α such that $\alpha(i) = -1, \alpha(\omega) = 1$	$\frac{N+3}{2}$	$N-1$	$\frac{N(N-1)}{2}$
W_α such that $\alpha(i) = -1, \alpha(\omega) \neq 1$	$\frac{N+3}{2}$	$N+2$	$\frac{N(N-1)}{2}$
X_β	$\frac{N-1}{2}$	$N-1$	$\frac{N(N-1)}{2}$
V	$\frac{N-1}{2}$	$N-1$	$\frac{N(N-1)}{2}$

$N \equiv 17 \pmod{24}$	$Ind_{H_1}^G \theta_1$	$Ind_{H_2}^G(\theta_2 + 2\theta_2^2)$	$Ind_{H_3}^G(\sum_{\ell=1}^{N-1} \ell\theta_3^\ell)$
$\mathbf{1}_G$	0	0	0
W'	$\frac{N-1}{4}$	$\frac{N+1}{2}$	$S_{\mathcal{Q}}$
W''	$\frac{N-1}{4}$	$\frac{N+1}{2}$	$S_{\mathcal{N}}$
W_α such that $\alpha(i) = 1$	$\frac{N-1}{2}$	$N+1$	$\frac{N(N-1)}{2}$
W_α such that $\alpha(i) = -1$	$\frac{N+3}{2}$	$N+1$	$\frac{N(N-1)}{2}$
X_β such that $\beta(\omega) = 1$	$\frac{N-1}{2}$	$N+1$	$\frac{N(N-1)}{2}$
X_β such that $\beta(\omega) \neq 1$	$\frac{N-1}{2}$	$N-2$	$\frac{N(N-1)}{2}$
V	$\frac{N-1}{2}$	$N+1$	$\frac{N(N-1)}{2}$

$N \equiv 19 \pmod{24}$	$Ind_{H_1}^G \theta_1$	$Ind_{H_2}^G (\theta_2 + 2\theta_2^2)$	$Ind_{H_3}^G (\sum_{\ell=1}^{N-1} \ell\theta_3^\ell)$
$\mathbf{1}_G$	0	0	0
W_α such that $\alpha(\omega) = 1$	$\frac{N+1}{2}$	$N-1$	$\frac{N(N-1)}{2}$
W_α such that $\alpha(\omega) \neq 1$	$\frac{N+1}{2}$	$N+2$	$\frac{N(N-1)}{2}$
X'	$\frac{N-3}{4}$	$\frac{N-1}{2}$	$S_{\mathcal{Q}}$
X''	$\frac{N-3}{4}$	$\frac{N-1}{2}$	$S_{\mathcal{N}}$
X_β such that $\beta(i) = 1$	$\frac{N+1}{2}$	$N-1$	$\frac{N(N-1)}{2}$
X_β such that $\beta(i) = -1$	$\frac{N-3}{2}$	$N-1$	$\frac{N(N-1)}{2}$
V	$\frac{N+1}{2}$	$N-1$	$\frac{N(N-1)}{2}$

$N \equiv 23 \pmod{24}$	$Ind_{H_1}^G \theta_1$	$Ind_{H_2}^G (\theta_2 + 2\theta_2^2)$	$Ind_{H_3}^G (\sum_{\ell=1}^{N-1} \ell\theta_3^\ell)$
$\mathbf{1}_G$	0	0	0
W_α	$\frac{N+1}{2}$	$N+1$	$\frac{N(N-1)}{2}$
X'	$\frac{N+1}{4}$	$\frac{N+1}{2}$	$S_{\mathcal{Q}}$
X''	$\frac{N+1}{4}$	$\frac{N+1}{2}$	$S_{\mathcal{N}}$
X_β such that $\beta(i) = \beta(\omega) = 1$	$\frac{N+1}{2}$	$N+1$	$\frac{N(N-1)}{2}$
X_β such that $\beta(i) = 1, \beta(\omega) \neq 1$	$\frac{N+1}{2}$	$N-2$	$\frac{N(N-1)}{2}$
X_β such that $\beta(i) = -1, \beta(\omega) = 1$	$\frac{N-3}{2}$	$N+1$	$\frac{N(N-1)}{2}$
X_β such that $\beta(i) = -1, \beta(\omega) \neq 1$	$\frac{N-3}{2}$	$N-2$	$\frac{N(N-1)}{2}$
V	$\frac{N+1}{2}$	$N+1$	$\frac{N(N-1)}{2}$

11. Appendix B: Ramification modules and GAP code

We used GAP to compute the ramification module for many small values of N. Here are the first results:

N	G-module structure of $\tilde{\Gamma_G}$ by definition	G-module structure of $\tilde{\Gamma_G}$ using JK formula
5	[0, 3, 3, 4, 5]	[0, 3, 3, 4, 5]
7	[0, 3, 4, 6, 7, 8]	[0, 7/2, 7/2, 6, 7, 8]
11	[0, 5, 6, 11, 10, 12, 12, 12]	[0, 11/2, 11/2, 11, 10, 12, 12, 12]
13	[0, 7, 7, 13, 13, 13, 13, 14, 15]	[0, 7, 7, 13, 13, 13, 13, 14, 15]
17	[0, 9, 9, 18, 17, 17, 17, 18, ...]	[0, 9, 9, 18, 17, 17, 17, 18, ...]
19	[0, 9, 10, 20, 20, 19, 19, 20, ...]	[0, 19/2, 19/2, 20, 20, 19, 19, 20, ...]
23	[0, 11, 14, 24, 24, 24, 23, 23, ...]	[0, 25/2, 25/2, 24, 24, 24, 23, 23, ...]
29	[0, 16, 16, 30, 31, 31, 30, ...]	[0, 16, 16, 30, 31, 31, 30, ...]

GAP was able to go up to and including $N = 277$ (in which case G is order about 20 million).

Here is the GAP code we used to create the table above.

```
ram_module_X:=function(p) ## p is a prime ## output is (m1,...,mn)

## where n = # conj classes of G=PSL(2,p)##
and mi = mult of pi_i in ram mod of G ##

local G,i,j,n,H,G1,H_chars,CG,G_chars,w,m,theta,pi_theta_to_the;
  G:=PSL(2,p);
  H:=[];
  H_chars:=[];
  CG:=ConjugacyClassesSubgroups(G);
  H[1]:=Representative(CG[2]); # size 2
  H[2]:=Representative(CG[3]); # size 3
  n :=Size(CG);
  for i in [1..n] do
    if Size(Representative(CG[i]))=p then
      H[3]:=Representative(CG[i]);
    fi;
  od;
  for i in [1..Size(H)] do
   H_chars[i]:=Irr(H[i]);
  od;
  G_chars:=Irr(G);
  m:=[];
  m[1]:=[];m[2]:=[];m[3]:=[];
  theta:=List([1..3],i->H_chars[i][2]);
  pi_theta_to_the:=List([1..3],i->Sum([1..(Size(H[i])-1)],
      j->j*InducedClassFunction(theta[i]^(j),G)));
  for i in [1..3] do
    m[i]:=List(G_chars, pi->ScalarProduct(pi_theta_to_the[i],pi))/Size(H[i]);
  od;
Print("\n\n m = ",m,"\n\n"); return Sum(m); end;
```

```
ram_module_X_JK:=function(p) ## p is a prime ## output is (m1,...,mn)

## where n = # conj classes of G=PSL(2,p) ## and
mi = "mult of pi_i in ram mod of G"

using JK formula ## local G,i,j,n,H,G1,H_chars,CG,G_chars,A,B,C,D,pi;
  G:=PSL(2,p);
  H:=[];
  H_chars:=[];
  CG:=ConjugacyClassesSubgroups(G);
  H[1]:=Representative(CG[2]); # size 2
  H[2]:=Representative(CG[3]); # size 3
  n:=Size(CG);
  for i in [1..n] do
    if Size(Representative(CG[i]))=p then
      H[3]:=Representative(CG[i]);
    fi;
  od;
  for i in [1..Size(H)] do
   H_chars[i]:=Irr(H[i]);
  od;
  G_chars:=Irr(G);
  n:=Length(G_chars);
  A:=[];
  for i in [1..n] do
    pi:=G_chars[i];
    A[i]:=ScalarProduct(H_chars[1][1],RestrictedClassFunction(pi,H[1]));
  od;
  B:=[];
  for i in [1..n] do
    pi:=G_chars[i];
    B[i]:=ScalarProduct(H_chars[2][1],RestrictedClassFunction(pi,H[2]));
  od;
  C:=[];
  for i in [1..n] do
    pi:=G_chars[i];
    C[i]:=ScalarProduct(H_chars[3][1],RestrictedClassFunction(pi,H[3]));
  od;
  D:=[];
  for i in [1..n] do
    pi:=G_chars[i];
    D[i]:=DegreeOfCharacter(pi);
  od;
 return (1/2)*(3*D-A-B-C);
end;


equiv_deg_module_X:=function(p,ii,r) ## p is a prime ##
ii = 1 for H[1], size 2 ## ii = 2 for H[2], size 3 ##
ii = 3 for H[3], size p ## output is (m1,...,mn) ##
```

```
where n = # conj classes of G=PSL(2,p) ## and mi
= mult of pi_i in deg_equiv module of G ## local
G,i,j,n,H,G1,H_chars,CG,G_chars,w,m,theta,pi_theta,pi_theta_to_the;

G:=PSL(2,p); H:=[]; # 3 decomp gps H_chars:=[];
CG:=ConjugacyClassesSubgroups(G);
H[1]:=Representative(CG[2]); # size 2 H[2]:=Representative(CG[3]);
# size 3 n:=Size(CG);

for i in [1..n] do
  if Size(Representative(CG[i]))=p then
    H[3]:=Representative(CG[i]);
  fi;
od; for i in [1..3] do
 H_chars[i]:=Irr(H[i]);
od;
G_chars:=Irr(G); m:=[]; m[1]:=[];m[2]:=[];m[3]:=[];
theta:=List([1..3],i->H_chars[i][2]);
pi_theta_to_the:=List([1..3],i->List([1..r],j->
                   InducedClassFunction(theta[i]^(-j),G)));
for i in [1..3] do
 for j in [1..r] do
  m[i][j]:=List(G_chars, pi->ScalarProduct(pi_theta_to_the[i][j],pi));
 od;
od; return Sum(m[ii]); end;
```

References

1. A. Adler, "The Mathieu group M_{11} and the modular curve X_{11}," Proc. London Math Society, 74(1997)1-28.
   See also the preprint "$X(11)$ and M_{11}," available at **http://swiss.csail.mit.edu/~adler/ARTICLES/**
2. ——, "Some integral representations of $PSL_2(\mathbb{F}_p)$ and their applications," J. Algebra 72(1981)115-145.
3. P. Bending, A. Camina, R. Guralnick "Automorphisms of the modular curve $X(p)$ in positive characteristic," 2003 preprint.
4. I. Berkovich, E. Zhmud, **Characters of Finite Groups, vol 2**. AMS, 1999.
5. N. Borne, "Une formule de Riemann-Roch equivariante pour des courbes," thesis, Univ. Bordeaux, 1999. Available on the web at **http://www.dm.unibo.it/~borne/**
6. N. Elkies, "The Klein quartic in number theory," in **The eightfold way**, MSRI Publ. vol 38, 1998.
7. W. Fulton and J. Harris, **Representation theory: a first course**, Springer-Verlag, 1991.
8. The GAP Group, **GAP – Groups, Algorithms, and Programming, Version 4.4**; 2002, (http://www.gap-system.org).
9. R. Hartshorne, **Algebraic geometry**, Springer-Verlag, 1977.

10. G. Janusz, "Simple components of $\mathbb{Q}[SL(2,q)]$," Comm. Alg. 1(1974)1-22.
11. D. Joyner and A. Ksir, Representations of finite groups on Riemann-Roch spaces, II, preprint 2003, available at `http://front.math.ucdavis.edu/math.AG/0312383`
12. D. Joyner and W. Traves, "Representations of finite groups on Riemann-Roch spaces," preprint, 2002 available at `http://front.math.ucdavis.edu/math.AG/0210408` (revised version posted April, 2004)
13. C. Khare and D. Prasad, "Extending local representations to global representations," Kyoto Journal of Mathematics, 36(1996)471-480
14. W. Bosma, J. Cannon, C. Playoust, "The MAGMA algebra system, I: The user language," J. Symb. Comp., 24(1997)235-265. (See also the MAGMA homepage at `http://www.maths.usyd.edu.au:8000/u/magma/`).
15. H. Martin, J. Migliore, S. Nollet, "Degrees of generators of ideals defining curves in projective space," Comm. Alg.26 (1998)1209-1231.
16. S. Nakajima, "Galois module structure of cohomology groups for tamely ramified coverings of algebraic varieties," J. Number Theory 22 (1986) 115-123.
17. C. Ritzenthaler, "Problèmes arithmétiques relatifs à certaines familles de courbes sur les corps finis," Thesis, Univ. Paris 7, 2003.
18. ——, "Action du groupe de Mathieu M_{11} sur la courbe modulaire $X(11)$ en caractéristique 3," Masters thesis, Univ. Paris 6, 1998.
19. ——, Automorphismes des courbes modulaires $X(n)$ en caractristique p,Manuscripta Math. 109(2002)49 - 62 .
20. C. Schoen, "On certain modular representations in the cohomology of algebraic curves," J. Alg. 135(1990)1-18.
21. G. Shimura, **Arithmetic theory of automorphic functions**, Princeton Univ. Press, 1971.
22. H. Stichtenoth, **Algebraic function fields and codes**, Springer-Verlag, 1993.
23. M. A. Tsfasman and S. G. Vladut, **Algebraic-geometric codes**, Mathematics and its Applications, Kluwer Academic Publishers, Dordrechet 1991.
24. J. Velu, "Courbes elliptiques munies dun sous-groupe $\mathbb{Z}/n\mathbb{Z} \times \mu_n$," Bull. Soc. Math. France, Memoire, 1978.

Genus two curves covering elliptic curves: a computational approach

T. Shaska*

Department of Mathematics
Oakland University
Rochester, MI, 48309-4485.
shaska@oakland.edu

A genus 2 curve C has an elliptic subcover if there exists a degree n maximal covering $\psi : C \to E$ to an elliptic curve E. Degree n elliptic subcovers occur in pairs (E, E'). The Jacobian J_C of C is isogenous of degree n^2 to the product $E \times E'$. We say that J_C is (n, n)-split. The locus of C, denoted by $\mathcal{L}_n$, is an algebraic subvariety of the moduli space $\mathcal{M}_2$. The space $\mathcal{L}_2$ was studied in Shaska/Völklein [32] and Gaudry/Schost [10]. The space $\mathcal{L}_3$ was studied in [26] were an algebraic description was given as sublocus of $\mathcal{M}_2$.

In this survey we give a brief description of the spaces $\mathcal{L}_n$ for a general n and then focus on small n. We describe some of the computational details which were skipped in [32] and [26]. Further we explicitly describe the relation between the elliptic subcovers E and E'. We have implemented most of these relations in computer programs which check easily whether a genus 2 curve has $(2, 2)$ or $(3, 3)$ split Jacobian. In each case the elliptic subcovers can be explicitly computed.

1. Introduction

Let C be a genus 2 curve defined over an algebraically closed field k, of characteristic zero. Let $\psi : C \to E$ be a degree n maximal covering (i.e. does not factor through an isogeny) to an elliptic curve E defined over k. We say that C has a *degree n elliptic subcover*. Degree n elliptic subcovers occur in pairs. Let (E, E') be such a pair. It is well known that there is an isogeny of degree n^2 between the Jacobian J_C of C and the product $E \times E'$. We say that C has **(n,n)-split Jacobian**. The locus of such C, denoted

*Supported by the NSA grant R1-05-0129

by $\mathcal{L}_n$, is a 2-dimensional algebraic subvariety of the moduli space $\mathcal{M}_2$ of genus two curves.

In this survey we study the genus 2 curves with (n, n)-split Jacobian for small n. While such curves have been studied by many authors, our approach is simply computational. Some of the results have appeared in previous articles of the author.

Curves of genus 2 with elliptic subcovers go back to Legendre and Jacobi. Legendre, in his *Théorie des fonctions elliptiques*, gave the first example of a genus 2 curve with degree 2 elliptic subcovers. In a review of Legendre's work, Jacobi (1832) gives a complete description for $n = 2$. The case $n = 3$ was studied during the 19th century from Hermite, Goursat, Burkhardt, Brioschi, and Bolza. For a history and background of the 19th century work see Krazer [19, pg. 479]. Cases when $n > 3$ are more difficult to handle. Frey and Kani note the difficulty to get explicit examples, see Frey [8] and Frey, Kani [9].

In §2 we give a brief description of genus 2 curves and their isomorphism classes which are classified by the absolute invariants of binary sextics. Further, we display the list of groups that occur as full automorphism groups of genus 2 curves defined over a field of characteristic $\neq 2$.

In §3 we study degree n covers $\mathcal{C} \to E$ from a genus 2 curve to an elliptic curve. Such covers induce a degree n covering $\phi : \mathbb{P}^1 \to \mathbb{P}^1$. A careful study of such covers leads to determining an equation for the curves $\mathcal{C}$. The covering $\phi : \mathbb{P}^1 \to \mathbb{P}^1$ could have different ramification structure. All such structures are described in section 3.

The moduli space of coverings $\phi : \mathbb{P}^1 \to \mathbb{P}^1$ with fixed ramification structure is a Hurwitz space. The irreducibility of such space, dimension, and the genus (in the case 1-dimensional spaces) can be computed via the braid action. For n an odd integer we display such results in section 4. There is a natural morphism between the Hurwitz space and the locus $\mathcal{L}_n$ (cf. §4). In the second part of section 4 we describe the correspondence between the points of $\mathcal{L}_n$ and the Humbert space of discriminant n^2 which we denote by $\mathcal{H}_{n^2}$.

In section 5 we study genus 2 curves with degree 2 elliptic subcovers. Jacobi [4] gives a general form of such curves: $Y^2 = X^6 - s_1 X^4 + s_2 X^2 - 1$, and a description of $\mathcal{L}_2$ in terms of the cross ratios of the roots $\alpha_1, \ldots, \alpha_6$ of the sextic:

$$\frac{\alpha_3 - \alpha_1}{\alpha_3 - \alpha_2} : \frac{\alpha_4 - \alpha_1}{\alpha_4 - \alpha_2} = \frac{\alpha_5 - \alpha_1}{\alpha_5 - \alpha_2} : \frac{\alpha_6 - \alpha_1}{\alpha_6 - \alpha_2}$$

Thus, $\mathcal{L}_2$ is parameterized by the pair $(s_1, s_2) \in k^2$. We note that this parametrization of $\mathcal{L}_2$ factors through a ramified Galois covering: $k^2 \longrightarrow k^2$, $(s_1, s_2) \to (u, v)$, where $u = s_1 s_2$ and $v = s_1^3 + s_2^3$. This induces a birational parametrization of $\mathcal{L}_2$ by the pairs (u, v). All our computations use these coordinates (u, v). We use this to compute an equation for $\mathcal{L}_2$ in terms of the classical invariants. We give a general relation between the j-invariants of degree 2 elliptic subfields of K. This improves [10], where each isomorphism type of G is treated separately. We determine conditions when degree 2 elliptic subfields of K are 2 or 3-isogenous. For a generalization of such invariants u, and v see Remark 9 or Gutierrez/Shaska [10].

In section 6, we study the case $n = 3$. We show that every genus 2 curve with a degree 3 elliptic subcover can be written in the form

$$Y^2 = (X^3 + aX^2 + bX + 1)(4X^3 + b^2X^2 + 2bX + 1)$$

for $a, b \in k$. So $\mathcal{L}_3$ is parameterized by the pairs $(a, b) \in k^2$. The invariants of the two cubics r_1, r_2 give a birational parametrization of $\mathcal{L}_3$. This parametrization of $\mathcal{L}_3$ factors through ramified Galois coverings of degree 3 (resp. 2)

$$\begin{gathered} k^2 \to k^2 \to k^2 \\ (a, b) \to (u, v) \to (r_1, r_2) \end{gathered} \tag{1}$$

where $ab = u$ and $b^3 = v$. The equation of $\mathcal{L}_3$ is computed in terms of the absolute invariants and is displayed in [26, Appendix A]. If $\mathcal{C} \in \mathcal{L}_3$ then $Aut(\mathcal{C})$ is isomorphic to $\mathbb{Z}_2, V_4$, D_4 or D_6. Moreover, there are exactly six genus 2 curves with automorphism group D_4 or D_6. The rational models of these 12 curves and rational points on them were studied in [29]. We determine the j-invariants of the elliptic subcovers and show that they satisfy the Fricke polynomial of level 2.

In the last section we give information on computer programs that we have made available for such computations.

2. Preliminaries

Let k be an algebraically closed field of characteristic zero and C a genus 2 curve defined over k. Then C can be described as a double cover of $\mathbb{P}^1(k)$ ramified in 6 places $w_1, \dots, w_6$. This sets up a bijection between isomorphism classes of genus 2 curves and unordered distinct 6-tuples $w_1, \dots, w_6 \in \mathbb{P}^1(k)$ modulo automorphisms of $\mathbb{P}^1(k)$. An unordered 6-tuple $\{w_i\}_{i=1}^6$ can be described by a binary sextic (i.e. a homogenous equation $f(X,Z)$ of degree 6). Let $\mathcal{M}_2$ denote the moduli space of genus 2 curves. To describe $\mathcal{M}_2$ we need to find polynomial functions of the coefficients of a binary sextic $f(X,Z)$ invariant under linear substitutions in X, Z of determinant one. These invariants were worked out by Clebsch and Bolza in the case of zero characteristic and generalized by Igusa for any characteristic different from 2; see [3], [15], or [32] for a more modern treatment.

Consider a binary sextic, i.e. a homogeneous polynomial $f(X,Z)$ in $k[X,Z]$ of degree 6:

$$f(X,Z) = a_6X^6 + a_5X^5Z + \cdots + a_0Z^6.$$

Igusa J-invariants $\{J_{2i}\}$ of $f(X,Z)$ are homogeneous polynomials of degree $2i$ in $k[a_0, \dots, a_6]$, for $i = 1, 2, 3, 5$; see [15], [32] for their definitions. Here J_{10} is simply the discriminant of $f(X,Z)$. It vanishes if and only if the binary sextic has a multiple linear factor. These J_{2i} are invariant under the natural action of $SL_2(k)$ on sextics. Dividing such an invariant by another one of the same degree gives an invariant under $GL_2(k)$ action.

Two genus 2 fields K (resp., curves) in the standard form $Y^2 = f(X,1)$ are isomorphic if and only if the corresponding sextics are $GL_2(k)$ conjugate. Thus if I is a $GL_2(k)$ invariant (resp., homogeneous $SL_2(k)$ invariant), then the expression $I(K)$ (resp., the condition $I(K) = 0$) is well defined. Thus the $GL_2(k)$ invariants are functions on the moduli space $\mathcal{M}_2$ of genus 2 curves. This $\mathcal{M}_2$ is an affine variety with coordinate ring

$$k[\mathcal{M}_2] = k[a_0, \dots, a_6, J_{10}^{-1}]^{GL_2(k)}$$

which is the subring of degree 0 elements in $k[J_2, \dots, J_{10}, J_{10}^{-1}]$. The *absolute invariants*

$$i_1 := 144\frac{J_4}{J_2^2}, \quad i_2 := -1728\frac{J_2J_4 - 3J_6}{J_2^3}, \quad i_3 := 486\frac{J_{10}}{J_2^5},$$

are even $GL_2(k)$-invariants. Two genus 2 curves with $J_2 \neq 0$ are isomorphic if and only if they have the same absolute invariants. If $J_2 = 0$ then we can define new invariants as in [27]. For the rest of this paper if we say "there is a genus 2 curve $\mathcal{C}$ defined over k" we will mean the k-isomorphism class of $\mathcal{C}$. We have the following; see [32, Theorem 2].

Lemma 1: *The automorphism group G of a genus 2 curve $\mathcal{C}$ in characteristic $\neq 2$ is isomorphic to $\mathbb{Z}_2$, $\mathbb{Z}_{10}$, V_4, D_8, D_{12}, $\mathbb{Z}_3 \rtimes D_8$, $GL_2(3)$, or 2^+S_5. The case when $G \cong 2^+S_5$ occurs only in characteristic 5. If $G \cong \mathbb{Z}_3 \rtimes D_8$ (resp., $GL_2(3)$) then $\mathcal{C}$ has equation $Y^2 = X^6 - 1$ (resp., $Y^2 = X(X^4 - 1)$). If $G \cong \mathbb{Z}_{10}$ then $\mathcal{C}$ has equation $Y^2 = X^6 - X$.*

3. Curves of genus 2 with split Jacobians

Let C and E be curves of genus 2 and 1, respectively. Both are smooth, projective curves defined over k, $char(k) = 0$. Let $\psi : C \longrightarrow E$ be a covering of degree n. From the Riemann-Hurwitz formula, $\sum_{P \in C} (e_\psi(P) - 1) = 2$ where $e_\psi(P)$ is the ramification index of points $P \in C$, under ψ. Thus, we have two points of ramification index 2 or one point of ramification index 3. The two points of ramification index 2 can be in the same fiber or in different fibers. Therefore, we have the following cases of the covering ψ:

Case I: There are $P_1, P_2 \in C$, such that $e_\psi(P_1) = e_\psi(P_2) = 2$, $\psi(P_1) \neq \psi(P_2)$, and $\forall P \in C \setminus \{P_1, P_2\}$, $e_\psi(P) = 1$.

Case II: There are $P_1, P_2 \in C$, such that $e_\psi(P_1) = e_\psi(P_2) = 2$, $\psi(P_1) = \psi(P_2)$, and $\forall P \in C \setminus \{P_1, P_2\}$, $e_\psi(P) = 1$.

Case III: There is $P_1 \in C$ such that $e_\psi(P_1) = 3$, and $\forall P \in C \setminus \{P_1\}$, $e_\psi(P) = 1$.

In case I (resp. II, III) the cover ψ has 2 (resp. 1) branch points in E.

Denote the hyperelliptic involution of C by w. We choose $\mathcal{O}$ in E such that w restricted to E is the hyperelliptic involution on E. We denote the restriction of w on E by v, $v(P) = -P$. Thus, $\psi \circ w = v \circ \psi$. E[2] denotes the group of 2-torsion points of the elliptic curve E, which are the points fixed by v. The proof of the following two lemmas is straightforward and will be omitted.

Lemma 2: *a) If $Q \in E$, then $\forall P \in \psi^{-1}(Q)$, $w(P) \in \psi^{-1}(-Q)$.*

b) For all $P \in C$, $e_\psi(P) = e_\psi(w(P))$.

Let W be the set of points in C fixed by w. Every curve of genus 2 is given, up to isomorphism, by a binary sextic, so there are 6 points fixed by the hyperelliptic involution w, namely the Weierstrass points of C. The following lemma determines the distribution of the Weierstrass points in fibers of 2-torsion points.

Lemma 3: *The following hold:*

(1) $\psi(W) \subset E[2]$
(2) If n *is an odd number then i)* $\psi(W) = E[2]$ *ii) If* $Q \in E[2]$ *then* $\#(\psi^{-1}(Q) \cap W) = 1 \mod (2)$
(3) If n *is an even number then for all* $Q \in E[2]$, $\#(\psi^{-1}(Q) \cap W) = 0 \mod (2)$

Let $\pi_C : C \longrightarrow \mathbb{P}^1$ and $\pi_E : E \longrightarrow \mathbb{P}^1$ be the natural degree 2 projections. The hyperelliptic involution permutes the points in the fibers of π_C and π_E. The ramified points of π_C, π_E are respectively points in W and $E[2]$ and their ramification index is 2. There is $\phi : \mathbb{P}^1 \longrightarrow \mathbb{P}^1$ such that the diagram commutes.

$$\begin{array}{ccc} C & \xrightarrow{\pi_C} & \mathbb{P}^1 \\ \psi \downarrow & & \downarrow \phi \\ E & \xrightarrow{\pi_E} & \mathbb{P}^1 \end{array} \qquad (2)$$

Next, we will determine the ramification of induced coverings $\phi : \mathbb{P}^1 \longrightarrow \mathbb{P}^1$. First we fix some notation. For a given branch point we will denote the ramification of points in its fiber as follows. Any point P of ramification index m is denoted by (m). If there are k such points then we write $(m)^k$. We omit writing symbols for unramified points, in other words $(1)^k$ will not be written. Ramification data between two branch points will be separated by commas. We denote by $\pi_E(E[2]) = \{q_1, \dots, q_4\}$ and $\pi_C(W) = \{w_1, \dots, w_6\}$.

3.0.1. *The Case When n is Odd*

The following theorem classifies the ramification types for the induced coverings $\phi : \mathbb{P}^1 \longrightarrow \mathbb{P}^1$ when the degree n is odd.

Theorem 4: Let $\psi : C \longrightarrow E$ be a covering of odd degree n and $\phi : \mathbb{P}^1 \longrightarrow \mathbb{P}^1$ be the induced covering induced by ψ. This induces a partitioning of the set of 6 Weierstrass points of C into two sets $W^{(1)} = W^{(1)}(C, E)$ and $W^{(2)} = W^{(2)}(C, E)$, each of cardinality 3 such that $|\phi(W^{(1)})| = 1$ and $|\phi(W^{(2)})| = 3$. Then the ramification structure of ϕ is as follows.

Case I (the generic case)
$\left((2)^{\frac{n-1}{2}}, (2)^{\frac{n-1}{2}}, (2)^{\frac{n-1}{2}}, (2)^{\frac{n-3}{2}}, (2)^1\right)$
Or the following degenerate cases:

Case II (the 4-cycle case and the dihedral case)

i) $\left((2)^{\frac{n-1}{2}},(2)^{\frac{n-1}{2}},(2)^{\frac{n-1}{2}},(4)^1(2)^{\frac{n-7}{2}}\right)$

ii) $\left((2)^{\frac{n-1}{2}},(2)^{\frac{n-1}{2}},(2)^{\frac{n-1}{2}},(2)^{\frac{n-1}{2}}\right)$

iii) $\left((2)^{\frac{n-1}{2}},(2)^{\frac{n-1}{2}},(4)^1(2)^{\frac{n-5}{2}},(2)^{\frac{n-3}{2}}\right)$

Case III (the 3-cycle case)

i) $\left((2)^{\frac{n-1}{2}},(2)^{\frac{n-1}{2}},(2)^{\frac{n-1}{2}},(3)^1(2)^{\frac{n-5}{2}}\right)$

ii) $\left((2)^{\frac{n-1}{2}},(2)^{\frac{n-1}{2}},(3)^1(2)^{\frac{n-3}{2}},(2)^{\frac{n-3}{2}}\right)$

3.0.2. *The Case When n is Even*

Let us assume now that $deg(\psi) = n$ is an even number. The following theorem classifies the induced coverings in this case.

Theorem 5: If n is an even number then the generic case for $\psi : C \longrightarrow E$ induce the following three cases for $\phi : \mathbb{P}^1 \longrightarrow \mathbb{P}^1$:

I $\left((2)^{\frac{n-2}{2}},(2)^{\frac{n-2}{2}},(2)^{\frac{n-2}{2}},(2)^{\frac{n}{2}},(2)\right)$

II $\left((2)^{\frac{n-4}{2}},(2)^{\frac{n-2}{2}},(2)^{\frac{n}{2}},(2)^{\frac{n}{2}},(2)\right)$

III $\left((2)^{\frac{n-6}{2}},(2)^{\frac{n}{2}},(2)^{\frac{n}{2}},(2)^{\frac{n}{2}},(2)\right)$

Each of the above cases has the following degenerations (two of the branch points collapse to one)

I (1) $\left((2)^{\frac{n}{2}},(2)^{\frac{n-2}{2}},(2)^{\frac{n-2}{2}},(2)^{\frac{n}{2}}\right)$

(2) $\left((2)^{\frac{n-2}{2}},(2)^{\frac{n-2}{2}},(4)(2)^{\frac{n-6}{2}},(2)^{\frac{n}{2}}\right)$

(3) $\left((2)^{\frac{n-2}{2}},(2)^{\frac{n-2}{2}},(2)^{\frac{n-2}{2}},(4)(2)^{\frac{n-4}{2}}\right)$

(4) $\left((3)(2)^{\frac{n-4}{2}},(2)^{\frac{n-2}{2}},(2)^{\frac{n-2}{2}},(2)^{\frac{n}{2}}\right)$

II (1) $\left((2)^{\frac{n-2}{2}},(2)^{\frac{n-2}{2}},(2)^{\frac{n}{2}},(2)^{\frac{n}{2}}\right)$

(2) $\left((2)^{\frac{n-4}{2}},(2)^{\frac{n}{2}},(2)^{\frac{n}{2}},(2)^{\frac{n}{2}}\right)$

(3) $\left((4)(2)^{\frac{n-8}{2}},(2)^{\frac{n-2}{2}},(2)^{\frac{n}{2}},(2)^{\frac{n}{2}}\right)$

(4) $\left((2)^{\frac{n-4}{2}},(4)(2)^{\frac{n-6}{2}},(2)^{\frac{n}{2}},(2)^{\frac{n}{2}}\right)$

(5) $\left((2)^{\frac{n-4}{2}},(2)^{\frac{n-2}{2}},(2)^{\frac{n-4}{2}},(2)^{\frac{n}{2}}\right)$

(6) $\left((3)(2)^{\frac{n-6}{2}},(2)^{\frac{n-2}{2}},(4)(2)^{\frac{n}{2}},(2)^{\frac{n}{2}}\right)$

(7) $\left((2)^{\frac{n-4}{2}}, (3)(2)^{\frac{n-4}{2}}, (2)^{\frac{n}{2}}, (2)^{\frac{n}{2}}\right)$

III (1) $\left((2)^{\frac{n-4}{2}}, (2)^{\frac{n}{2}}, (2)^{\frac{n}{2}}, (4)(2)^{\frac{n}{2}}\right)$

(2) $\left((2)^{\frac{n-6}{2}}, (4)(2)^{\frac{n-4}{2}}, (2)^{\frac{n}{2}}, (2)^{\frac{n}{2}}\right)$

(3) $\left((2)^{\frac{n}{2}}, (2)^{\frac{n}{2}}, (2)^{\frac{n}{2}}, (4)(2)^{\frac{n-10}{2}}\right)$

(4) $\left((3)(2)^{\frac{n-8}{2}}, (2)^{\frac{n}{2}}, (2)^{\frac{n}{2}}, (2)^{\frac{n}{2}}\right)$

3.1. *Maximal coverings* $\psi : C \longrightarrow E$.

Let $\psi_1 : C \longrightarrow E_1$ be a covering of degree n from a curve of genus 2 to an elliptic curve. The covering $\psi_1 : C \longrightarrow E_1$ is called a **maximal covering** if it does not factor through a nontrivial isogeny. A map of algebraic curves $f : X \to Y$ induces maps between their Jacobians $f^* : J_Y \to J_X$ and $f_* : J_X \to J_Y$. When f is maximal then f^* is injective and $ker(f_*)$ is connected, see [30] for details.

Let $\psi_1 : C \longrightarrow E_1$ be a covering as above which is maximal. Then ${\psi^*}_1 : E_1 \to J_C$ is injective and the kernel of $\psi_{1,*} : J_C \to E_1$ is an elliptic curve which we denote by E_2; see [9] or [21]. For a fixed Weierstrass point $P \in C$, we can embed C to its Jacobian via

$$\begin{aligned} i_P : C &\longrightarrow J_C \\ x &\to [(x) - (P)] \end{aligned} \tag{3}$$

Let $g : E_2 \to J_C$ be the natural embedding of E_2 in J_C, then there exists $g_* : J_C \to E_2$. Define $\psi_2 = g_* \circ i_P : C \to E_2$. So we have the following exact sequence

$$0 \to E_2 \xrightarrow{g} J_C \xrightarrow{\psi_{1,*}} E_1 \to 0$$

The dual sequence is also exact

$$0 \to E_1 \xrightarrow{\psi_1^*} J_C \xrightarrow{g_*} E_2 \to 0$$

If $deg(\psi_1)$ is an odd number then the maximal covering $\psi_2 : C \to E_2$ is unique (up to isomorphism of elliptic curves), see Kuhn [21]. If the cover $\psi_1 : C \longrightarrow E_1$ is given, and therefore ϕ_1, we want to determine $\psi_2 : C \longrightarrow E_2$ and ϕ_2. The study of the relation between the ramification structures of ϕ_1 and ϕ_2 provides information in this direction. The following lemma (see [9, pg. 160]) answers this question for the set of Weierstrass points $W = \{P_1, \ldots, P_6\}$ of C when the degree of the cover is odd.

Lemma 6: *Let $\psi_1 : C \longrightarrow E_1$, be maximal of degree n. Then, the map $\psi_2 : C \to E_2$ is a maximal covering of degree n. Moreover,*

i) if n is odd and $\mathcal{O}_i \in E_i[2]$, $i = 1, 2$ are the places such that $\#(\psi_i^{-1}(\mathcal{O}_i) \cap W) = 3$, then $\psi_1^{-1}(\mathcal{O}_1) \cap W$ and $\psi_2^{-1}(\mathcal{O}_2) \cap W$ form a disjoint union of W.

ii) if n is even and $Q \in E[2]$, then $\#\left(\psi^{-1}(Q)\right) = 0$ or 2.

The above lemma says that if ψ is maximal of even degree then the corresponding induced covering can have only type **I** ramification, see theorem 5.

4. The locus of genus two curves with (n, n) split Jacobians

In this section we will discuss the Hurwitz spaces of coverings with ramification as in the previous section and the Humbert spaces of discriminant n^2.

4.1. *Hurwitz spaces of covers* $\phi : \mathbb{P}^1 \to \mathbb{P}^1$

Two covers $f : X \to \mathbb{P}^1$ and $f' : X' \to \mathbb{P}^1$ are called **weakly equivalent** if there is a homeomorphism $h : X \to X'$ and an analytic automorphism g of $\mathbb{P}^1$ (i.e., a Moebius transformation) such that $g \circ f = f' \circ h$. The covers f and f' are called **equivalent** if the above holds with $g = 1$.

Consider a cover $f : X \to \mathbb{P}^1$ of degree n, with branch points $p_1, ..., p_r \in \mathbb{P}^1$. Pick $p \in \mathbb{P}^1 \setminus \{p_1, ..., p_r\}$, and choose loops γ_i around p_i such that $\gamma_1, ..., \gamma_r$ is a standard generating system of the fundamental group $\Gamma := \pi_1(\mathbb{P}^1 \setminus \{p_1, ..., p_r\}, p)$, in particular, we have $\gamma_1 \cdots \gamma_r = 1$. Such a system $\gamma_1, ..., \gamma_r$ is called a homotopy basis of $\mathbb{P}^1 \setminus \{p_1, ..., p_r\}$. The group Γ acts on the fiber $f^{-1}(p)$ by path lifting, inducing a transitive subgroup G of the symmetric group S_n (determined by f up to conjugacy in S_n). It is called the **monodromy group** of f. The images of $\gamma_1, ..., \gamma_r$ in S_n form a tuple of permutations $\sigma = (\sigma_1, ..., \sigma_r)$ called a tuple of **branch cycles** of f.

We say a cover $f : X \to \mathbb{P}^1$ of degree n is of type σ if it has σ as tuple of branch cycles relative to some homotopy basis of $\mathbb{P}^1$ minus the branch points of f. Let $\mathcal{H}_\sigma$ be the set of weak equivalence classes of covers of type σ. The **Hurwitz space** $\mathcal{H}_\sigma$ carries a natural structure of an quasiprojective variety.

We have $\mathcal{H}_\sigma = \mathcal{H}_\tau$ if and only if the tuples σ, τ are in the same **braid orbit** $\mathcal{O}_\tau = \mathcal{O}_\sigma$. In the case of the covers $\phi : \mathbb{P}^1 \to \mathbb{P}^1$ from above, the corresponding braid orbit consists of all tuples in S_n whose cycle type matches the ramification structure of ϕ.

This and the genus of $\mathcal{H}_\sigma$ in the degenerate cases (see the following table) has been computed in GAP by the BRAID PACKAGE written by K. Magaard.

deg	Case	cycle type of σ	$\#(\mathcal{O}_\sigma)$	G	$\dim \mathcal{H}_\sigma$	genus of $\mathcal{H}_\sigma$
3		$(2^2, 2^2, 2^2, 2, 2)$	40	S_3	2	–
	1	$(2^2, 2^2, 4, 2)$	8	S_5	1	0
	2	$(2^2, 2^2, 2 \cdot 3, 2)$	6	S_5	1	0
	3	$(2^2, 2^2, 2^2, 3)$	9	A_5	1	1
	4					
5		$(2^2, 2^2, 2^2, 2, 2)$	40	S_5	2	–
	1	$(2^2, 2^2, 4, 2)$	8	S_5	1	0
	2	$(2^2, 2^2, 2 \cdot 3, 2)$	6	S_5	1	0
	3	$(2^2, 2^2, 2^2, 3)$	9	A_5	1	1
	4					
7		$(2^2, 2^2, 2^2, 2, 2)$	168	S_7	2	–

4.2. *Humbert surfaces*

Let $\mathcal{A}_2$ denote the moduli space of principally polarized abelian surfaces. It is well known that $\mathcal{A}_2$ is the quotient of the Siegel upper half space $\mathfrak{H}_2$ of symmetric complex 2×2 matrices with positive definite imaginary part by the action of the symplectic group $Sp_4(\mathbb{Z})$; see [11, p. 211].

Let Δ be a fixed positive integer and N_Δ be the set of matrices

$$\tau = \begin{pmatrix} z_1 & z_2 \\ z_2 & z_3 \end{pmatrix} \in \mathfrak{H}_2$$

such that there exist nonzero integers a, b, c, d, e with the following properties:

$$\begin{aligned} & a z_1 + b z_2 + c z_3 + d(z_2^2 - z_1 z_3) + e = 0 \\ & \Delta = b^2 - 4ac - 4de \end{aligned} \tag{4}$$

The *Humbert surface* $\mathcal{H}_\Delta$ of discriminant Δ is called the image of N_Δ under the canonical map

$$\mathfrak{H}_2 \to \mathcal{A}_2 := Sp_4(\mathbb{Z}) \setminus \mathfrak{H}_2,$$

see [2, 14, 24] for details. It is known that $\mathcal{H}_\Delta \neq \emptyset$ if and only if $\Delta > 0$ and $\Delta \equiv 0$ *or* $1 \mod 4$. Humbert (1900) studied the zero loci in Eq. (4) and discovered certain relations between points in these spaces and certain plane configurations of six lines; see [14] for more details.

For a genus 2 curve C defined over $\mathbb{C}$, $[C]$ belongs too $\mathcal{L}_n$ if and only if the isomorphism class $[J_C] \in \mathcal{A}_2$ of its (principally polarized) Jacobian J_C belongs to the Humbert surface $\mathcal{H}_{n^2}$, viewed as a subset of the moduli space $\mathcal{A}_2$ of principally polarized abelian surfaces; see [24, Theorem 1, p. 125] for the proof of this statement. In [24] is shown that there is a one to one correspondence between the points in $\mathcal{L}_n$ and points in $\mathcal{H}_{n^2}$. Thus, we have the map:

$$\begin{array}{ccccc} \mathcal{H}_\sigma & \longrightarrow & \mathcal{L}_n & \longrightarrow & \mathcal{H}_{n^2} \\ ([f],(p_1,\dots,p_r) & \to & [\mathcal{X}] & \to & [J_\mathcal{X}] \end{array} \tag{5}$$

In particular, every point in $\mathcal{H}_{n^2}$ can be represented by an element of $\mathfrak{H}_2$ of the form

$$\tau = \begin{pmatrix} z_1 & \frac{1}{n} \\ \frac{1}{n} & z_2 \end{pmatrix}, \qquad z_1,\, z_2 \in \mathfrak{H}.$$

There have been many attempts to explicitly describe these Humbert surfaces. For some small discriminant this has been done by several authors; see [32], [26], [21]. Geometric characterizations of such spaces for $\Delta = 4, 8, 9,$ and 12 were given by Humbert (1900) in [14] and for $\Delta = 13, 16, 17, 20,$ 21 by Birkenhake/Wilhelm (2003) in [2].

5. Genus 2 curves with degree 2 elliptic subcovers

An **elliptic involution** of K is an involution in G which is different from z_0 (the hyperelliptic involution). Thus the elliptic involutions of G are in 1-1 correspondence with the elliptic subfields of K of degree 2 (by the Riemann-Hurwitz formula).

If z_1 is an elliptic involution and z_0 the hyperelliptic one, then $z_2 := z_0\, z_1$ is another elliptic involution. So the elliptic involutions come naturally in pairs. This pairs also the elliptic subfields of K of degree 2. Two such subfields E_1 and E_2 are paired if and only if $E_1 \cap k(X) = E_2 \cap k(X)$. E_1

and E_2 are G-conjugate unless $G \cong D_6$ or $G \cong V_4$ (This can be checked from Lemma 1).

Theorem 7: Let K be a genus 2 field and $e_2(K)$ the number of $Aut(K)$-classes of elliptic subfields of K of degree 2. Suppose $e_2(K) \geq 1$. Then the classical invariants of K satisfy the equation,

$$
\begin{aligned}
-J_2^7 J_4^4 + 8748 J_{10} J_2^4 J_6^2 507384000 J_{10}^2 J_4^2 J_2 - 19245600 J_{10}^2 J_4 J_2^3 - 592272 J_{10} J_4^4 J_2^2 \\
-81 J_2^3 J_6^4 - 3499200 J_{10} J_2 J_6^3 + 4743360 J_{10} J_4^3 J_2 J_6 - 870912 J_{10} J_4^2 J_2^3 J_6 \\
+1332 J_2^4 J_4^4 J_6 - 125971200000 J_{10}^3 + 384 J_4^6 J_6 + 41472 J_{10} J_4^5 + 159 J_4^6 J_2^3 \\
-47952 J_2 J_4 J_6^4 + 104976000 J_{10}^2 J_2^2 J_6 - 1728 J_4^5 J_2^2 J_6 + 6048 J_4^4 J_2 J_6^2 + 108 J_2^4 J_4 J_6^3 \\
+12 J_2^6 J_4^3 J_6 + 29376 J_2^2 J_4^2 J_6^3 - 8910 J_2^3 J_4^3 J_6^2 - 2099520000 J_{10}^2 J_4 J_6 - 236196 J_{10}^2 J_2^5 \\
+31104 J_6^5 - 6912 J_4^3 J_6^3 4 + 972 J_{10} J_2^6 J_4^2 + 77436 J_{10} J_4^3 J_2^4 - 78 J_2^5 J_4^5 \\
+3090960 J_{10} J_4 J_2^2 J_6^2 - 5832 J_{10} J_2^5 J_4 J_6 - 80 J_4^7 J_2 - 54 J_2^5 J_4^2 J_6^2 - 9331200 J_{10} J_4^2 J_6^2 = 0
\end{aligned}
\tag{6}
$$

Further, $e_2(K) = 2$ unless $K = k(X, Y)$ with

$$Y^2 = X^5 - X$$

in which case $e_2(K) = 1$.

Proof: Since $e_2(K)$ is the number of conjugacy classes of elliptic involutions in G the claim about $e_2(K)$ follows from theorem 5. For the proof of the following lemma see [32].

Lemma 8: *Suppose z_1 is an elliptic involution of K. Let $z_2 = z_1 z_0$, where z_0 is the hyperelliptic involution. Let E_i be the fixed field of z_i for $i = 1, 2$. Then $K = k(X, Y)$ where*

$$Y^2 = X^6 - s_1 X^4 + s_2 X^2 - 1 \tag{7}$$

and $27 - 18 s_1 s_2 - s_1^2 s_2^2 + 4 s_1^3 + 4 s_2^3 \neq 0$. Further E_1 and E_2 are the subfields $k(X^2, Y)$ and $k(X^2, YX)$.

We need to determine to what extent the normalization in the above proof determines the coordinate X. The condition $z_1(X) = -X$ determines the coordinate X up to a coordinate change by some $\gamma \in \Gamma$ centralizing z_1. Such γ satisfies $\gamma(X) = mX$ or $\gamma(X) = \frac{m}{X}$, $m \in k \setminus \{0\}$. The additional condition $abc = 1$ forces $1 = -\gamma(\alpha_1) \dots \gamma(a_6)$, hence $m^6 = 1$. So X is determined up to a coordinate change by the subgroup $H \cong D_6$ of Γ generated by $\tau_1 : X \to \xi_6 X$, $\tau_2 : X \to \frac{1}{X}$, where ξ_6 is a primitive 6-th root of unity. Let $\xi_3 := \xi_6^2$. The coordinate change by τ_1 replaces s_1 by $\xi_3 s_2$ and s_2 by

$\xi_3^2 s_2$. The coordinate change by τ_2 switches s_1 and s_2. Invariants of this H-action are:

$$u := s_1 s_2, \quad v := s_1^3 + s_2^3 \tag{8}$$

Remark 9: Such invariants were quite important in simplifying computations for the locus $\mathcal{L}_2$. Later they have been used by Duursma and Kiyavash to show that genus 2 curves with extra involutions are suitable for the vector decomposition problem; see [7] for details. In this volume they are used again, see the paper by Cardona and Quer. They were later generalized to higher genus hyperelliptic curves and were called dihedral invariants; see [13].

Classical invariants of the field K given by lemma 5 are:

$$\begin{aligned} J_2 &= 240 + 16u \\ J_4 &= 48v + 4u^2 + 1620 - 504u \\ J_6 &= -20664u + 96v - 424u^2 + 24u^3 + 160uv + 119880 \\ J_{10} &= 64(27 - 18u - u^2 + 4v)^2 \end{aligned} \tag{9}$$

For $J_2 \neq 0$ we express the absolute invariants i_1, i_2, i_3 in terms of u and v. We can eliminate u and v and get the following equation of $\mathcal{L}_2$.

$$\begin{aligned} -27i_1^6 + 9i_1^7 + 161243136 i_3 i_1^3 - 12441600 i_3 i_2^3 + 2i_2^5 + 107495424 i_3 i_1^2 i_2 + 54 i_1^3 i_2^2 \\ -52254720 i_3 i_1 i_2^2 - 47278080 i_3 i_1^3 i_2 - 8294400 i_3 i_1^2 i_2^2 - 9459597312000 i_3^2 i_1^2 - 18 i_1^4 i_2^2 \\ -240734712102912 i_3^2 + 111451255603200 i_3^2 i_1 + 20639121408000 i_3^2 i_2 - 55240704 i_3 i_1^4 \\ +2i_1^6 i_2 - 4i_1^3 i_2^3 + 331776 i_3 i_1^5 - 27 i_2^4 - 2866544640000 i_3^2 i_1 i_2 + 161243136 i_3 i_2^2 + 9 i_1 i_2^4 \\ -264180754022400000 i_3^3 = 0 \end{aligned} \tag{10}$$

To get rid of the condition $J_2 \neq 0$ we multiply by J_2^5 to get the "projective" equation (6) of $\mathcal{L}_2$. This holds indeed for all $K \in \mathcal{L}_2$, as can be checked by substituting from (9). This completes the proof of Theorem 7.

□

The following proposition determines the group G in terms of u and v.

Proposition 10: *Let $\mathcal{C}$ be a genus 2 curve such that $G := Aut(\mathcal{C})$ has an elliptic involution and $J_2 \neq 0$. Then,*

a) $G \cong \mathbb{Z}_3 \rtimes D_4$ if and only if $(u,v) = (0,0)$ or $(u,v) = (225, 6750)$.

b) $G \cong W_1$ if and only if $u = 25$ and $v = -250$.

c) $G \cong D_6$ if and only if $4v - u^2 + 110u - 1125 = 0$, for $u \neq 9, 70 + 30\sqrt{5}, 25$.

Moreover, the classical invariants satisfy the equations,

$$\begin{aligned} -J_4J_2^4 + 12J_2^3J_6 - 52J_4^2J_2^2 + 80J_4^3 + 960J_2J_4J_6 - 3600J_6^2 = 0 \\ 864J_{10}J_2^5 + 3456000J_{10}J_4^2J_2 - 43200J_{10}J_4J_2^3 - 2332800000J_{10}^2 - J_4^2J_2^6 \\ -768J_4^4J_2^2 + 48J_4^3J_2^4 + 4096J_4^5 = 0 \end{aligned} \quad (11)$$

d) $G \cong D_4$ *if and only if* $v^2 - 4u^3 = 0$, *for* $u \neq 1, 9, 0, 25, 225$. *Cases* $u = 0, 225$ *and* $u = 25$ *are reduced to cases a),and b) respectively. Moreover, the classical invariants satisfy* (6) *and the following equation,*

$$1706J_4^2J_2^2 + 2560J_4^3 + 27J_4J_2^4 - 81J_2^3J_6 - 14880J_2J_4J_6 + 28800J_6^2 = 0 \quad (12)$$

Proposition 11: *The mapping*

$$A : (u, v) \longrightarrow (i_1, i_2, i_3)$$

gives a birational parametrization of $\mathcal{L}_2$. *The fibers of* A *of cardinality* > 1 *correspond to those curves* $\mathcal{C}$ *with* $|Aut(\mathcal{C})| > 4$.

Proof: See [32] for the details. □

5.1. *Elliptic subcovers*

Let j_1 and j_2 denote the j-invariants of the elliptic curves E_1 and E_2 from lemma 5. The invariants j_1 and j_2 and are roots of the quadratic

$$j^2 + 256\frac{(2u^3 - 54u^2 + 9uv - v^2 + 27v)}{(u^2 + 18u - 4v - 27)}j + 65536\frac{(u^2 + 9u - 3v)}{(u^2 + 18u - 4v - 27)^2} = 0 \quad (13)$$

5.1.1. *Isomorphic elliptic subcovers*

The elliptic curves E_1 and E_2 are isomorphic when equation (13) has a double root. The discriminant of the quadratic is zero for

$$(v^2 - 4u^3)(v - 9u + 27) = 0$$

Remark 12: From lemma 5, $v^2 = 4u^3$ if and only if $Aut(\mathcal{C}) \cong D_4$. So for $\mathcal{C}$ such that $Aut(\mathcal{C}) \cong D_4$, E_1 is isomorphic to E_2. It is easily checked that z_1 and $z_2 = z_0z_1$ are conjugate when $G \cong D_4$. So they fix isomorphic subfields.

If $v = 9(u - 3)$ then the locus of these curves is given by,

$$\begin{aligned} 4i_1^5 - 9i_1^4 + 73728i_1^2i_3 - 150994944i_3^2 = 0 \\ 289i_1^3 - 729i_1^2 + 54i_1i_2 - i_2^2 = 0 \end{aligned} \quad (14)$$

For $(u,v)=(\frac{9}{4},-\frac{27}{4})$ the curve has $Aut(\mathcal{C})\cong D_4$ and for $(u,v)=(137,1206)$ it has $Aut(\mathcal{C})\cong D_6$. All other curves with $v=9(u-3)$ belong to the general case, so $Aut(\mathcal{C})\cong V_4$. The j-invariants of elliptic curves are $j_1=j_2=256(9-u)$. Thus, these genus 2 curves are parameterized by the j-invariant of the elliptic subcover.

Remark 13: This embeds the moduli space $\mathcal{M}_1$ into $\mathcal{M}_2$ in a functorial way.

5.2. *Isogenous degree 2 elliptic subfields*

In this section we study pairs of degree 2 elliptic subfields of K which are 2 or 3-isogenous. We denote by $\Phi_n(x,y)$ the n-th modular polynomial (see Blake et al. [1] for the formal definitions. Two elliptic curves with j-invariants j_1 and j_2 are n-isogenous if and only if $\Phi_n(j_1,j_2)=0$.

5.2.1. *3-Isogeny.*

Suppose E_1 and E_2 are 3-isogenous. Then, from equation (13) and $\Phi_3(j_1,j_2)=0$ we eliminate j_1 and j_2. Then,

$$(4v-u^2+110u-1125)\cdot g_1(u,v)\cdot g_2(u,v)=0 \tag{15}$$

where g_1 and g_2 are

$$\begin{aligned} g_1 &= -27008u^6+256u^7-2432u^5v+v^4+7296u^3v^2-6692v^3u-1755067500u \\ &+2419308v^3-34553439u^4+127753092vu^2+16274844vu^3-1720730u^2v^2 \\ &-1941120u^5+381631500v+1018668150u^2-116158860u^3+52621974v^2 \\ &+387712u^4v-483963660vu-33416676v^2u+922640625 \end{aligned} \tag{16}$$

$$\begin{aligned} g_2 &= 291350448u^6-v^4u^2-998848u^6v-3456u^7v+4749840u^4v^2+17032u^5v^2 \\ &+4v^5+80368u^8+256u^9+6848224u^7-10535040v^3u^2-35872v^3u^3+26478v^4u \\ &-77908736u^5v+9516699v^4+307234984u^3v^2-419583744v^3u-826436736v^3 \\ &+27502903296u^4+28808773632vu^2-23429955456vu^3+5455334016u^2v^2 \\ &-41278242816v+82556485632u^2-108737593344u^3-12123095040v^2 \\ &+41278242816vu+3503554560v^2u+5341019904u^5-2454612480u^4v \end{aligned} \tag{17}$$

Thus, there is a isogeny of degree 3 between E_1 and E_2 if and only if u and v satisfy equation (15). The vanishing of the first factor is equivalent to $G \cong D_6$. So, if $Aut(\mathcal{C}) \cong D_6$ then E_1 and E_2 are isogenous of degree 3. This was also noted by Gaudry and Schost [10].

5.2.2. *2-Isogeny*

Below we give the modular 2-polynomial.

$$\Phi_2 = x^3 - x^2y^2 + y^3 + 1488xy(x+y) + 40773375xy - 162000(x^2 - y^2) + 8748000000(x+y) - 157464000000000 \tag{18}$$

Suppose E_1 and E_2 are isogenous of degree 2. Substituting j_1 and j_2 in Φ_2 we get

$$f_1(u,v) \cdot f_2(u,v) = 0 \tag{19}$$

where f_1 and f_2 are

$$f_1 = -16v^3 - 81216v^2 - 892296v - 2460375 + 3312uv^2 + 707616vu + 3805380u + 18360vu^2 - 1296162u^2 - 1744u^3v - 140076u^3 + 801u^4 + 256u^5 \tag{20}$$

$$\begin{aligned} f_2 = {} & 4096u^7 + 256016u^6 - 45824u^5v + 4736016u^5 - 2126736vu^4 + 23158143u^4 \\ & - 25451712u^3v - 119745540u^3 + 5291136v^2u^2 - 48166488vu^2 - 2390500350u^2 \\ & - 179712uv^3 + 35831808uv^2 + 1113270480vu + 9300217500u - 4036608v^3 \\ & - 1791153000v - 8303765625 - 1024v^4 + 163840u^3v^2 - 122250384v^2 + 256u^2v^3 \end{aligned} \tag{21}$$

5.2.3. *Other isogenies between elliptic subcovers*

If $G \cong D_4$, then z_1 and z_2 are in the same conjugacy class. There are again two conjugacy classes of elliptic involutions in G. Thus, there are two degree 2 elliptic subfields (up to isomorphism) of K. One of them is determined by double root j of the equation (13), for $v^2 - 4u^3 = 0$. Next, we determine the j-invariant j' of the other degree 2 elliptic subfield and see how it is related to j.

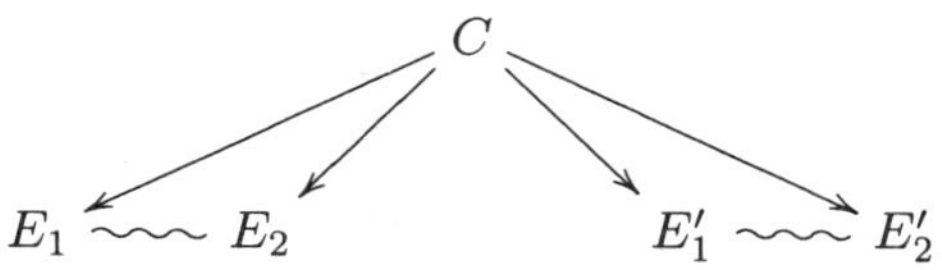

If $v^2 - 4u^3 = 0$ then $\mathbf{G} \cong V_4$ and $\mathcal{P} = \{\pm 1, \pm\sqrt{a}, \pm\sqrt{b}\}$. Then, $s_1 = a + \frac{1}{a} + 1 = s_2$. Involutions of $\mathcal{C}$ are $\tau_1 : X \to -X$, $\tau_2 : X \to \frac{1}{X}$, $\tau_3 : X \to -\frac{1}{X}$. Since τ_1 and τ_3 fix no points of $\mathcal{P}$ the they lift to involutions in G. They each determine a pair of isomorphic elliptic subfields. The j-invariant of elliptic subfield fixed by τ_1 is the double root of equation (13), namely

$$j = -256\frac{v^3}{v+1}$$

To find the j-invariant of the elliptic subfields fixed by τ_3 we look at the degree 2 covering $\phi : \mathbb{P}^1 \to \mathbb{P}^1$, such that $\phi(\pm 1) = 0$, $\phi(a) = \phi(-\frac{1}{a}) = 1$, $\phi(-a) = \phi(\frac{1}{a}) = -1$, and $\phi(0) = \phi(\infty) = \infty$. This covering is, $\phi(X) = \frac{\sqrt{a}}{a-1}\frac{X^2-1}{X}$. The branch points of ϕ are $q_i = \pm\frac{2i\sqrt{a}}{\sqrt{a-1}}$. From lemma 5 the elliptic subfields E_1' and E_2' have 2-torsion points $\{0, 1, -1, q_i\}$. The j-invariants of E_1' and E_2' are

$$j' = -16\frac{(v-15)^3}{(v+1)^2}$$

Then $\Phi_2(j, j') = 0$, so E_1 and E_1' are isogenous of degree 2. Thus, τ_1 and τ_3 determine degree 2 elliptic subfields which are 2-isogenous.

6. Genus 2 curves with degree 3 elliptic subcovers

This case was studied in detail in [26]. The main theorem was:

Theorem 14: Let K be a genus 2 field and $e_3(K)$ the number of $Aut(K/k)$-classes of elliptic subfields of K of degree 3. Then;

i) $e_3(K) = 0, 1, 2$, or 4

ii) $e_3(K) \geq 1$ if and only if the classical invariants of K satisfy the irreducible equation $F(J_2, J_4, J_6, J_{10}) = 0$ displayed in [26, Appendix A].

There are exactly two genus 2 curves (up to isomorphism) with $e_3(K) = 4$. The case $e_3(K) = 1$ (resp., 2) occurs for a 1-dimensional (resp., 2-dimensional) family of genus 2 curves, see [26].

Lemma 15: *Let K be a genus 2 field and E an elliptic subfield of degree 3.*

i) Then $K = k(X, Y)$ such that

$$Y^2 = (4X^3 + b^2X^2 + 2bX + 1)(X^3 + aX^2 + bX + 1) \tag{22}$$

for $a, b \in k$ such that

$$(4a^3 + 27 - 18ab - a^2b^2 + 4b^3)(b^3 - 27) \neq 0 \tag{23}$$

The roots of the first (resp. second) cubic correspond to $W^{(1)}(K,E)$, *(resp.* $W^{(2)}(K,E)$*) in the coordinates* X, Y, *(see theorem 4).*

ii) $E = k(U,V)$ *where*

$$U = \frac{X^2}{X^3 + aX^2 + bX + 1}$$

and

$$V^2 = U^3 + 2\frac{ab^2 - 6a^2 + 9b}{R}U^2 + \frac{12a - b^2}{R}U - \frac{4}{R} \tag{24}$$

where $R = 4a^3 + 27 - 18ab - a^2b^2 + 4b^3 \neq 0$.

iii) Define

$$u := ab, \quad v := b^3$$

Let K' *be a genus 2 field and* $E' \subset K'$ *a degree 3 elliptic subfield. Let* a', b' *be the associated parameters as above and* $u' := a'b'$, $v = (b')^3$. *Then, there is a* k*-isomorphism* $K \to K'$ *mapping* $E \to E'$ *if and only if exists a third root of unity* $\xi \in k$ *with* $a' = \xi a$ *and* $b' = \xi^2 b$. *If* $b \neq 0$ *then such* ξ *exists if and only if* $v = v'$ *and* $u = u'$.

iv) The classical invariants of K *satisfy equation [26, Appendix A].*

Let

$$\begin{aligned} F(X) &:= X^3 + aX^2 + bX + 1 \\ G(X) &:= 4X^3 + b^2X^2 + 2bX + 1 \end{aligned} \tag{25}$$

Denote by $R = 4a^3 + 27 - 18ab - a^2b^2 + 4b^3$ the resultant of F and G. Then we have the following lemma.

Lemma 16: *Let* $a, b \in k$ *satisfy equation* (23). *Then equation* (22) *defines a genus 2 field* $K = k(X,Y)$. *It has elliptic subfields of degree 3,* $E_i = k(U_i, V_i)$, $i = 1, 2$, *where* U_i, *and* V_i *are as follows:*

$$U_1 = \frac{X^2}{F(X)}, \quad V_1 = Y\frac{X^3 - bX - 2}{F(X)^2}$$

$$U_2 = \begin{cases} \dfrac{(X-s)^2(X-t)}{G(X)} & if \quad b(b^3 - 4ba + 9) \neq 0 \\ \dfrac{(3X - a)}{3(4X^3 + 1)} & if \quad b = 0 \\ \dfrac{(bX + 3)^2}{b^2G(X)} & if \quad (b^3 - 4ba + 9) = 0 \end{cases} \tag{26}$$

where

$$s = -\frac{3}{b}, \quad t = \frac{3a - b^2}{b^3 - 4ab + 9}$$

$$V_2 = \begin{cases} \frac{\sqrt{27 - b^3}Y}{G(X)^2}((4ab - 8 - b^3)X^3 - (b^2 - 4ab)X^2 + bX + 1) & if \quad b(b^3 - 4ba + 9) \neq 0 \\ Y\frac{8X^3 - 4aX^2 - 1}{(4X^3 + 1)^2} & if \quad b = 0 \\ \frac{8}{b}\sqrt{b}\frac{Y}{G(X)}(bX^3 + 9X^2 + b^2X + b) & if \quad (b^3 - 4ba + 9) = 0 \end{cases} \tag{27}$$

Proof: We skip the details of the proof. □

6.1. *Function field of* $\mathcal{L}_3$

The absolute invariants i_1, i_2, and i_3 are expressed in terms of u, v. Let u, v be independent transcendentals over k and $i_1, i_2, i_3 \in k(u, v)$. Further elements $r_1, r_2 \in k(u, v)$ are defined below; see § 6.1.1.

From the resultants of equations if i_1, i_2, i_3 in terms of u, v, we determine that $[k(v) : k(i_1, i_2)] = 16$, $[k(v) : k(i_2, i_3)] = 40$, and $[k(v) : k(i_1, i_3)] = 26$. We also can show that $u \in k(i_1, i_2, i_3, v)$, the expression is large and we display it on [26, Appendix A]. Thus, $[k(u, v) : k(i_1, i_2, i_3)] \leq 2$, see figure 1.

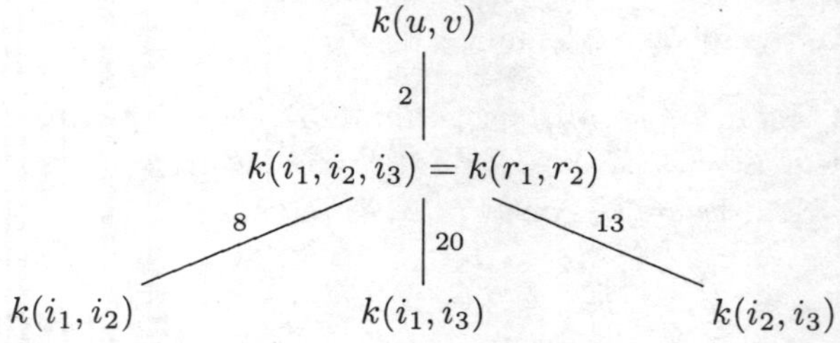

Fig. 1.

Computing the equation [26, Appendix A] directly from the equations of i_1, i_2, i_3 in terms of u, v, exceeds available computer power. We use additional invariants r_1, r_2 to overcome this problem.

6.1.1. *Invariants of Two Cubics*

We define the following invariants of two cubic polynomials. For $F(X) = a_3X^3 + a_2X^2 + a_1X + a_0$ and $G(X) = b_3X^3 + b_2X^2 + b_1X + b_0$ define

$$H(F,G) := a_3b_0 - \frac{1}{3}a_2b_1 + \frac{1}{3}a_1b_2 - a_0b_3$$

We denote by $R(F,G)$ the resultant of F and G and by $D(F)$ the discriminant of F. Also,

$$r_1(F,G) = \frac{H(F,G)^3}{R(F,G)}, \quad r_2(F,G) = \frac{H(F,G)^4}{D(F)\,D(G)}$$

Remark 17: Note that $D(FG) = D(F) \cdot D(G) \cdot R^2(F,G)$.

For

$$F(X) = X^3 + aX^2 + bX + 1, \quad G(X) = 4X^3 + b^2X^2 + 2bX + 1$$

from lemma 15 we have

$$\begin{aligned} r_1(F,G) &= 27\frac{v(v-9-2u)^3}{4v^2 - 18uv + 27v - u^2v + 4u^3} \\ r_2(F,G) &= -1296\frac{v(v-9-2u)^4}{(v-27)(4v^2 - 18uv + 27v - u^2v + 4u^3)} \end{aligned} \tag{28}$$

Remark 18: Note that r_1, r_2 are defined for any u, v by (23).

Taking the resultants from the above equations we get the following equations for u and v over $k(r_1, r_2)$:

$$\begin{aligned} &65536r_1r_2^3u^2 + (42467328r_2^4 + 21233664r_2^4r_1 + 480r_2r_1^4 + 2r_1^5 + 41472r_2^2r_1^3 \\ &+1548288r_2^3r_1^2 - 294912r_2^3r_1)u - 382205952r_2^4 + 238878720r_2^4r_1 - 2654208r_2^3r_1 \\ &+13934592r_2^3r_1^2 + 285696r_2^2r_1^3 + 2400r_2r_1^4 + 7r_1^5 = 0 \end{aligned} \tag{29}$$

$$\begin{aligned} &16384v^2r_2^3 + (221184r_2^3r_1 + r_1^4 + 11520r_2^2r_1^2 - 442368r_2^3 + 192r_2r_1^3)v \\ &-5971968r_2^3r_1 - 864r_2r_1^3 - 124416r_2^2r_1^2 - 2r_1^4 = 0 \end{aligned} \tag{30}$$

In equation (29) express r_1 and r_2 in terms of u and v. Roots of this equation are u and $\nu(u)$ where,

$$\nu(u) = \frac{(v-3u)(324u^2 + 15u^2v - 378uv - 4uv^2 + 243v + 72v^2)}{(v-27)(4u^3 + 27v - 18uv - u^2v + 4v^2)} \tag{31}$$

Similarly for v we get

$$\nu(v) = -\frac{4(v-3u)^3}{4u^3 + 27v - 18uv - u^2v + 4v^2} \tag{32}$$

Define a ring homomorphism

$$\nu : k[u,v] \to k(u,v)$$

$$u \to \nu(u), \quad v \to \nu(v)$$

Then, we compute $\nu^2 = 1$. Thus, ν extends to an involutory automorphism of $k(u,v)$ which we again denote by ν. Since,

$$\tau : k(u,v) \to k(u,v)$$

$$(u,v) \to (u, \nu(v))$$

is not involutory, then $[k(u,v) : k(r_1,r_2)] = 2$ and $Gal_{k(u,v)/k(r_1,r_2)} = \langle \nu \rangle$.

Lemma 19: *The fields $k(i_1,i_2,i_3) = k(r_1,r_2)$ are the same.*

Remark 20: To find the equation in [26, Appendix A] we eliminate r_1 and r_2 from the three equations of the above lemma. This equation has degree 8, 13, and 20 in i_1, i_2, i_3 respectively.

Proof: (Theorem 14) The map

$$\theta : (u,v) \to (i_1, i_2, i_3)$$

generically has degree 2, by previous section. Denote the minors of the Jacobian matrix of θ by $M_1(u,v), M_2(u,v), M_3(u,v)$. The system

$$\begin{cases} M_1(u,v) = 0 \\ M_2(u,v) = 0 \\ M_3(u,v) = 0 \end{cases} \tag{33}$$

has solutions

$$8v^3 + 27v^2 - 54uv^2 - u^2v^2 + 108u^2v + 4u^3v - 108u^3 = 0 \tag{34}$$

and 7 further solutions which we display in the following table together with the corresponding values (i_1, i_2, i_3) and properties of the corresponding genus 2 field K.

(u,v)	(i_1,i_2,i_3)	$Aut(K)$	$e_3(K)$
$(-\frac{7}{2},2)$	$J_{10}=0$, no associated genus 2 field K		
$(-\frac{775}{8},\frac{125}{96})$, $(\frac{25}{2},\frac{250}{9})$	$-\frac{8019}{20},-\frac{1240029}{200},\frac{531441}{100000}$	D_4	2
$(27-\frac{77}{2}\sqrt{-1},23+\frac{77}{9}\sqrt{-1})$, $(27+\frac{77}{2}\sqrt{-1},23-\frac{77}{9}\sqrt{-1})$	$(\frac{729}{2116},\frac{1240029}{97336},\frac{531441}{13181630464}$	D_4	2
$(-15+\frac{35}{8}\sqrt{5},\frac{25}{2}+\frac{35}{6}\sqrt{5})$, $(-15-\frac{35}{8}\sqrt{5},\frac{25}{2}-\frac{35}{6}\sqrt{5})$	$81,-\frac{5103}{25},-\frac{729}{12500}$	D_6	2

Fig. 2. Corresponding (u,v) for which the Jacobian matrix of θ is 0

Assume that equation (34) holds for some $(u,v)\in k^2$. Then the corresponding quantities J_{2i}, $i=1,2,3,5$ satisfy the equation

$$F(J_2,J_4,J_6,J_{10})=0 \tag{35}$$

where $F(J_2,J_4,J_6,J_{10})$ is displayed in [26]. This is obtained by taking the resultants of equations of i_1,i_2,i_3 and (34). We define $J_{48}:=F(J_2,J_4,J_6,J_{10})$. By previous section θ is generically a covering of degree 2. So exists a Zariski open subset $\mathcal{U}$ of k^2 with the following properties: Firstly, θ is defined everywhere on $\mathcal{U}$ and is a covering of degree 2 from $\mathcal{U}$ to $\theta(\mathcal{U})$. Further, if $\mathrm{u}\in\mathcal{U}$ then all $\mathrm{u}'\in k^2$ with θ defined at u' and $\theta(\mathrm{u}')=\theta(\mathrm{u})$ also lie in $\mathcal{U}$. Suppose $\underline{i}\in k^3$ such that $|\theta^{-1}(\underline{i})|>2$ and $det(Jac(\theta))$ does not vanish at any point of $\theta^{-1}(\underline{i})$. Then by implicit function theorem, there is an open ball B around each element of $\theta^{-1}(\underline{i})$ such that each point in $\theta(B)$ has >2 inverse images under θ. But B has to intersect the Zariski open set $\mathcal{U}$. This is a contradiction. Thus, if $\underline{i}\in k^3$ and $|\theta^{-1}(\underline{i})|>2$, then $det(Jac(\theta))=0$ at some point of $\theta^{-1}(\underline{i})$ and so J_{48} vanishes.

Let $e_3(K)>1$ and $J_2\neq 0$, $J_{48}\neq 0$. Then i_1,i_2,i_3 are defined and by previous paragraph $|\theta^{-1}(i_1,i_2,i_3)|\leq 2$. Thus, by lemma 15 part iii) $e_3(K)\leq 2$. This completes the proof of theorem 14. □

6.2. *Elliptic subcovers*

We express the j-invariants j_i of the elliptic subfields E_i of K, from lemma 16, in terms of u and v as follows:

$$\begin{aligned} j_1 &= 16v\frac{(vu^2+216u^2-126vu-972u+12v^2+405v)^3}{(v-27)^3(4v^2+27v+4u^3-18vu-vu^2)^2} \\ j_2 &= -256\frac{(u^2-3v)^3}{v(4v^2+27v+4u^3-18vu-vu^2)} \end{aligned} \tag{36}$$

where $v \neq 0, 27$.

Remark 21: The automorphism $\nu \in Gal_{k(u,v)/k(r_1,r_2)}$ permutes the elliptic subfields. One can easily check that:

$$\nu(j_1) = j_2, \quad \nu(j_2) = j_1$$

Define T and N as follows;

$$\begin{aligned} T = &\frac{1}{16777216r_2^3r_1^8}(1712282664960r_2^3r_1^6 + 1528823808r_2^4r_1^6 + 49941577728r_2^4r_1^5 \\ &- 38928384r_2^5r_1^5 - 258048r_2^6r_1^4 + 12386304r_2^6r_1^3 + 901736973729792r_2r_1^{10} \\ &+ 966131712r_2^5r_1^4 + 16231265527136256r_1^{10} + 480r_2^8r_1 + 101376r_2^7r_1^2 + 479047767293952r_2r_1^8 \\ &+ 7247757312r_2^3r_1^8 + 7827577896960r_2^2r_1^9 + 2705210921189376r_1^9 + 619683250176r_2^3r_1^7 \\ &+ 21641687369515008r_1^{12} + 32462531054272512r_1^{11} + r_2^9 + 37572373905408r_2^2r_1^7 \\ &+ 1408964021452800r_2r_1^9 + 45595641249792r_2^2r_1^8) \\ N = &-\frac{1}{68719476736r_1^12r_2^3}(84934656r_1^5 + 1179648r_1^4r_2 - 5308416r_1^4 - 442368r_1^3r_2 \\ &- 13824r_1^2r_2^2 - 192r_1r_2^3 - r_2^4)^3 \end{aligned} \tag{37}$$

Lemma 22: *The j-invariants of the elliptic subfields satisfy the following quadratic equations over $k(r_1, r_2)$;*

$$j^2 - T\,j + N = 0 \tag{38}$$

Proof: Substitute j_1 and j_2 as in (36) in equation (38). □

6.2.1. *Isomorphic Elliptic Subfields*

Suppose that $E_1 \cong E_2$. Then, $j_1 = j_2$ implies that

$$8v^3 + 27v^2 - 54uv^2 - u^2v^2 + 108u^2v + 4u^3v - 108u^3 = 0 \tag{39}$$

or

$$
\begin{aligned}
&324v^4u^2 - 5832v^4u + 37908v^4 - 314928v^3u - 81v^3u^4 + 255879v^3 + 30618v^3u^2 \\
&- 864v^3u^3 - 6377292uv^2 + 8503056v^2 - 324u^5v^2 + 2125764u^2v^2 - 215784u^3v^2 \\
&+ 14580u^4v^2 + 16u^6v^2 + 78732u^3v + 8748u^5v - 864u^6v - 157464u^4v + 11664u^6 = 0
\end{aligned} \tag{40}
$$

The former equation is the condition that $det(Jac(\theta)) = 0$ see (35). From equation 35 and expressions of i_1, i_2, i_3 we can express u as a rational function in i_1, i_2, and v. This is displayed in [26, Appendix B]. Also, $[k(v) : k(i_1)] = 8$ and $[k(v) : k(i_2)] = 12$. Eliminating v we get a curve in i_1 and i_2 which has degree 8 and 12 respectively. Thus, $k(u, v) = k(i_1, i_2)$. Hence, $e_3(K) = 1$ for any K such that the associated u and v satisfy equation (35).

6.2.2. *The Degenerate Case*

We assume now that one of the extensions K/E_i from lemma 16 is degenerate, i.e. has only one branch point. The following lemma determines a relation between j_1 and j_2.

Lemma 23: *Suppose that K/E_2 has only one branch point. Then,*

$$729 j_1 j_2 - (j_2 - 432)^3 = 0$$

Making the substitution $T = -27 j_1$ we get

$$j_1 = F_2(T) = \frac{(T+16)^3}{T}$$

where $F_2(T)$ is the Fricke polynomial of level 2.

If both K/E_1 and K/E_2 are degenerate then

$$
\begin{cases}
729 j_1 j_2 - (j_1 - 432)^3 = 0 \\
729 j_1 j_2 - (j_2 - 432)^3 = 0
\end{cases} \tag{41}
$$

There are 7 solutions to the above system. Three of which give isomorphic elliptic curves

$$j_1 = j_2 = 1728, \quad j_1 = j_2 = \frac{1}{2}(297 \pm 81\sqrt{-15})$$

The other 4 solutions are given by:

$$
\begin{cases}
729 j_1 j_2 - (j_1 - 432)^3 = 0 \\
j_1^2 + j_2^2 - 1296(j_1 + j_2) + j_1 j_2 + 559872 = 0
\end{cases} \tag{42}
$$

This corrects [21] where it is claimed there is only one solution $j_1 = j_2 = 1728$.

7. Further remarks

If $e_3(\mathcal{C}) \geq 1$ then the automorphism group of $\mathcal{C}$ is one of the following: $\mathbb{Z}_2, V_4, D_4$, or D_6. Moreover; there are exactly 6 curves $\mathcal{C} \in \mathcal{L}_3$ with automorphism group D_4 and six curves $\mathcal{C} \in \mathcal{L}_3$ with automorphism group D_6. They are listed in [29] where rational points of such curves are found.

Genus 2 curves with degree 5 elliptic subcovers are studied in [33] where a description of the space $\mathcal{L}_5$ is given and all its degenerate loci. The case of degree 7 is the first case when all possible degenerate loci occur.

We have organized the results of this paper in a Maple package which determines if a genus 2 curve has degree $n = 2, 3$ elliptic subcovers. Further, all its elliptic subcovers are determined explicitly. We intend to implement the results for $n = 5$ and the degenerate cases for $n = 7$.

References

1. I. Blake, G. Seroussi and N. Smart, Elliptic Curves in Cryptography, LMS, 265, (1999).
2. C. Birkenhake, H. Wilhelm, Humbert surfaces and the Kummer plane. Trans. Amer. Math. Soc. 355 (2003), no. 5, 1819–1841.
3. O. Bolza, On binary sextics with linear transformations into themselves. *Amer. J. Math.* **10**, 47-70.
4. R. Brandt, Über Die Automorphismengruppen von algebraischen Funktionenkörpern, (unpublished) PhD thesis. Universität-Gesamthochschule Essen, 1988.
5. R. Brandt and H. Stichtenoth, Die Automorphismengruppen hyperelliptischer Kurven, *Man. Math.* 55, 83-92, 1986.
6. A. Clebsch, Theorie der Binären Algebraischen Formen, Verlag von B.G. Teubner, Leipzig, 1872.
7. I. Duursma and N. Kiyavash, The Vector Decomposition Problem for Elliptic and Hyperelliptic Curves, (preprint)
8. G. Frey, On elliptic curves with isomorphic torsion structures and corresponding curves of genus 2. *Elliptic curves, modular forms, and Fermat's last theorem (Hong Kong, 1993)*, 79-98, Ser. Number Theory, I, *Internat. Press, Cambridge, MA*, 1995.
9. G. Frey and E. Kani, Curves of genus 2 covering elliptic curves and an arithmetic application. *Arithmetic algebraic geometry (Texel, 1989)*, 153-176, *Progr. Math.*, 89, Birkhäuser Boston, MA, 1991.
10. P. Gaudry and E. Schost, Invariants des quotients de la Jacobienne d'une courbe de genre 2, (in press)
11. G. van der Geer, Hilbert modular surfaces, Springer, Berlin, 1987.
12. Geyer W. Geyer, Invarianten binärer Formen, *Lecture Notes in Math.*, Springer, New York, (1972).
13. J. Gutierrez and T. Shaska, Hyperelliptic curves with extra involutions, *LMS J. of Comput. Math.*, 8 (2005), 102-115.

14. G. HUMBERT Sur les fonctionnes abliennes singulires. I, II, III. J. Math. Pures Appl. serie 5, t. V, 233–350 (1899); t. VI, 279–386 (1900); t. VII, 97–123 (1901).
15. J. IGUSA, Arithmetic Variety Moduli for genus 2. *Ann. of Math.* (2), 72, 612-649, 1960.
16. C. JACOBI, Review of Legendre, Théorie des fonctions elliptiques. Troiseme supplém ent. 1832. J. reine angew. Math. 8, 413-417.
17. E, KANI AND W. SCHANZ, Diagonal quotient surfaces. Manuscripta Math. 93, no. 1, 67–108, 1997.
18. E. KANI, The number of curves of genus two with elliptic differentials. J. Reine Angew. Math. 485 (1997), 93–121.
19. A. KRAZER, Lehrbuch der Thetafunctionen, Chelsea, New York, 1970.
20. V. KRISHNAMORTHY, T. SHASKA, H. VÖLKLEIN, Invariants of binary forms , *Developments in Mathematics*, Vol. 12, Springer 2005, pg. 101-122.
21. M. R. KUHN, Curves of genus 2 with split Jacobian. *Trans. Amer. Math. Soc* **307**, 41-49, 1988.
22. H. LANGE, Über die Modulvarietät der Kurven vom Geschlecht 2. *J. Reine Angew. Math.*, 281, 80-96, 1976.
23. K. MAGAARD, T. SHASKA, S. SHPECTOROV, AND H. VÖLKLEIN, The locus of curves with prescribed automorphism group. *Communications in arithmetic fundamental groups* (Kyoto, 1999/2001). Sūrikaisekikenkyūsho Kōkyūroku No. 1267 (2002), 112–141.
24. N. MURABAYASHI, The moduli space of curves of genus two covering elliptic curves. Manuscripta Math. 84 (1994), no. 2, 125–133.
25. D. SEVILLA, T. SHASKA, Hyperelliptic curves with reduced automorphism group A_5, (submitted).
26. T. SHASKA, Genus 2 curves with degree 3 elliptic subcovers, *Forum. Math.*, vol. **16**, 2, pg. 263-280, 2004.
27. T. SHASKA, Some special families of hyperelliptic curves, *J. Algebra Appl.*, vol **3**, No. 1 (2004), 75-89.
28. T. SHASKA, Computational algebra and algebraic curves, ACM, *SIGSAM Bulletin, Comm. Comp. Alg.*, Vol. **37**, No. 4, 117-124, 2003.
29. T. SHASKA, Genus 2 curves with (3,3)-split Jacobian and large automorphism group, Algorithmic Number Theory (Sydney, 2002), **6**, 205-218, *Lect. Not. in Comp. Sci.*, 2369, Springer, Berlin, 2002.
30. T. SHASKA, Curves of genus 2 with (n, n)-decomposable Jacobians, *J. Symbolic Comput.* 31 (2001), no. 5, 603–617.
31. T. SHASKA AND J. L. THOMPSON, On the generic curves of genus 3, *Affine Algebraic Geometry*, Cont. Math., American Mathematical Society, (2004).
32. T. SHASKA AND H. VÖLKLEIN, Elliptic subfields and automorphisms of genus two fields, *Algebra, Arithmetic and Geometry with Applications*, pg. 687 - 707, Springer (2004).
33. T. SHASKA, H. VÖLKLEIN, Genus 2 curves with degree 5 elliptic subcovers, (preprint).
34. G. TAMME, Ein Satz über hyperelliptische Funktionenkörper. J. Reine Angew. Math. 257, 217–220, 1972.

A Question about Pic(X) as a G-module

D. Goldstein, R. Guralnick, & D. Joyner*

IDA/CCR,
4320 Westerra Court,
San Diego, CA 92121.
dgoldste@ccrwest.org

Department of Mathematics,
University of Southern California,
Los Angeles, CA 90089-2532.
guralnic@usc.edu

Department of Mathematics,
United States Naval Academy,
Annapolis, MD 21402
wdj@usna.edu

Let G be a finite group acting faithfully on an irreducible non-singular projective curve. Does every G-invariant divisor class contain a G-invariant divisor? We show that the answer depends only on G and not on the curve. We address the same question for degree-0 divisor classes.

1. Introduction

This paper addresses the following question: Let X be an irreducible non-singular projective variety defined over a field F, and let G be a finite subgroup of $\mathrm{Aut}_F(X)$.

*The authors are thankful to Amy Ksir, David Saltman, Joe Wetherell, and David Zelinsky for stimulating discussions. We thank Bernard Köch for the reference to [2]. The second author gratefully acknowledge the support of the NSF grant DMS-0140578.

Question 1: Is it true that every G-invariant divisor class on X is the class of a G-invariant divisor on X?

A special case of Question 1 was posed by Lonsted in [2]. He shows that the question has an affirmative answer if F is an algebraically closed field whose characteristic does not divide the order of the cyclic group G. We settle the question if F is algebraically closed and X is a curve (Corollary 9). We show that the answer depends only on the group and the characteristic of the field. We show by example that the answer may be negative.

Lonsted shows [2, Proposition 2.9] that a G-invariant divisor class on X is the class of a G-invariant divisor if and only if the corresponding line bundle $\mathcal{E}$ on X has a *G-linearization*, which is, loosely speaking, a family of isomorphisms $g^*\mathcal{E} \cong \mathcal{E}$, one for each g in G, that is compatible in a natural sense. In view of this result, Question 1 is equivalent to the question: Does every G-invariant line bundle on X have a G-linearization?

The basic idea is to use group cohomology to attack Question 1. A basic reference for Section 2 is Serre [3, §VII]. In Section 3 we treat a curve over an algebraically closed field. In Section 4 we address an analogous question for degree-0 divisors. In Section 5 we give some examples.

2. Cohomology

Throughout this section, X will be denote a non-singular projective variety defined over a field F. Let $F(X)$ be the field of functions of X. Corresponding to any nonzero function f in $F(X)$ is the divisor (f). By definition (f) is the zero divisor of f minus the polar divisor of f. Thus $(f) = 0$ in the divisor group $\mathrm{Div}(X)$ precisely when f has no zeros and no poles. Since X is projective, such an f is constant. Therefore there is a short exact sequence

$$1 \to F^\times \to F(X)^\times \to \mathrm{Prin}(X) \to 0, \qquad (1)$$

where a divisor on X is principal if it has the form (f) for some nonzero $f \in F(X)$ and where $\mathrm{Prin}(X)$ is the group of principal divisors on X.

Now let G be a finite subgroup of $\mathrm{Aut}_F(X)$. Then we have from the long exact sequence in cohomology:

$$H^1(G, F(X)^\times) \to H^1(G, \mathrm{Prin}(X)) \to H^2(G, F^\times) \to H^2(G, F(X)^\times)$$

Lemma 2: *The abelian group $H^1(G, \mathrm{Prin}(X))$ is isomorphic to the kernel of the map $H^2(G, F^\times) \to H^2(G, F(X)^\times)$.*

Proof: By Hilbert's Theorem 90 we have that $H^1(G, F(X)^\times) = 0$. The lemma now follows from the sequence above. □

Next consider the short exact sequence

$$0 \to \mathrm{Prin}(X) \to \mathrm{Div}(X) \to \mathrm{Pic}(X) \to 0, \tag{2}$$

where by definition $\mathrm{Pic}(X)$ is the quotient. We get from the long exact sequence in cohomology:

$$H^0(G, \mathrm{Div}(X)) \to H^0(G, \mathrm{Pic}(X)) \to H^1(G, \mathrm{Prin}(X)) \to H^1(G, \mathrm{Div}(X))$$

Of course $H^0(G, \mathrm{Div}(X)) = \mathrm{Div}(X)^G$ and $H^0(G, \mathrm{Pic}(X)) = \mathrm{Pic}(X)^G$.

Proposition 3: $H^1(G, \mathrm{Div}(X)) = 0$.

Proof: For each orbit $\mathcal{O}$ of G on X, set $L_{\mathcal{O}} = \oplus_{P \in \mathcal{O}} \mathbb{Z}[P]$. Then $\mathrm{Div}(X)$ is the direct sum of the $L_{\mathcal{O}}$ as $\mathcal{O}$ varies over all orbits of G on X. So it suffices to show that $H^1(G, L_{\mathcal{O}}) = 0$ for each orbit $\mathcal{O}$.

Now choose $P \in \mathcal{O}$ and set $H = \{g \in G \mid gP = P\}$. Then $L_{\mathcal{O}}$ is equal to the induced module $\mathrm{Ind}_H^G(\mathbb{Z})$, where H acts trivially on $\mathbb{Z}$. By Shapiro's Lemma, $H^1(G, \mathrm{Ind}_H^G(\mathbb{Z})) = H^1(H, \mathbb{Z})$. Now $H^1(H, Z) = \mathrm{Hom}(H, \mathbb{Z})$ since the action is trivial, and $\mathrm{Hom}(H, \mathbb{Z}) = 0$ since H is finite. □

Corollary 4: *The sequence*

$$\mathrm{Div}(X)^G \to \mathrm{Pic}(X)^G \to H^1(G, \mathrm{Prin}(X)) \to 0$$

is exact.

Next we combine Lemma 2 and Corollary 4.

Lemma 5: *The sequence*

$$\mathrm{Div}(X)^G \to \mathrm{Pic}(X)^G \to \ker\left(H^2(G, F^\times) \to H^2(G, F(X)^\times)\right) \to 0$$

is exact. In particular, the map $\mathrm{Div}(X)^G \to \mathrm{Pic}(X)^G$ *is surjective whenever the natural map from* $H^2(G, F^\times)$ *to* $H^2(G, F(X)^\times)$ *is injective.*

The next proposition generalizes Proposition 2.6 of [2] and follows from Lemma 5.

Proposition 6: *If* $H^2(G, F^\times) = 0$*, then the map* $\mathrm{Div}(X)^G \to \mathrm{Pic}(X)^G$ *is surjective.*

If the Sylow ℓ-subgroup of G is cyclic, then the ℓ part of $H^2(G, F^\times) = 1$. Indeed, the restriction map to the Sylow ℓ-subgroup is an injection on the ℓ part of cohomology. If all Sylow subgroups of G are cyclic, then $H^2(G, F^\times) = 1$. Such groups are well-known to be metacyclic. This yields:

Corollary 7: *Suppose that for every prime ℓ the ℓ-Sylow subgroup of G is cyclic. Then the map* $\mathrm{Pic}(X)^G \to \mathrm{Div}(X)^G$ *is surjective. In particular, the map is surjective if G is cyclic.*

3. Curves

Throughout this section and the next, X will be denote a non-singular projective curve defined over a field F and G will be a finite subgroup of $\mathrm{Aut}_F(X)$. The field F is assumed to be algebraically closed except briefly in Example 11.

Corollary 8: $H^1(G, \mathrm{Prin}(X)) = H^2(G, F^\times)$. *The cokernel of the map* $\mathrm{Div}(X)^G \to \mathrm{Pic}(X)^G$ *is equal to* $H^2(G, F^\times)$.

Proof: Since G acts faithfully we have $H^2(G, F(X)^\times) = 0$ by Tsen's theorem. The first statement now follows from Lemma 2. The second statement follows from Lemma 5. □

The following is now immediate:

Let p be the characteristic of the field F. If $p = 0$ then $H^2(G, F^\times)$ is the Schur multiplier of G. If $p > 0$ then $H^2(G, F^\times)$ is the prime-to-p-part of the Schur multiplier of G.

Corollary 9: *Let p be the characteristic of the field F.*

(1) If $p = 0$, the cokernel of the map $\mathrm{Div}(X)^G \to \mathrm{Pic}(X)^G$ *is equal to the Schur multiplier of G.*
(2) If $p > 0$, the cokernel of the map $\mathrm{Div}(X)^G \to \mathrm{Pic}(X)^G$ *is equal to the prime-to-p-part of the Schur multiplier of G.*

Remark 10: It follows from Corollary 9 that the cokernel of the map $\mathrm{Div}(X)^G \to \mathrm{Pic}(X)^G$ (and in particular its surjectivity) depends only on the group G and the characteristic of the field F, and not on the curve X.

Example 11: Somewhat more generally, we can consider the following setup. Let F/k be a finite Galois extension, and let the non-singular projective curve Y be defined over k. Set $X = Y \otimes_F k$. Now let G be a finite subgroup of the arithmetic automorphism group $A = \mathrm{Aut}_k(Y \otimes_F k)$. Note that A is the middle term of the short exact sequence

$$1 \longrightarrow \mathrm{Aut}_F(X) \longrightarrow \mathrm{Aut}_k(X) \xrightarrow{\pi} \mathrm{Gal}(F/k) \longrightarrow 1$$

In this context, the map $\mathrm{Div}(X)^G \to \mathrm{Pic}(X)^G$ is surjective if (1) π is injective on G and (2) the Brauer group of $F^{\pi G}$ is trivial. Note that condition (2) holds if F is a finite field.

We give an easy corollary that has no cohomology in the statement.

Corollary 12: *We have:*

(1) Let the finite group G act faithfully on the non-singular projective curve X defined over the field of complex numbers. Suppose that $\mathrm{Div}(X)^G$ *surjects onto* $\mathrm{Pic}(X)^G$. *Then* $\mathrm{Div}(Y)^G$ *surjects onto* $\mathrm{Pic}(Y)^G$ *for any curve Y (over any algebraically closed field) on which G acts faithfully.*

(2) Let the finite group G act faithfully on the curve X_i over the algebraically closed field K_i for $i = 1, 2$. Suppose that $\mathrm{Div}(X_i)^G$ *surjects onto* $\mathrm{Pic}(X_i)^G$ *for $i = 1, 2$, and suppose that the characteristics of K_1 and K_2 are distinct. Then* $\mathrm{Div}(Y)^G$ *surjects onto* $\mathrm{Pic}(Y)^G$ *surjects onto* $\mathrm{Pic}(Y)^G$ *for any curve Y (over any algebraically closed field) on which G acts faithfully.*

4. Degree zero

Throughout this section, X will be denote a non-singular projective curve defined over an algebraically closed field F and G will be a finite subgroup of $\mathrm{Aut}_F(X)$.

We next want to consider $\mathrm{Pic}_0(X)$ (the Jacobian of the non-singular curve X) and discuss the surjectivity of the map $\phi : \mathrm{Div}_0(X)^G \to \mathrm{Pic}_0(X)^G$.

Consider the degree map deg on $\mathrm{Div}(X)$ and also on $\mathrm{Pic}(X)$. Let B be $\deg(\mathrm{Div}(X)^G)$. We identify B with a subgroup of $\mathrm{Pic}(X)/\mathrm{Pic}_0(X) = \mathbb{Z}$. Clearly, B contains $|G|\mathbb{Z}$. It may be bigger. As a substitute for Proposition 3, we have the following result. Let $\bar{F}$ denote an algebraic closure of the field F.

Lemma 13: *Let b be the* gcd *of the indices $|G : \mathrm{Stab}_G(x)|$ as x ranges over the $\bar{F}$-points of X. Then (1) $B = b\mathbb{Z}$ and (2) $H^1(G, \mathrm{Div}_0(X)) \cong \mathbb{Z}/B$.*

Proof: For $\mathcal{O}$ a G-orbit on X, let $\Sigma_{\mathcal{O}}$ be the sum of the points in $\mathcal{O}$. Observe that $\mathrm{Div}(X)^G$ is spanned by the set of $\Sigma_{\mathcal{O}}$ as $\mathcal{O}$ varies over the G-orbits on X. If I is the stabilizer of a point in the orbit, then the degree of the divisor $\Sigma_{\mathcal{O}}$ is $|G : I|$. Thus $\deg(\mathrm{Div}(X)^G) = b\mathbb{Z}$, where b is the gcd of the indices $|G : I|$. The first statement is proved.

Consider the sequence $0 \to \mathrm{Div}_0(X) \to \mathrm{Div}(X) \to \mathbb{Z} \to 0$. The map from Div($X$) to $\mathbb{Z}$ is deg. Using Proposition 3 yields:

$$\mathrm{Div}(X)^G \to \mathbb{Z} \to H^1(G, \mathrm{Div}_0(X)) \to 0.$$

The second statement is proved. □

Now consider the short exact sequence

$$0 \to \mathrm{Prin}(X) \to \mathrm{Div}_0(X) \to \mathrm{Pic}_0(X) \to 0.$$

Taking fixed points leads to the long exact sequence for cohomology, part of which is:

$$\mathrm{Div}_0(X)^G \xrightarrow{\phi} \mathrm{Pic}_0(X)^G \to H^1(G, \mathrm{Prin}(X)) \xrightarrow{\beta} H^1(G, \mathrm{Div}_0(X)) \qquad (3)$$

Here ϕ denotes the natural map from Div(X) to Pic(X). We note :

Proposition 14: *The image of* $H^1(G, \mathrm{Prin}(X))$ *in* $H^1(G, \mathrm{Div}_0(X))$ *is isomorphic to* A/B *where* $A = \deg(\mathrm{Pic}(X)^G)$ *and* $B = \deg(\mathrm{Div}(X)^G)$.

Proof: Let W denote the kernel of the mapping β from $H^1(G, \mathrm{Prin}(X))$ to $H^1(G, \mathrm{Div}_0(X))$. We have already seen that the rows of the diagram:

$$\begin{array}{ccccccccc} 1 & \longrightarrow & \phi(\mathrm{Div}_0(X)^G) & \longrightarrow & \mathrm{Pic}_0(X)^G & \longrightarrow & W & \longrightarrow & 1 \\ & & \downarrow{\scriptstyle \alpha_1} & & \downarrow{\scriptstyle \alpha_2} & & \downarrow{\scriptstyle \alpha_3} & & \\ 1 & \longrightarrow & \phi(\mathrm{Div}(X)^G) & \longrightarrow & \mathrm{Pic}(X)^G & \longrightarrow & H^1(G, \mathrm{Prin}(X)) & \longrightarrow & 1 \end{array}$$

are exact. It is clear that the diagram commutes, where the vertical maps α_i are the natural inclusions.

By the snake lemma the sequence

$$1 \to \mathrm{coker}(\alpha_1) \to \mathrm{coker}(\alpha_2) \to \mathrm{coker}(\alpha_3) \to 1$$

is exact. By definition $\mathrm{coker}(\alpha_1) = B$ and $\mathrm{coker}(\alpha_2) = A$, so it follows that $\mathrm{coker}(\alpha_3) = A/B$, as desired. □

The exact sequence (3), together with and Lemmas 8 and 13 yield the following result.

Theorem 15: The sequence

$$\mathrm{Div}_0(X)^G \to \mathrm{Pic}_0(X)^G \to H^2(G, F^\times) \to \mathbb{Z}/b\mathbb{Z}$$

is exact. In particular, if $H^2(G, F^\times) = 1$ then each G-invariant divisor class of degree-0 is the class of a degree-0 G-invariant divisor.

5. Examples

Now $H^1(G, \mathrm{Prin}(X)) \cong H^2(G, F^\times)$ can be essentially any abelian group one wishes (by taking direct products for example). On the other hand, we saw that $H^1(G, \mathrm{Div}_0(X))$ is a finite cyclic group. So there are certainly examples where this kernel is nontrivial (and as large as one wishes). We now give some examples:

Example 16: Let X be a non-singular projective curve. Let the finite group G act faithfully on X. Clearly the canonical class K in $\mathrm{Pic}(X)$ is G-invariant. Let f be any G-invariant function on X such that $df \neq 0$. Then $K = [df]$. Thus in this case Question 1 has an affirmative answer. (To see that such an f exists, let $\Omega_0 = \{f \mid df = 0\}$ and let $\Omega_1 = \{f \mid gf = f \text{ for all } g \text{ in } G\}$. Then the extension $[F(X) : \Omega_0]$ is infinite while the extension $[F(X) : \Omega_1]$ is finite. Therefore $\Omega_1 \setminus \Omega_0$ is nonempty.)

Example 17: We work over the field of complex numbers. If $X = \mathbb{P}^1$ then $\mathrm{Pic}(X) = \mathbb{Z}$ and $H^0(G, \mathrm{Pic}(X)) = \mathbb{Z}$ as well. Since $\mathrm{Pic}(X)_0 = 0$, the answer to the question is "yes" if and only if there is a fixed divisor of any given degree (since there is only one class with a given degree). We show that this can fail. Fix an embedding of $G = A_5$ into $PGL(2, \mathbb{C})$ (to get an action of G on $\mathbb{P}^1$). In this case, inertia groups (i.e. stabilizers of points on the curve) are always cyclic, so the only possibilities are of order $1, 2, 3$ and 5. Thus the possible orbit sizes are $12, 20, 30$ and 60 and any G-fixed divisor has degree a multiple of the greatest common divisor of these numbers (i.e. 2). Now Corollary 9 implies that $H^2(A_5, \mathbb{C}^\times) = \mathrm{Pic}(X)^G/\mathrm{Div}(X)^G = \mathbb{Z}/2\mathbb{Z}$, which is well-known (see e.g. [6, pp. 245–246]).

Now let F be an algebraically closed field of characteristic 2. Then $H^2(A_5, F^\times) = 0$, since the $2'$ part of the Schur multiplier of A_5 is zero. Fix an embedding of A_5 into $PGL(2, F)$ so that A_5 acts on $X = \mathbb{P}^1$. It follows from Corollary 9 that $\mathrm{Div}(X)^G = \mathrm{Pic}(X)^G = \mathbb{Z}$. Indeed, the inertia groups are A_4 and cyclic of order 5, and the gcd of their indices is one.

Example 18: Let F be an algebraically closed field of characteristic not 2. Let G be an elementary abelian group of order 4 acting on the curve $\mathbb{P}^1(F)$. One can argue as in the previous example that $H^2(G, F^\times) = \mathbb{Z}/2\mathbb{Z}$.

Example 19: Let F be an algebraically closed field of characteristic not 2. Let G be an elementary abelian group of order 2^r acting on the curve X. If G acts fixed point freely on X, then $B = 2^r\mathbb{Z}$ in the notation above. Since A/B is cyclic and embeds in $H^2(G, F^\times)$, a group of exponent 2, it follows that every G-invariant divisor class on X has degree a multiple of 2^{r-1}. We show below that there may be such a class or there may not be. Suppose that G does not act fixed point freely on X. Since G acts tamely on X, all inertia groups are cyclic and so have index a multiple of 2^{r-1}, it follows that $B = 2^{r-1}\mathbb{Z}$ and so every G-invariant divisor class has degree a multiple of 2^{r-2}.

Example 20: Assume that F is algebraically closed. We show by example that it is possible that both the map $\mathrm{Div}(X)^G \to \mathrm{Pic}(X)^G$ is surjective and $H^2(G, F^\times) \neq 1$. Of course it is necessary to have $\dim(X) > 1$. Let r be a prime and let G be an elementary abelian group of order r^2. Suppose that the characteristic of the algebraically closed field is not equal to r. Let $X = \mathbb{P}^2$. Then $\mathrm{Pic}(X) \cong \mathbb{Z}$. We can write $F(X) = F(x, y)$.

Let G act faithfully on X in such a way that x and y are nontrivial eigenvalues with independent characters. The hyperplane $x = 0$ is an invariant divisor of degree 1 and so $\mathrm{Div}(X)^G \to \mathrm{Pic}(X)^G$ is surjective.

It follows that the map $H^2(G, F^\times) \to H^2(G, F(X)^\times)$ is injective. Since $H^2(G, F^\times)$ is cyclic of order r, the image of a generator is nonzero. This says that a certain twisted group algebra $F(X)*[G]$ (twisted by both the cocycle and the action of G) is nontrivial in the Brauer group of $F(X)^G = F(x^p, y^p)$.

Now let X be a non-singular projective curve on which G acts faithfully. We have already observed that there are exact sequences:

$$\mathrm{Div}^G(X) \to \mathrm{Pic}^G(X) \to H^2(G, F^\times) \to 0 \text{ and} \tag{4}$$

$$\mathrm{Div}_0^G(X) \to \mathrm{Pic}_0^G(X) \to H^2(G, F^\times) \to A/B \to 0, \tag{5}$$

where $A = \deg(\mathrm{Pic}(X)^G)$ and $B = \deg(\mathrm{Div}(X)^G)$. Note that B is generated by the gcd of $|G : \mathrm{Stab}_G(x)|$ as x ranges over X.

It is convenient to define $C = C(X, G) = A/B$. We summarize some properties of this group.

Lemma 21: *We have:*

(1) The group $C = A/B$ is finite and cyclic.
(2) $|C|$ divides the gcd of the $|G|/|I|$ as I ranges over the inertia groups of G.
(3) $C = 1$ if there is a totally ramified point in X.

Only the third statement deserves comment. Indeed, if the point $P \in X$ is totally ramified, then $[P]$ is a G-invariant degree-one divisor.

David Saltman has shown us how to explicitly compute the last map $H^2(G, F^\times) \to A/B$ in equation (5). He has kindly allowed us to include his observations here. Let n be an integer coprime to the characteristic of F. Let c be a 2-cocycle in $H^2(G, \mu_n)$. Then $c = \delta(d_g)$, where d_g is a coboundary with values in $F(X)^\times$ by Tsen's theorem. Since d_g^n is trivial in H^1, it follows that

$$d_g^n = g\theta/\theta \tag{6}$$

for some function θ on X. Now $F(X)(\theta^{1/n})/F(X)^G$ is a Galois cover that realizes c. The class of d_g vanishes in $H^1(G, \mathrm{Div}(X))$ and so we can write (switching now from multiplicative to additive notation)

$$(d_g) = gD - D$$

for a divisor D. Now multiply by n and use (6) to get $(g\theta) - (\theta) = ngD - D$. Hence

$$(\theta) - D \quad \text{is a } G\text{-invariant divisor.}$$

Now D is not well-defined – but its degree is well-defined modulo b. The map in question $H^2(G, F^\times) \to A/B$ sends (the class of) c to the degree of D via the construction above.

Note that the question of how many G-invariant divisor classes are not in the image of a G-invariant divisor depends only on the group. We show that this is not the case for the analogous question for degree-0 divisors and divisor classes. The answer depends on the curve even if we assume that $X \to X/G$ is unramified.

We have already set $C(X, G)$ to be $\deg(\mathrm{Div}(X)^G)/\deg(\mathrm{Pic}(X)^G)$. By definition, $C(X, G)$ equals A/B. We already observed that the term $H^2(G, F^\times)$ in (4) does not depend on the curve X. So it seems natural to ask whether the term $C = C(X, G)$ only depends on G and the ramification data of X. We next show Theorem 24 that indeed, this fails over the complex field even for G equals the Klein four group acting without fixed points.

We will need two lemmas to prove the theorem.

Lemma 22: *Let the finite group G act on the curve X. Let N be a normal subgroup of G that acts on X without fixed points. Let Q be the quotient group and let Y be the quotient curve. Then $|C(Y, Q)|$ divides $|C(X, G)|$.*

Proof: A Q-invariant divisor class on Y of degree d lifts to a G-invariant divisor class of degree $|N|d$ on X. However, for divisors, you can also go down: a G-invariant divisor of degree $|N|d$ on X always comes from a Q-invariant divisor on Y of degree d. □

Lemma 23: *Let X be a curve over the algebraically closed field F of characteristic not 2. Let G be an elementary abelian 2-group of order 2^r with $r \geq 3$. Then the cokernel of* $\mathrm{Div}(X)^G \to \mathrm{Pic}(X)^G$ *is an elementary abelian group of order 2^m where $m = r(r-1)/2$.*

Proof: The lemma follows from Corollary 9 and from the fact that $H^2(G, F^\times)$ equals $\wedge^2(G)$. □

Theorem 24: Let K be the Klein four group.

(1) There exists a curve E over the field of complex numbers such that K acts without fixed points on E and $C(E, K) = \mathbb{Z}/2\mathbb{Z}$.
(2) There exists a curve Y over the field of complex numbers such that K acts without fixed points on Y and $C(Y, K) = 1$.

Proof: Let E be an elliptic curve over $\mathbb{C}$ and let K act by translations on E. Note that indeed K acts without fixed points. In particular, any K-invariant divisor on E has degree 4.

Let $\{0, P, Q, R,\}$ be the of two-torsion of $E(\mathbb{C})$. Using the group law on E we have $0 + P = Q + R$, which implies that as divisors $(0) + (P) \sim (Q) + (R)$. Note further $E(\mathbb{C})$ and in particular K acts trivially on $\mathrm{Pic}_0(E)$. Thus there is an invariant divisor class of degree 2. Thus $C(E, K) = \mathbb{Z}/2\mathbb{Z}$ and the first part is proved.

Let $r \geq 3$ be a positive integer. Let J be the affine group $AGL(r, 2)$. The subgroup of translations G is elementary abelian of order 2^r. Note that G is normal in J. Now J acts on some curve X with the cover $X \to X/J$ unramified (in characteristic 0, this follows from the description of the fundamental group – in positive characteristic, this follows from [4]. Now we note that it follows from our proofs above that the map from $H^1(G, \mathrm{Prin}(X)) \to H^1(G, \mathrm{Div}_0(X))$ is J-equivariant. Since $H^1(G, \mathrm{Prin}(X)) \cong \wedge^2(G)$ is J-irreducible and not cyclic (since $r \geq 3$), it follows that $C(X, G)$ is trivial. Now let N be a subgroup of G of index 4. We have $K = G/N$. Set $Y = X/N$. Now lemma 22 implies that $C(Y, K)$ is trivial. So we have constructed two curves with the same group acting without fixed points such that the cokernels of $\mathrm{Div}_0(X)^G \to \mathrm{Pic}_0(X)^G$ are different. The theorem is proved. □

Remark 25: One can choose the two curves to have the same genus.

References

1. G. Karpilovsky, **The Schur multiplier**, Oxford Univ. Press, 1987.
2. K. Lonsted, On G-linebundles and $K_G(X)$, J. Math. Kyoto Univ. 23(1983), 775-793.
3. J.-P. Serre, **Local fields**, Springer-Verlag, Berlin, 1979.
4. K. Stevenson, Galois groups of unramified covers of projective curves in characteristic p, J. Algebra 182 (1996), 770–804.

Galois groups of prime degree polynomials with nonreal roots

A. Bialostocki and T. Shaska*

Department of Mathematics
University of Idaho
Moscow, ID, 83843.
arieb@uidaho.edu

Department of Mathematics
Oakland University
Rochester, MI, 48309-4485.
shaska@oakland.edu

In the process of computing the Galois group of a prime degree polynomial $f(x)$ over $\mathbb{Q}$ we suggest a preliminary checking for the existence of non-real roots. If $f(x)$ has non-real roots, then combining a 1871 result of Jordan and the classification of transitive groups of prime degree which follows from CFSG we get that the Galois group of $f(x)$ contains A_p or is one of a short list. Let $f(x) \in \mathbb{Q}[x]$ be an irreducible polynomial of prime degree $p \geq 5$ and $r = 2s$ be the number of non-real roots of $f(x)$. We show that if s satisfies $s\,(s \log s + 2 \log s + 3) \leq p$ then $Gal(f) = A_p, S_p$.

1. Introduction

Solving algebraic equations is one of the oldest and most fundamental problems in mathematics. The problem was put on a firm basis with the contribution of E. Galois and was the main motivation for the development of modern algebra. It is now considered basic knowledge that a polynomial equation with rational coefficients can be solved by radicals if and only if its Galois groups is solvable. Hence, the problem of determining Galois groups is important and has been considered in many areas of mathematics

*Supported by the NSA grant R1-05-0129

as number theory, group theory, algebraic geometry, and differential equations.

However, computing Galois groups is still a difficult task. Even with the development of new computer algebra systems this remains a challenge and can be accomplished only for small degree polynomials. For example, Maple 9 can only handle polynomials of degree ≤ 9 and Kant up to degree 15. Other computer algebra packages can handle polynomials whose degree is in the same range.

Let $f(x) = a_0 + a_1 x + \cdots + a_n x^n = 0$ be an algebraic equation. Suppose that $a_i = \pm 1$; then Littlewood and Offord proved that almost all these equations have less than $25(\log n^2)$ real roots; see [3]. If the a_i have normal distribution with density $e^{-u^2}/\pi^{\frac{1}{2}}$; then Kac (1944) showed that the mean value of the number of real roots of $f(x)$ equals

$$(4/\pi)\int_0^1 \frac{[1-n^2[x^2(1-x^2)/(1-x^{2n})]^2]^{\frac{1}{2}}}{1-x^2}\,dx \sim (2/\pi)\log n,$$

see [7]. If the coefficients are random variables with distribution function $\sigma(u)$ then in a second paper Kac (1949) gives an expansion for the average number of roots of the above equation in the interval (a, b). He then considers the special case when $\sigma(u) = \frac{1}{2}u$, and deduces from the general formula that in this case the average number of real roots is asymptotic to $(2\pi)^{-1}\log n$. Consequently, polynomials in general have plenty of non-real roots.

The existence of non-real roots of a polynomial makes the computation of its Galois group much easier. Computing the Galois group in this case for polynomials of prime degree p will be the focus of this short note. Checking whether a polynomial has non-real roots is very efficient since numerical methods can be used. Once the existence of non-real roots is established then from a theorem of Jordan (1871) it follows that if their number is "small" enough with respect to the degree p of the polynomial, then the Galois group is A_p or S_p. Furthermore, using the classification of finite simple groups we know nowadays the complete classification of transitive groups of prime degree. This enables us to provide a complete list of possible Galois groups for every polynomial of prime degree p which has non-real roots.

In section 2 we briefly describe the existing techniques used to compute the Galois group of a polynomial $f(x) \in \mathbb{Q}[x]$ of degree n. These techniques are mainly based on the Dedekind's theorem and knowledge of the list of all transitive subgroups of S_n and the cycle structure of their elements. First, a factorization mod p (i.e., good prime p) of $f(x)$ is required to obtain information on the cycle structure of the group elements. Such information enables us to eliminate groups from the list of transitive subgroups of S_n. For as long as the group is not uniquely determined we repeat the procedure with a different prime p. This is a rather expensive technique since algorithms of factorizing polynomials are not very efficient. Furthermore, many primes p might be needed in the process.

In section 3 we study polynomials of prime degree p with non-real roots. We describe the transitive groups of S_p and compute all such groups for $n \leq 30$. Using a theorem of Jordan on permutation groups we show that for a fixed number of non-real roots $r = 2s$ and

$$p \geq N(r) := [s\,(s \log s + 2 \log s + 3)]$$

the Galois group is A_p or S_p. Furthermore, we classify all groups that occur for $p < N(r)$ and provide an algorithm that computes the Galois group of a polynomial with non-real roots of degree prime p. We conclude with some final remarks in section 4.

Notation: Throughout this paper the ground field $k = \mathbb{Q}$. All polynomials are assumed to be irreducible over $\mathbb{Q}$. The group D_n denotes the dihedral group of n elements, M_{11} and M_{23} are respectively Mathew groups of degree 11 and 23. For all other groups we use the GAP notation.

2. The Galois group of a polynomial

Let $f(x) \in k[x]$ be a degree n polynomial. A *splitting field* of f is a field extension of k of the form $k(\alpha_1, \ldots, \alpha_n)$ where $f(x) = (x - \alpha_1) \cdots (x - \alpha_n)$. Any two splitting fields of f are isomorphic under an isomorphism trivial on k. Thus we normally speak of *the splitting field* of f.

If the discriminant Δ_f of $f(x)$ is nonzero then $f(x)$ has n-distinct roots $\alpha_1, \ldots, \alpha_n$ in the splitting field E_f of f. E_f/k is a Galois extension. The *Galois group* of f over k, denoted by $G_k(f)$, is the group $G(E_f/k)$, viewed as a permutation group of the roots $\alpha_1, \ldots, \alpha_n$. Thus $G_k(f)$ is a subgroup of S_n, determined up to conjugacy by f. The following is elementary and we avoid the proof.

Proposition 1: *Let $f(x) \in k[x]$ be a degree n polynomial and $G = G_k(f)$. (i) Let $H = G \cap A_n$. Then $H = G(E_f/k(\sqrt{\Delta_f}))$. In particular, G is contained in the alternating group A_n if and only if the discriminant Δ_f is a square in k.*

(ii) The irreducible factors of f in $k[x]$ correspond to the orbits of G. In particular, G is a transitive subgroup of S_n if and only if f is irreducible.

Remark 2: Recall that for a degree n irreducible polynomial $f(x)$ over k with splitting field E_f we have $n \mid [E_f : k]$.

Hence, for a given polynomial $f(x)$ of degree n its Galois group must satisfy:
i) G is isomorphic to a transitive subgroup of S_n.
ii) n divides $|G|$,
iii) G is a subgroup of A_n if and only if Δ_f is a square in $\mathbb{Q}$.

These conditions narrow down the possible choices of groups that can be Galois groups of $f(x)$. In order to determine precisely the group G we need to determine the type of cycles in G.

2.1. *Reduction mod p*

The reduction method uses the fact that every polynomial with rational coefficients can be transformed into a monic polynomial with integer coefficients without changing the splitting field.

Let $f(x) \in \mathbb{Q}[x]$ be given by

$$f(x) = x^n + a_{n-1}x^{n-1} + \cdots + a_1 x + a_0$$

Let d be the common denominator of all coefficients $a_0, \cdots, a_{n-1}$. Then $g(x) := d \cdot f(\frac{x}{d})$ is a monic polynomial with integer coefficients. Clearly the splitting field of $f(x)$ is the same as the splitting field of $g(x)$. Thus, without loss of generality we can assume that $f(x)$ is a monic polynomial with integer coefficients. The following theorem gives information on the cycle shape of permutations of Gal (f).

Theorem 3: (Dedekind) Let $f(x) \in \mathbb{Z}[x]$ be a monic polynomial such that $\deg f = n$, $\mathrm{Gal}_{\mathbb{Q}}(f) = G$, and p a prime such that $p \nmid \Delta_f$. If $f_p := f(x) \mod p$ factors in $\mathbb{Z}_p[x]$ as a product of irreducible factors of degree

$$n_1, n_2, n_3, \cdots, n_k,$$

then G contains a permutation of type

$$(n_1)(n_2)\cdots(n_k)$$

From the Chebotarev density theorem we know that if $p \to \infty$ then the distribution of factor degrees approaches the distribution of the cycle shapes in the group. Assuming that we can compute all transitive subgroups of S_n and their cycle shapes then the above two theorems give a basis of an algorithm to determine $Gal(f)$. The transitive subgroups of S_n can be computed in GAP for all $n \leq 30$.

Remark 4: The above technique doesn't determine the Galois group uniquely in all cases. For example there are two non-isomorphic degree 8 groups with the same cycle structure. In such cases other methods such as invariants of groups are used to determine the group uniquely.

2.2. *Transitive subgroups of S_n*

By *degree of a permutation group* G we mean the number of points moved by G. The *degree of a permutation* $\alpha \in S_n$ is the number of points moved by $\langle \alpha \rangle$. The *minimal degree* of G, denoted by $m(G)$, is the smallest of degrees of elements $\alpha \neq 1$ in G.

In the case n is a prime $n = p$ then a non-solvable transitive subgroup G of S_p is doubly transitive (Burnside theorem); see [1, pg. 431] or [12, Theorem 11.7]. Using the classification of simple groups one gets a complete list of G:

Theorem 5: Let G be a doubly transitive subgroup of S_p, for a prime p. Then G isomorphic to one of the following:
(i) A_p, S_p,
(ii) $p = 11, G = L_2(11)$ or M_{11},
(iii) $p = 23,\ G = M_{23}$,
(iv) $p = \frac{(q^k-1)}{(q-1)}$ and $L_k(q) \leq G \leq Aut(L_k(q))$.

Proof: One can check Prop. 4.4.1 in [9] for all these statements. □

Thus, the above theorem classifies all transitive non-solvable subgroups of S_p. For the solvable cases we use GAP to compute them in each case for ≤ 29. In the Table 1, we display the number of transitive subgroups and the number of non-solvable transitive subgroups of S_n for all $n \leq 30$.

3. Polynomials with non-real roots

Let $f(x) \in \mathbb{Q}[x]$ be an irreducible polynomial of degree $n > 5$. Denote by r the number of non-real roots of $f(x)$. Since the complex conjugation permutes the roots then r is even, say $r = 2s$. By a reordering of the roots we may assume that if $f(x)$ has r non-real roots then

$$\alpha := (1,2)(3,4)\cdots(r-1,r) \in Gal(f).$$

Deg.	Nr. of trans. groups	Nr. of unsolvable groups
5	5	2
6	16	4
7	7	3
8	50	5
9	34	4
10	45	21
11	8	4
12	301	36
13	9	3
14	63	27
15	104	40
16	1954	49
17	10	5
18	983	91
19	8	2
20	1117	358
21	164	56
22	59	27
23	7	3
24	25000	807
25	211	79
26	96	26
27	2392	64
28	1854	617
29	8	2
30	5712	1896

Since determining the number of non-real roots can be very fast, we would like to know to what extent the number of non-real roots of $f(x)$ determines $Gal(f)$. The complex conjugation assures that $m(G) \leq r$. The existence of α can narrow down the list of candidates for $Gal(f)$. However, it is unlikely that the group can be determined only on this information unless

p is "large" enough. In this case the number of non-real roots of $f(x)$ can almost determine the Galois group of $f(x)$, as we will see in the next section. Nevertheless, the test is worth running for all p since it is very fast and improves the algorithm overall.

3.1. *Polynomials of prime degree*

An approach of computing Galois groups of polynomials or solving for roots is to check whether or not the polynomial is decomposable. The polynomial decomposition problem can be stated as follows: given a degree n polynomial $f \in k[x]$, determine whether there exist polynomials f_1, f_2 of degree greater than one such that $f = f_1 \circ f_2 = f_1(f_2(x))$, and in the affirmative case to compute them. From the classical Lüroth's theorem this problem is equivalent to deciding if there exists a proper intermediate field in the finite algebraic extension $k(f) \subset k(x)$. From the computational point of view, there are several polynomial time algorithms for decomposing polynomials. The computation of $f_1(x)$ and $f_2(x)$ only requires $O(n^2)$ arithmetic operations in the ground field k; see for instance [2]. A polynomial $f(x) \in F[x]$ is indecomposable over the subfield $F \subset k$ if and only if $f(x)$ is indecomposable over k. There are fast algorithms to compute the decomposition of polynomials; see [2]. However, for indecomposable polynomials we would like to have better methods of determining the Galois group and possible roots of the polynomial. Hence, polynomials of prime degree are of interest since they are, of course, indecomposable.

The next theorem determines the Galois group of a prime degree polynomial $f(x)$ with r non-real roots when the degree of $f(x)$ is large enough with respect to r.

Theorem 6: Let $f(x) \in \mathbb{Q}[x]$ be an irreducible polynomial of prime degree $p \geq 5$ and $r = 2s$ be the number of non-real roots of $f(x)$. If s satisfies

$$s\,(s \log s + 2 \log s + 3) \leq p$$

then $Gal(f) = A_p, S_p$.

Proof: Since p is prime then every transitive subgroup of S_p is primitive. Let G denote the Galois group of $f(x)$ and $m(G)$ its minimal degree. By reordering the roots we can assume that

$$(1,2)(3,4)\cdots(r-1,r) \in Gal(f).$$

Hence, $m := m(G) \leq r$. From a theorem of Jordan [5] we have that if

$$\frac{m^2}{4} \log \frac{m}{2} + m \left(\log \frac{m}{2} + \frac{3}{2} \right) \leq p$$

then $G = A_p$ or S_p. Hence, if

$$s\,(s \log s + 2 \log s + 3) \leq p$$

then $Gal(f) = A_p$ or S_p. □

Remark 7: For a modern view of Jordan's theorem and its implications to number theory, Galois representations, and topology see [10].

For a fixed p the above bound is not sharp as we will see below. However, the above theorem can be used successfully if s is fixed. We denote the above bound on p by

$$N(r) := [s\,(s \log s + 2 \log s + 3)]$$

for $r = 2s$. Hence, for a fixed number of non-real roots, for $p \geq N(r)$ the Galois group is always A_p or S_p.

Corollary 8: *Let a polynomial of prime degree p have r non-real roots. If one of the following holds:*

(i) $r = 4$ *and* $p > 7$,
(ii) $r = 6$ *and* $p > 13$,
(iii) $r = 8$ *and* $p > 23$,
(iv) $r = 10$ *and* $p > 37$,

then $Gal(f) = A_p$ *or* S_p.

Remark 9: The above results gives a very quick way of determining the Galois group for polynomials with non-real roots. Whether or not the discriminant is a complete square can be used to distinguish between A_p and S_p.

If $p < N(r)$ then some exceptional cases occur. Next theorem determines these exceptional cases for polynomials of degree up to 29. The computations were made using GAP.

Theorem 10: Let $f(x) \in \mathbb{Q}[x]$ be an irreducible polynomial of prime degree $p > 5$. Let r be the number of complex roots of $f(x)$. If $r > 0$ then $Gal(f)$ is A_p, S_p or one of the groups as in the following Table 2.

Table 2. Galois groups (other then A_p, S_p) of polynomials with non-real roots.

p	Solv.	Sign.	Nonsol.	Sign.
7	D_7	$(2)^3, (7)$	$L(7)$	$(2)^2, (4)(2), (3)^2, (7)$
	$(7,4)$	$(2)^3, (3)^2, (7)$		
11	D_{11}	$(2)^5, (11)$	$L(11)$	$(2)^4, (3)^3, (5)^2, (2)(6)(3), (11)$
	$(11,4)$	$(2)^5, (5)^2, (10), (11)$	M_{11}	$(2)^4, (2)(6)(3), (2)(8), (3)^3, (4)^2, (5)^2, (11)$
13	D_{13}	$(2)^6, (13)$	$L(13)$	$(2)^4, (3)^3, (3)^4, (4)^2(2)^2, (6)(3)(2), (8)(4), (13)$
	$(13,4)$	$(2)^6, (4)^3, (13)$		
	$(13,5)$	$(2)^6, (3)^4, (6)^2, (13)$		
	$(13,6)$	$(2)^6, (3)^4, (4)^3, (6)^2, (12), (13)$		
17	D_{17}	$(2)^8, (17)$	$PSL_2(16)$	$(2)^8, (3)^5, (5)^3, (15), (17)$
	$(17,3)$	$(2)^8, (4)^4, (17)$	$(17,7)$	$(2)^6, (2)^8, (3)^5, (4)^4, (5)^3, (6)^2(3), (5)(10)(2), (15), (17)$
	$(17,4)$	$(2)^8, (4)^4, (8)^2, (17)$	$(17,8)$	$(2)(5)(10), (2)(4), (2)(4)^3, (2)^6, (2)^8, (3)(6)^2, (3)^5, (3)^2(12), (4)^3, (5)^3, (8)^2, (15), (17)$
	$(17,5)$	$(2)^8, (4)^4, (8)^2, (16), (17)$		
19	D_{19}	$(2)^9, (19)$		
	$(19,4)$	$(2)^9, (3)^6, (6)^3, (19)$		
	$(19,6)$	$(2)^9, (3)^6, (6)^3, (9)^2, (18), (19)$		
23	D_{23}	$(2)^{11}, (23)$	M_{23}	$(2)^8, (2)^2(4)^4, (2)(7)(14), (2)(4)(8)^2, (2)^2(3)^2(6)^2, (3)(5)(15), (5)^3, (5)^4, (7)^3, (11)^2, (23)$
	$(23,4)$	$(2)^{11}, (11)^2, (22), (23)$		
29	D_{29}	$(2)^{14}, (29)$		
	$(29,3)$	$(2)^{14}, (4)^7, (29)$		
	$(29,5)$	$(2)^{14}, (7)^4, (14)^2, (29)$		
	(29.6)	$(2)^{14}, (4)^7, (7)^4, (14)^2, (28), (29)$		

Proof: The proof is computational and follows from the tables of transitive subgroups of S_p. It is easy to decide which ones of these groups are nonsolvable and compute their cycle types. □

Remark 11: We used in Table 2 notations which we considered standard as D_n, M_{11}, and $L(p)$, otherwise we used the GAP notation (p,i) which is the i-th group in the list of transitive groups of degree p. These groups can be generated in GAP simple by typing `TransitiveGroup(n,i);`. The group M_{23} is not realized as a Galois groups over $\mathbb{Q}$.

Notice that no two groups of Table 2 have the same cycle structure. Hence the Galois group can be determined uniquely by reduction mod p for all polynomials of prime degree ≤ 29 with non-real roots.

Example 12: Let $f(x)$ be the polynomial

$$f(x) = x^{11} + 5x^7 - 4x^6 - 20x^5 + 4x^4 + 20x^3 + 1$$

This polynomial is irreducible over $\mathbb{Q}$ and has exactly 10 non-real roots. The reader can easily check these facts in Maple using the commands:
`factor(f(x)); realroot(f(x));`
From the above theorem, its Galois group is A_{11} or S_{11}. Its discriminant is

$$\Delta_f = -59 \cdot 139121293609142912303 3$$

which is not a square in $\mathbb{Q}$. Hence the Galois group of $f(x)$ is S_{11}. Maple can not compute the Galois group of this polynomial since its degree is > 8.

Combining the above results we have the following algorithm for computing the Galois group of prime degree polynomials with non-real roots. `D(f)` denotes the discriminant of $f(x)$ and `A_p`, `S_p` the alternating and symmetric group of p letters. Note that in the case $p < N(r)$ the we know that a permutation of the type $(2)^{\frac{r}{2}}$ is in the group. Hence, the list of transitive subgroups is much shorter than in general. This information was obtained by computing the number of real roots other then by some factorization modulo p. Thus, even in this case the algorithm is improved.

ALGORITHM: Computing the Galois group of prime degree polynomials with non-real roots.

Input: An irreducible monic polynomial $f(x) \in \mathbb{Q}[x]$ of prime degree p.
Output: Galois group $Gal(f)$ of $f(x)$ over $\mathbb{Q}$

```
begin

r :=NumberOfRealRoots(f(x));

If p > N(r) then
   if  D(f) is a complete square then
```

```
        Gal (f)=A_p; else Gal (f) = S_p;
    endif;
else ReductionMethod(f(x)); endif;

end;
```

3.2. *Polynomials of prime degree p with Galois group A_p*

Let $f(x)$ be a polynomial in $\mathbb{Q}(t)$ as below

$$f(x) = (n-1)x^n - nx^{n-1} + t.$$

The discriminant of $f(x)$ with respect to x is

$$\Delta_f = (-1)^{\frac{n(n-1)}{2}} n^n (n-1)^{n-1} t^{n-2}(t-1).$$

Δ_f is a complete square in $\mathbb{Q}$ if $(-1)^{\frac{(n-1)}{2}} nt(t-1)$ is a complete square; see [11] (pg. 44) for more on this family of polynomials. Let $n = 23$. Then

$$\Delta_f = -2^{22} \cdot 11^{22} \cdot 23^{23} \cdot t^{21}(t-1).$$

Hence, Δ_f is a complete square in $\mathbb{Q}$ if $G(t) = -23t(t-1)$ is a complete square. In other words, for all those rational points on the curve

$$y^2 = G(t).$$

This is a genus 0 curve and can be parametrized as follows:

$$(y,t) = \left(-\frac{23m}{m^2+23}, \frac{23}{(m^2+23}\right)$$

Consider $f(x)$ for $t = \frac{23}{(m^2+23}$. Since we prefer to work with polynomials with integer coefficients then take

$$f(x) = (22m^2 + 506)x^{23} - (23m^2 + 529)x^{22} + 23.$$

It is easily checked that $f(x)$ is irreducible over $\mathbb{Q}$ and its discriminant is

$$\Delta_f = 2^{22} \cdot 11^{22} \cdot 23^{44} \cdot m^2 (23 + m^2)^{22}$$

which is a complete square in $\mathbb{Q}$. Thus, $Gal(f)$ is inside A_{23}. It is an simple calculus exercise to show that the number of real roots of these polynomials is ≤ 3. Hence, the Galois group is A_{23}.

We conclude with the following open problem:

Problem: *Find a degree 23 polynomial $f(x) \in \mathbb{Q}[x]$ with exactly 7 real roots such that Δ_f is a complete square in $\mathbb{Q}$ but $Gal(f)$ is not isomorphic to A_{23}.*

The reader is probably aware that the above is an open problem of the inverse Galois problem, see [10]. Its solution would realize M_{23} as a Galois group over $\mathbb{Q}$.

4. Concluding remarks

The algorithm suggested in this paper works very well with prime degree polynomials which have nonreal roots. Since most polynomials have such roots this is an effective test to be implemented on all algorithms computing Galois groups. Most computer algebra packages have already algorithms implemented to find the number of real roots of a polynomial. For example in Maple the user can easily check using the command `realroot( f(x) );`. One can also generalize this algorithm to any degree n polynomial. However, in this case a more detailed analysis is required.

References

1. W. Burnside, Theory of groups of finite order. 2d ed. Dover Publications, Inc., New York, 1955.
2. J. Gutierrez, A polynomial decomposition algorithm over factorial domains, Comptes Rendues Mathematiques, de Ac. de Sciences, 13 (1991), 81-86.
3. J. E. Littlewood, A. C. Offord, On the number of real roots of a random algebraic equation. III. Rec. Math. [Mat. Sbornik] N.S. 12(54), (1943). 277–286.
4. Janusz, Gerald J. Permutation groups generated by a transposition and another element. Enseign. Math. (2) 38 (1992), no. 1-2, 41–53.
5. C. Jordan, Recherches sur les substitutions, J. Liouville 17 (1872), 351–367.
6. Kochetkov, Yu. Yu. Permutation groups generated by a $(2,2)$-cycle and an n-cycle. (Russian) Funktsional. Anal. i Prilozhen. 31 (1997), no. 2, 76–79; translation in Funct. Anal. Appl. 31 (1997), no. 2, 133–135
7. M. Kac, On the average number of real roots of a random algebraic equation. Bull. Amer. Math. Soc. 49, (1943). 314–320.
8. M. Kac, On the average number of real roots of a random algebraic equation. II. Proc. London Math. Soc. (2) 50, (1949). 390–408.
9. A. Kostrikin, P. Tiep, Orthogonal decompositions and integral lattices. de Gruyter Expositions in Mathematics, 15. Walter de Gruyter & Co., Berlin, 1994. x+535 pp. ISBN: 3-11-013783-6.
10. J. P. Serre, On a theorem of Jordan. Bull. Amer. Math. Soc. (N.S.) 40 (2003), no. 4, 429–440
11. J. P. Serre, Topics in Galois theory. Research Notes in Mathematics, 1. Jones and Bartlett Publishers, Boston, MA, 1992. xvi+117 pp. ISBN 0-86720-210-6.
12. H. Wielandt, Finite permutation groups. Academic Press, New York-London, 1964.

13. H. Völklein, *Groups as Galois groups. An introduction.* Cambridge Studies in Advanced Mathematics, 53. Cambridge University Press, Cambridge, 1996.

Counting generating systems of a finite group from given conjugacy classes

R. Staszewski, H. Völklein, & G. Wiesend

Institute for Experimental Mathematics
Universität Essen,
Ellernstr. 29, 45326 Essen,
Germany

Let G be a finite group with conjugacy classes C_1, ..., C_r. We describe an algorithm (and its implementation in GAP and MAGMA) for computing the number of tuples $(g_1, ..., g_r)$ generating G such that $g_1 \cdots g_r = 1$ and $g_i \in C_i$. This is a new ingredient of the BRAID package, a collection of programs for computing braid orbits of such tuples.

1. Introduction

It is well-known that the compact Riemann surfaces of fixed genus g are parametrized (up to isomorphism) by an irreducible variety — the moduli space $\mathcal{M}_g$. We can re-phrase this by saying that the genus is the finest discrete (continuous) invariant of a compact Riemann surface. Similarly, there is a finest discrete invariant of a cover of the Riemann sphere: It is the braid orbit consisting of all associated "branch cycle tuples". The corresponding moduli space for covers of the Riemann sphere is called a **Hurwitz space** (see [2], [10]).

These braid orbits are group-theoretic objects of considerable complexity. Only in very few cases it is possible to give a theoretical description. In general one needs computational group theory to compute braid orbits and associated information. The BRAID package of Magaard, Shpectorov and Völklein [8] has been devised to perform these computations. A previously missing ingredient is supplied by the algorithm described in the present paper.

Let $C_1, ..., C_r$ be non-trivial conjugacy classes of the finite group G. The set of generating systems $(g_1, ..., g_r)$ of G with $g_1 \cdots g_r = 1$ and such that there is a permutation $\pi \in S_r$ with $g_i \in C_{\pi(i)}$ for $i = 1, ..., r$ is called a **Nielsen class**. Each Nielsen class is the disjoint union of **braid orbits,** which are defined as the smallest subsets of the Nielsen class closed under the **braid operations**

$$(1) \qquad (g_1, ..., g_r)^{Q_i} = (g_1, \ldots, g_{i+1}, g_{i+1}^{-1} g_i g_{i+1}, \ldots, g_r)$$

for $i = 1, ..., r-1$.

The intersection of a Nielsen class (resp., braid orbit) with the set $C_1 \times \cdots \times C_r$ (tuples where the classes C_i occur in the given order) is called a **pure Nielsen class** (resp., **pure braid orbit**).

Existing parts of the BRAID package compute the pure braid orbit of a given tuple $(g_1, ..., g_r)$ (and associated information of geometrical relevance). The tuples are found by a random search. It was a problem to decide when the entire (pure) Nielsen class has been exhausted because there was no independent way of computing its cardinality. We achieve the latter in a two step algorithm described in the next two sections. It is usually much faster than the actual computation of the braid orbits.

Viability of the algorithm depends on the following two observations. Firstly, that it is possible to find all subgroups U of G generated by tuples in $C_1 \times \cdots \times C_r$ with product 1, **without** actually computing all these tuples. This is Step 1 of the algorithm. Secondly, that the well-known character formula for finding the number of such tuples in each U can be re-written to avoid summing over all combinations of U-classes contained in $C_1, ..., C_r$.

In the last section of the paper we describe an application for tuples in alternating groups consisting only of three-cycles and double transpositions. This is motivated by theoretical results of Clebsch and Fried on braid orbits of tuples in S_n and A_n.

Reliability of results: Correctness of the results of the computation is checked automatically in a very satisfying way. Indeed, the new algorithm for computing the cardinality of a Nielsen class is completely independent from the algorithm that partitions the Nielsen class into braid orbits (see [8, §2.3]. If the numbers match, one can be practically certain that the result is correct.

Implementation in GAP and MAGMA: We have implemented the above algorithm in both GAP [4] and MAGMA [6]. It became part of the BRAID PACKAGE of Magaard, Shpectorov and Völklein [8], which was originally written in GAP and translated to MAGMA by R. Staszewski. All these programs are freely available from H. Völklein at voelkle@iem.uni-due.de.

Notation: Let $C_1, ..., C_r$ be non-trivial conjugacy classes of the finite group G. Let U be a subgroup of G.

2. First step of the algorithm

Find all subgroups of G generated by tuples $(g_1, ..., g_r) \in C_1 \times \cdots \times C_r$ *with* $g_1 \cdots g_r = 1$.

We compute all pairs $(U^{(j)}, u^{(j)})$ (up to G-conjugacy) consisting of a subgroup $U^{(j)}$ of G generated by a tuple $(g_1, ..., g_j)$ with $g_1 \cdots g_j = u^{(j)}$ and $g_i \in C_i$, for $1 \leq j \leq r$. In order to pass from j to $j+1$, we only need the information given by the pairs $(U^{(j)}, u^{(j)})$ (up to G-conjugacy), but **NOT** the totality of all generating systems $(g_1, ..., g_j)$ of U. This makes the algorithm suitably effective.

3. Second step of the algorithm

We begin with the well-known character formula of Frobenius counting tuples from $C_1 \times \cdots \times C_r$ with product 1 (see [5, Th. 1.1.12]. It is implemented in GAP under the command `ClassStructureCharTable`. Let Irr(G) be the set of irreducible (complex) characters of G.

Lemma 1: *(Frobenius) The number of tuples* $(g_1, ..., g_r)$ *in* $C_1 \times \cdots \times C_r$ *with* $g_1 \cdots g_r = 1$ *equals*

$$\frac{1}{|G|} \sum_{\chi \in \mathrm{Irr(G)}} \frac{1}{\chi(1)^{r-2}} \prod_{i=1}^{r} |C_i| \, \chi(C_i)$$

In order to get the number of **generating** tuples, we need to subtract the number of tuples in proper subgroups U of G. Thus we need to apply Lemma 1 to all subgroups U of G found in the first step of the algorithm. At first, it appears there is the following problem with this approach: For certain (e.g., abelian) U there is a large number of combinations of classes of U fusing into $C_1, ..., C_r$. Simply summing over all these combinations would take too long. Miraculously, this sum can be re-written using the distributive law which turns it into a product of simpler terms. A further step to shorten the calculations is to replace U by its normalizer in G. We summarize this discussion in the following Lemma:

Lemma 2: *Let N be the normalizer of U in G. The number of tuples $(g_1, ..., g_r)$ in $(C_1 \cap U) \times \cdots \times (C_r \cap U)$ with $g_1 \cdots g_r = 1$ equals*

$$\frac{1}{|N|} \sum_{\chi \in \mathrm{Irr(N)}} \frac{1}{\chi(1)^{r-2}} \prod_{i=1}^{r} \Big(\sum_j |D_i^{(j)}| \, \chi(D_i^{(j)}) \Big)$$

The inner sum is over all classes $D_i^{(j)}$ of N contained in $C_i \cap U$.

4. An application: tuples of index 2 permutations

4.1. *The results of Clebsch and Fried*

It is a classical result of Clebsch (see [10, Lemma 10.15] that the generating tuples of r transpositions in S_n (with product 1) form a single braid orbit for any even $r \geq 2(n-1)$. An analogous result for alternating groups was proved by Fried [1]: Let $n \geq 5$. The tuples of r three-cycles (with product 1) generating A_n form a single braid orbit (resp., exactly two braid orbits) if $r = n-1$ (resp., $r > n-1$).

These results have important consequences for algebraic curves (see [1], [9], [10]) which we do not discuss in this paper. Not many other theoretical results on braid orbits are known.

The **index** of a permutation from S_n is n minus the number of orbits (or cycles) of the permutation. This notion of permutation index occurs prominently in the group-theoretic formulation of the Riemann-Hurwitz formula (see [1], [3], [7] or [8]).

Transpositions are the only elements of S_n of index 1. The only elements of S_n of index 2 are three-cycles and double transpositions (i.e., elements conjugate to $(1,2)(3,4)$). The double transpositions play an important role in [7] (which deals with problems on generic curves). This leads to the question how Fried's result extends to tuples consisting only of three-cycles and double transpositions. In the following we collect some experimental evidence on the this question.

4.2. *Braid orbits of (r, s)-tuples*

Call a tuple that generates A_n and has product 1 an **(r, s, n)-tuple** if it consists of s double transpositions and $r - s$ three-cycles. By the Riemann-Hurwitz formula, such tuples can only exist if $r \geq n-1$. We computed the braid orbits of such tuples for $n = 5, 6, 7$ and $r \leq 7$. The computations were carried out using MAGMA 2.10.18 under linux with an AMD Athlon 64 Processor 3400+.

Each braid orbit of tuples that generate a group G is a union of Inn(G)-classes of tuples (where Inn(G) acts componentwise on the tuples, see [10], Lemma 9.4). However, braid orbits of (r, s, n)-tuples may be permuted non-trivially by S_n/A_n. Thus we actually list braid orbits of S_n-classes of (r, s, n)-tuples. For fixed (r, s, n), if there are braid orbits of lengths $N_1, N_2, ...$ on (r, s, n)-tuples then we list them in the tables as $N_1 + N_2 + ...$; then the sum $N = N_1 + N_2 + ...$ is the total number of (r, s, n)-tuples mod S_n. Note that the number N was computed by the algorithm introduced in this paper, whereas the numbers $N_1, N_2...$ were computed by the braid program. The numbers matched in each case, which guarantees correctness of the results (as indicated in the Introduction).

Findings: For the values of n, r, s that we considered, there were at most two braid orbits in each case. For $s = 0$, it is Fried's lifting invariant (see [1] and [9]) that separates the two braid orbits. However, we also get two braid orbits in some cases where $s = r$; we have no explanation for that.

Another observation is that the numbers N increase (resp., decrease) with s when $n > 5$ (resp., $n = 5$). Again we don't know an explanation of this curious phenomenon.

Table 1. The case $n = 5$

r	s	lengths of braid orbits of $(r, s, 5)$-tuples mod S_5	CPU-time in sec.
4	0	9	< 1
4	1	12	< 1
4	2	9	< 1
4	3	9	< 1
4	4	0	< 1
5	0	216 + 126	< 1
5	1	288	< 1
5	2	234	< 1
5	3	180	< 1
5	4	135	< 1
5	5	96	< 1
6	0	4545 + 3456	24
6	1	6336	18
6	2	4860	15
6	3	3708	11
6	4	2781	8
6	5	2106	6
6	6	1440	4
7	0	90720 + 80010	800
7	1	130560	541
7	2	99000	431
7	3	74592	320
7	4	56106	242
7	5	42120	180
7	6	31590	139
7	7	23520	103

Table 2. The case $n = 6$

r	s	lengths of braid orbits of $(r, s, 6)$-tuples mod S_6	CPU-time in sec.
5	0	96	< 1
5	1	192	< 1
5	2	252	< 1
5	3	360	1
5	4	432	1
5	5	432	2
6	0	5940 + 3040	30
6	1	12160	39
6	2	15336	54
6	3	18720	66
6	4	22248	79
6	5	25920	95
6	6	19440 + 8280	113
7	0	275520 + 207900	2277
7	1	593920	2768
7	2	706320	3508
7	3	827712	4446
7	4	955260	4892
7	5	1095120	5579
7	6	1244160	6234
7	7	952560 + 433440	7593

Table 3. The case $n = 7$

r	s	lengths of braid orbits of $(r, s, 7)$-tuples mod S_7	CPU-time in sec.
6	0	1435	8
6	1	3640	13
6	2	6882	28
6	3	13105	62
6	4	22846	92
6	5	39780	159
6	6	45360 + 19600	271
7	0	176760 + 84765	1345
7	1	494485	2537
7	2	859730	4621
7	3	1452372	8090
7	4	2382805	12736
7	5	3845160	20900
7	6	6078780	41317
7	7	6298560 + 3115680	50218.24

References

1. M. FRIED, Alternating groups and lifting invariants, Preprint as of 07/01/96 (1996), 1–34.
2. M. FRIED AND H. VÖLKLEIN, The inverse Galois problem and rational points on moduli spaces, *Math. Annalen* **290** (1991), 771-800.a
3. D. FROHARDT AND K. MAGAARD, Composition factors of monodromy groups, Annals of Math. **154** (2001),1-19.
4. THE GAP GROUP, *GAP – Groups, Algorithms, and Programming*, Version 4.2; 2000. (`http://www.gap-system.org`)
5. S.K. LANDO AND A.K. ZVONKIN, Graphs on surfaces and their applications, Encyclopedia of Math. Sciences 141, Springer 2004
6. Computational Algebra Group at the University of Sydney, *MAGMA*, Version 2.10–18; 2004. (`http://magma.maths.usyd.edu.au/magma`)
7. K. MAGAARD AND H. VÖLKLEIN, The monodromy group of a function on a general curve Israel J. Math. 141 (2004), 355–368.
8. K. MAGAARD, S. SPHECTOROV AND H. VÖLKLEIN, A GAP package for braid orbit computation, and applications, Experimental Math. 12 (2003), 385–393.
9. J-P. SERRE, Relèvements dans $\tilde{A}_n$, C.R. Acad. Sci. Paris, serie I, **311** (1990), 477–482.
10. H. VÖLKLEIN, Groups as Galois Groups – an Introduction, Cambr. Studies in Adv. Math. 53, Cambridge Univ. Press 1996.

Group action on genus 3 curves and their Weierstrass points

H. Babu & P. Venkataraman

Department Of Mathematics
358 Little Hall, University Of Florida,
Gainesville, Florida 32611

We study the locus of genus 3 curves with prescribed automorphism group G, and inclusion relations between such loci. We supply a normal form for the curves in each locus, and generators for G. Furthermore, we determine the action of G on the Weierstrass points of the curve.

1. Introduction

We use the term "curve" to mean compact Riemann surface. There is a vast literature on automorphism groups of curves (see the survey in [12]). In particular there are a lot of papers classifying all these groups in genus 3, 4 and 5. For genus 3, for each group, equations for the corresponding curves have been found in several papers by Kuribayashi and various co-authors (see [6], [7], [8], [9]). A short and corrected account of that is given in [12], section 6. Here we reorganize this information in the nonhyperelliptic case so that the inclusion relations between the various Hurwitz loci are reflected in the equations. Furthermore, we supply explicit generators for the automorphism group G, and determine its action on the Weierstrass points.

The paper [12] replaces the ad hoc methods of the previous papers by the systematic use of group theory via Riemann's Existence Theorem. The latter implies that a finite group G acts on a genus g curve if and only if it has a genus g generating system (see [12], section 5 equation 8). Two such generating systems correspond to an equivalent group action if and only if they are in the same mapping class group orbit (respectively braid orbit in the case of orbit genus 0). See [12], [13] and [15]. This group theoretic setup also yields the information in Table 3 which displays the inclusion relations between all Hurwitz loci in $\mathcal{M}_3$. Here $\mathcal{M}_3$ is the moduli space of genus 3 curves. A Hurwitz locus in $\mathcal{M}_3$ is an irreducible component of the closure of the locus of curves with prescribed automorphism group. The locus of curves in $\mathcal{M}_3$ with nontrivial automorphisms is a closed subvariety of $\mathcal{M}_3$ which has three irreducible components. One component is the hyperelliptic locus (of dimension 5). The other two components are the nonhyperelliptic Hurwitz loci corresponding to C_2 and C_3. They can be described in other words as the locus of genus 3 curves which are double covers of elliptic curves (with dimension 4), and the locus of curves that admit a Galois cover to $\mathbb{P}^1$. These are the Hurwitz loci.

From now on, X_3 denotes a curve of genus 3. The group $Aut(X_3)$ is the group of analytic (equivalently algebraic) automorphisms of X_3. We say a group G acts on X_3 if there exists a homomorphism from G to $Aut(X_3)$.

We expand on the information in Section 6 of [12] which was derived from group-theoretic data computed by the BRAID program [13]. All groups occurring as $Aut(X_3)$ together with the corresponding signature and orbit genus are given in Tables 1, 2 and 3 of [12]. Furthermore, equations for the respective curves X_3 are given. The hyperelliptic case is very easy and we do not add anything here. For some of the other cases, we supply new equations that reflect the inclusion relations between the corresponding Hurwitz loci; i.e. if one Hurwitz locus is contained in another one, then the corresponding equation is obtained by specializing some of the parameters in the equation for the larger Hurwitz locus.

Our results are presented in Tables 1 and 2. For each group G occurring as $Aut(X_3)$ for non-hyperelliptic X_3 we supply the above mentioned information on the corresponding curves X_3. Recall that the signature of a G-action on X_3 is the collection of numbers $\frac{|G|}{|B|}$, where B runs over all the non-regular orbits of G on X_3. Furthermore, the stabilizer in G of each point of X_3 is cyclic. The equations for the curves depend on certain independent complex parameters $a_1, a_2, \ldots, a_7$, and the given data holds true for general values of these parameters. For certain special values of the parameters the curves may become singular or hyperelliptic, and the automorphism group becomes larger.

2. Group Action on Weierstrass Points

A genus 3 curve X_3 is either hyperelliptic or it is a non-singular plane quartic. In the hyperelliptic case X_3 has exactly 8 Weierstrass points which are the ramification points of the canonical map $X_3 \to \mathbb{P}^1$. The action of

$Aut(X_3)$ on the Weierstrass points can be easily deduced from Table 3 of [12].

From now on X_3 is a non-singular plane quartic. Its Weierstrass points are the inflection points. The weight of a Weierstrass point is the same as the multiplicity of the inflection point which is either 1 or 2. The weighted number of inflection points is 24. This and the facts collected in the following remark can be found in Chapter 2 of [1].

Remark 1: The inflection points are the intersection points of X_3 with its Hessian. The tangent to X_3 at the inflection point hits X_3 in N further points, where $N = 2$ (respectively 1) if the weight of the inflection point is 1 (respectively 2). This way the inflection points and their weights can be computed effectively from the equation of the curve.

The groups G occurring as $Aut(X_3)$ are denoted in [12] by their group ID from the GAP library of small groups. Here we use the following shorter notation: C_n, S_n, D_{2n}, V_4 denotes the cyclic group of order n, the symmetric group on n letters, the dihedral group of order $2n$, and the Klein-4 group respectively. Also, $L_3(2)$ ($= PGL_3(2)$) is the simple group of order 168. The groups of order 16, 48 and 96 are just denoted by their group order. Their group IDs are $(16, 13)$, $(48, 33)$ and $(96, 64)$ respectively. The group $(96, 64)$ is sometimes denoted $C_4^2.S_3$.

We show how to derive the information on the Weierstrass points given in Tables 1 and 2.

When $G = L_3(2)$ or $C_4^2.S_3$, then G has only one orbit of length ≤ 24 on X_3. Therefore this orbit has to consist of all Weierstrass points. For $G = L_3(2)$ they form one orbit of length 24 with a stabilizer of order 7 and they all have weight 1. For $G = C_4^2.S_3$ the Weierstrass points have weight 2 and form one orbit of length 12 with stabilizer of order 8.

From Table 3 we see that the $L_3(2)$-locus (a single point) is contained in each of the loci with groups S_4, D_8, S_3, V_4, C_2. Therefore the general X_3 in each of these loci has 24 distinct Weierstrass points of weight one. No element of order 2 or 3 in $L_3(2)$ fixes a Weierstrass point of the corresponding X_3, therefore the same holds for the general curve in the loci listed above. Therefore for all X_3 in these loci, the Weierstrass points have weight 1 and consist of regular orbits. This settles the corresponding entries of Table 1.

Remark 2: The specialization $a_1 = 3$ in the equation for the S_4-locus yields the only curve of genus 3, apart from the Fermat Curve, possessing exactly 12 Weierstrass points (see [5]).

When $G = (48, 33)$, we have three nonregular G-orbits of length 24, 16, and 4. Let S denote a Sylow 3-subgroup of G (of order 3). Then S stabilizes a point in the 4-orbit as well as in the 16-orbit (because these orbits have cyclic stabilizers of order 12 and 3 respectively). Let $N(S)$ denote the normalizer of S in G. Then the number of fixed points of S in the 16-orbit is $\mid \frac{N(S)}{S} \mid$. Since $N(S)$ contains the cyclic group of order 12 (point stabilizer of the 4-orbit) it follows that $\mid \frac{N(S)}{S} \mid \geq 4$. Thus S has at least 4 fixed points in the 16-orbit, and it fixes at least one point in the 4-orbit. Thus S has at least 5 fixed points. By [3, Theorem V.1.7] all fixed points of S are Weierstrass points. It follows all points in the 16-orbit and the 4-orbit are Weierstrass points. Thus the points in the 16-orbit have weight 1 and the points in the 4-orbit have weight 2.

When $G = (16, 13)$, there are four nonregular G-orbits of length 8, 8, 8, 4. The 4-orbit consists of the points $(0, 0, 1)$, $(1, 0, 1)$, $(1, 0, 0)$, $(t, 0, 1)$. We check by Maple [11] that the Hessian of the equation $y^4 = xz(x - z)(x - tz)$ is 0 at $(0, 0, 1)$ so this point is a Weierstrass point. It follows all points in the 4-orbit are Weierstrass points. Further, the 16-locus contains the 48-locus. Therefore by specialization, the general curve in the 16-locus has one regular orbit of Weierstrass points with weight 1. Thus the Weierstrass points of the 16-locus consist of one regular orbit with points of weight 1 and one 4-orbit with points of weight 2.

When $G = C_3$, C_6, C_9, the subgroup C_3 fixes five points: $(0, 0, 1)$, $(1, 0, 1)$, $(s, 0, 1)$, $(t, 0, 1)$, $(0, 1, 0)$. By [3, Theorem V.1.7] they are all Weierstrass points. Using Maple [11] and Remark 1, we compute the weight of these points, and get that the points $(0, 0, 1)$, $(1, 0, 1)$, $(s, 0, 1)$, $(t, 0, 1)$ have weight 1, and $(0, 1, 0)$ has weight 2.

Furthermore, we compute by Maple [11] the Hessian $h(x, y, z)$ of the C_9-equation $f(x, y, z)$. Since the Weierstrass points are the intersection of the Hessian with X_3, we set $z = 1$ and consider the system of equations $h(x, y, 1) = 0$ and $f(x, y, 1) = 0$. The resultant with respect to x of these two polynomials is a polynomial in y that has degree 18 and has nonzero discriminant. Thus the X_3 in the C_9-locus has 18 distinct Weierstrass points in addition to the 5 fixed points of the C_3-subgroup. Thus the Weierstrass points consist of 2 regular orbits with points of weight 1, plus the above 5 fixed points of C_3.

Now consider X_3 in the C_3-locus. By specialization to the C_9-locus we see that X_3 has 22 Weierstrass points of weight 1. This settles the C_3 case. The 48-group occurs also in Table 1, and so has already been dealt with.

Finally, we consider X_3 in the C_6-locus. The nonregular orbits consist of one fixed point and three additional orbits of length 3, 2, 2 with stabilizing subgroups of order 2, 3, 3 respectively. Now C_6 has only one subgroup of order 3 so C_3 fixes all points in the 2-orbits. Thus we obtain the 5 fixed (Weierstrass) points of C_3 whose weights we computed above.

By specialization to the 48-locus we see that X_3 has at least 16 distinct Weierstrass points of weight 1. Thus in addition to the 5 fixed points of C_3, the Weierstrass points must consist of either 3 regular orbits with points of weight 1, or 2 regular orbits with points of weight 1 and the 3-orbit with points of weight 2. To determine which, we compute the fixed points on X_3 of the involution N in C_6. They are the points $(1, y_i, 2)$, $1 \le i \le 3$, where the y_i are the roots of the equation $y^3 = \frac{(1-2t)^2}{2}$. We find, using Maple [11], that for $t \neq \frac{1}{2}, \frac{1+\zeta_4}{2}, \frac{1-\zeta_4}{2}$, the Hessian of X_3 is not 0 at these points. Thus for these values of t the Weierstrass points are a union of 3 regular orbits with points of weight 1, and the five fixed points of C_3.

3. Equations and Group Generators

We take $G \le \mathrm{PGL}_3(\mathbb{C})$. Note that if f is an irreducible homogeneous polynomial, then an element of $\mathrm{PGL}_3(\mathbb{C})$ induces an automorphism of the curve $f(x, y, z) = 0$ if and only if the corresponding linear substitution turns f into a scalar multiple of itself. The group $\mathrm{GL}_3(\mathbb{C})$ acts on such polynomials in the following way: $A \in \mathrm{GL}_3(\mathbb{C})$ maps $f(x, y, z)$ to $f((x, y, z)A^T)$. From now on let f denote a nonsingular quartic fixed by G up to scalar multiples. Note that every automorphism of the genus 3 curve $f = 0$ is induced by an element of $\mathrm{PGL}_3(\mathbb{C})$ (because the curve is given in its canonical embedding).

In Tables 1 and 2, we denote the diagonal matrix with diagonal entries d_1, d_2, d_3 by $\Delta(d_1, d_2, d_3)$.

3.1. *Table 1*

$G = C_2$: We may assume $G =< \Delta(1, -1, -1) >$. Thus f contains only even powers of x. (The other possibility is that f contains only odd powers of x but this would contradict irreducibility of f.) Let a_1, b and a_2 be the coefficients of x^2y^2, x^2yz, and x^2z^2 in the equation for f. Then via the substitution $y \to y - \frac{bz}{2a_1}$ we ensure that f contains no yz term. Thus we obtain the equation in Table 1.

Table 1: Non-hyperelliptic genus 3 curves arising as double covers of elliptic curves

G	Sign	equation	Number of G-orbits on W-pts of wt. 1	wt. 2	Generators of G
C_2	(2^4)	$x^4 + a_1x^2y^2 + a_2x^2z^2$ $+a_3y^2z^2 + a_4y^3z$ $+a_5y^4 + a_6yz^3 + a_7z^4 = 0$ either $a_6 = 1$ or $a_7 = 1$	12 reg. orb. (in general)		$\Delta(-1, 1, 1)$
V_4	(2^6)	$a_5 = a_7 = 1$, $a_4 = a_6 = 0$	6 reg. orb. (in general)		$\Delta(1,-1,1)$, $\Delta(-1,1,1)$
D_8	(2^5)	V_4-equation with $a_2 = a_3$: $x^4 + y^4 + z^4$ $+a_1x^2y^2 + a_2(x^2z^2 + y^2z^2) = 0$	3 reg. orb. (in general)		P_1, $\Delta(-1, 1, 1)$
S_4	$(2^3, 3)$	D_8-equation with $a_1 = a_2$: $x^4 + y^4 + z^4$ $+a_1(x^2y^2 + x^2z^2 + y^2z^2) = 0$	1 reg. orb. (in general)		$\Delta(-1, 1, 1)$, P_1, P_2
16	$(2^3, 4)$	D_8-equation with $a_2 = 0$: $x^4 + y^4 + z^4 + a_1x^2y^2 = 0$	1 reg. orb. (in general)	1 orb. of 4	P_1 $\Delta(1, 1, \zeta_4)$, $\Delta(-1, 1, 1)$
48	(2,3,12)	16-group equation with $a_1 = \sqrt{-12}$	1 orb. of 16	1 orb. of 4	$\Delta(1, 1, \zeta_4)$, $\Delta(-1, 1, 1)$, R
96	(2, 3, 8)	Fermat Curve S_4-equation and $a_1 = 0$: $x^4 + y^4 + z^4 = 0$		1 orb. of 12	$\Delta(1, 1, \zeta_4)$, P_1, P_2
$L_3(2)$	(2, 3, 7)	Klein Curve S_4-equation and $a_1 = \frac{-3+3\sqrt{-7}}{2}$	1 orb. of 24		$\Delta(-1, 1, 1)$, P_1, P_2, Q
S_3	$(2^4, 3)$	$a(x^4 + y^4 + z^4)+$ $b(x^2y^2 + x^2z^2 + y^2z^2)+$ $c(x^2yz + y^2xz + z^2xy) = 0$	4 reg. orb. (in general)		P_1, P_2

Remark 3: (*Singularities of the Hurwitz loci*) Suppose $H \leq G$. Then every G-action on X_3 induces an H-action by restriction. This yields an inclusion of the two corresponding Hurwitz loci. If H' is another subgroup of G, then the corresponding Hurwitz loci $\mathcal{H}$ and $\mathcal{H}'$ contain the Hurwitz locus $\mathcal{G}$ corresponding to G in their intersection. Now assume H and H' are not conjugate in G but nevertheless $\mathcal{H} = \mathcal{H}'$. In that case $\mathcal{H}$ is singular at the subvariety $\mathcal{G}$ because the two branches of $\mathcal{H}$ corresponding to H and H' come together at $\mathcal{G}$.

The C_2 locus is 4-dimensional and it is singular at various of the subloci by Remark 3. This explains why we need 7 parameters to describe it.

$G = V_4$, D_8, S_4, or $C_4^2.S_3$: The equations are as in [12]. It can be checked by GAP that the given generators for G leave the equation invariant and generate a group of the correct order.

$G = L_3(2)$: The specialization $a_1 = \frac{-3+3\sqrt{-7}}{2}$ which distinguishes the Klein curve among the above S_4-equations was found by Elkies, see [2]. He also computed the generators that we display in Table 1.

$G = 16, 48$: For the groups of order 16 and 48, we gave new equations which show the inclusion of the corresponding loci in the D_8-locus. Again, it can be checked by GAP that the given generators generate a group of the correct order and leave the equation invariant.

$G = S_3$: The S_3 case could not be fitted nicely into our tree of specializations, and so we just took the equation from [12].

3.2. *Table 2*

For $G = C_3$ and $G = C_6$ we took the equations from [12]. For $G = C_9$ and $G = 48$ we again computed the respective specializations of s and t. Thus we have provided two equations for the 48-locus: the equation in Table 1 reflects the containment of the 48-locus in the C_2 locus and the equation in Table 2 reflects the containment of the 48-locus in the C_3 locus. The result can be checked as before by verification of the group order.

Table 2: Non-hyperelliptic genus 3 curves that admit a Galois cover of degree 3 to $\mathbb{P}^1$

G	Sign	equation	Number of G-orbits on W-pts of		Generators of G
			wt. 1	wt. 2	
C_3	(3^5)	$y^3z = x(x-z)(x-sz)(x-tz)$	4 orb. of 1, 6 reg. orb. (in general)	1 orb. of 1	$\Delta(1, \zeta_3, 1)$
C_6	$(2, 3^2, 6)$	$s = 1 - t$	2 orb. of 2, 3 reg. orb. (in general)	1 orb. of 1	N
C_9	(3,9,9)	$s = \zeta_3$, $t = \zeta_3^{-1}$	1 orb. of 1, 1 orb of 3, 3 reg. orb.	1 orb. of 1	$\Delta(\zeta_3, \zeta_9, 1)$
48	(2,3,12)	$s = \frac{1-\zeta_4}{2}, t = \frac{1+\zeta_4}{2}$	1 orb. of 16	1 orb. of 4	M, $\Delta(1, \zeta_3, 1)$, $\Delta(\zeta_4, 1, 1)$

Table 4: Generating Matrices

$P_1 = \begin{pmatrix} 0 & 1 & 0 \\ 1 & 0 & 0 \\ 0 & 0 & 1 \end{pmatrix}$	$P_2 = \begin{pmatrix} 1 & 0 & 0 \\ 0 & 0 & 1 \\ 0 & 1 & 0 \end{pmatrix}$	$Q = \frac{1}{2}\begin{pmatrix} -1 & 1 & \bar{a} \\ a & a & 0 \\ -1 & 1 & -\bar{a} \end{pmatrix}$ where $a = \frac{-1+\sqrt{-7}}{2}$
$R = \begin{pmatrix} -d & d & 0 \\ d+1 & d+1 & 0 \\ 0 & 0 & -1 \end{pmatrix}$ where $d = \frac{-(1+\zeta_4)}{2}$	$M = \begin{pmatrix} 1 & \frac{-\sqrt{3}r}{3} & \frac{-1}{2} - \frac{\sqrt{3}}{6} \\ 0 & \frac{\sqrt{3}}{3} & \frac{-\sqrt{3}r^2}{6} \\ 0 & \frac{-2\sqrt{3}r}{3} & \frac{-\sqrt{3}}{3} \end{pmatrix}$ where $r = \sqrt[3]{2}$	$N = \begin{pmatrix} -1 & 0 & 1 \\ 0 & \zeta_3 & 0 \\ 0 & 0 & 1 \end{pmatrix}$

Table 3:

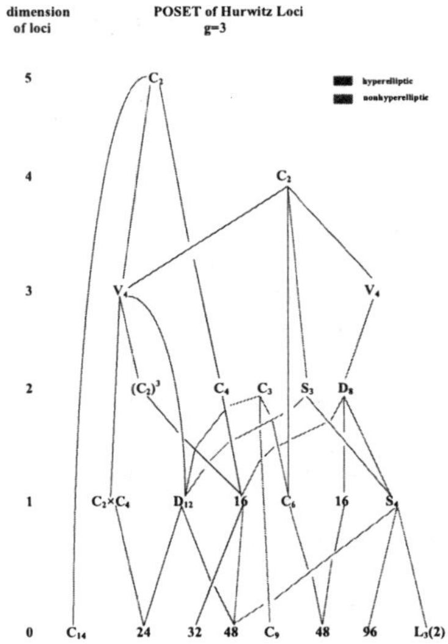

References

1. E. Brieskorn, *Plane Algebraic Curves*, Birkhäuser, 1986.
2. N. Elkies, The Klein quartic in Number theory. *Mathematical Sciences Research Institute Publications* **35**, Cambridge University Press, Cambridge, 1999.
3. M. Farkas and I. Kra, *Riemann Surfaces*, Springer-Verlag, 1992.
4. The GAP Group, *GAP-Groups, Algorithms, and Programming*, Version 4.2; 2000. (`http://www.gap-system.org`)
5. M. girard and P. Tzermias, Group generated by the Weierstrass points of a plane quartic. *Proc. of the American Mathematical Society* **130**.(no.3)(2001), 667–672.
6. A. Kuribayashi, K. Komiya, On Weierstrass points and automorphisms of curves of genus three. In: Algebraic geometry (Proc. Summer Meeting, Univ. Copenhagen, Copenhagen, 1978), pp. 253–299, *Lecture Notes in Math.* **732**, Springer, Berlin, 1979.
7. A. Kuribayashi, K. Komiya, On the structure of the automorphism group of a compact Riemann surface of genus 3. *Bull. Fac. Sci. Engrg. Chuo Univ.* **23** (1980), 1–34.
8. A. Kuribayashi and I. Kuribayashi, On subgroups of automorphism groups of hyperelliptic Riemann surfaces of genus three. *Bull. Fac. Sci. Engrg. Chuo Univ.* **24** (1981), 21–37.
9. A. Kuribayashi and I. Kuribayashi, On subgroups of automorphism groups of hyperelliptic Riemann surfaces of genus three. II. *Bull. Fac. Sci. Engrg. Chuo Univ.* **25** (1982), 5–43.
10. S. Levy, The eightfold way. The beauty of Klein's quartic curve. Ed. Silvio Levy. *Mathematical Sciences Research Institute Publications* **35**, Cambridge University Press, Cambridge, 1999.
11. MAPLE, Version 7; 2001.
12. K. Magaard, T. Shaska, S. Shpectorov and H. Völklein , The locus of curves with prescribed automorphism group RIMS Publication series **1267** (2002), 112–141 (Communications in Arithmetic Fundamental Groups, Proceedings of the RIMS workshop held at Kyoto University Oct. 01)
13. K. Magaard, S. Sphectorov and H. Völklein, A GAP Package for braid orbit computation, and applications. *Experimental Mathematics* **12**.(no.4)(2003), 385-393
14. H. Völklein, *Groups as Galois Groups – an Introduction*, Cambr. Studies in Adv. Math. **53**, Cambridge Univ. Press 1996.
15. H. Völklein, The image of a Hurwitz space under the moduli map, pp.135-150 in: "Progress in Galois Theory", edts. H. Völklein and T. Shaska, Developments in Math. **12**, Springer 2004.